1/5
8/08

REVISED
for the 17th Edition IEE
Wiring Regulations

Electrical Installations

LEVEL 3

2330 Technical Certificate and 2356

Dave Allen
John Blaus
Nigel Harman
Brian Tucker

www.heinemann.co.uk

✓ Free online support
✓ Useful weblinks
✓ 24 hour online ordering

01865 888118

Heinemann

Heinemann is an imprint of Pearson Education Limited, a company incorporated in England and Wales, having its registered office at Edinburgh Gate, Harlow, Essex, CM20 2JE. Registered company number: 872828

www.heinemann.co.uk

Heinemann is a registered trademark of Pearson Education Limited

Text © EAS and JTL 2008

First published 2008

12 11 10 09 08
10 9 8 7 6 5 4 3 2 1

British Library Cataloguing in Publication Data
A catalogue record for this book is available from the British Library

ISBN 978 0 435401 10 8

Typeset by HL Studios
Original illustrations © Pearson Education Limited 2008
Illustrated by Illustration Ltd. (Stuart Holmes), Ken Vail Graphic Design and HL Studios
Cover photo/illustration © Getty Images
Printed in the UK by Scotprint

The websites used in this book were correct and up-to-date at the time of publication. It is essential for tutors to preview each website before using it in class so as to ensure that the URL is still accurate, relevant and appropriate. We suggest that tutors bookmark useful websites and consider enabling students to access them through the school/college intranet.

Original material by JTL. Additional content supplied by Nigel Harman and Brian Tucker on behalf of Electrical Assessment Services UK Ltd.

JTL endorsement
The electrotechnical industry has determined that the minimum requirement for skilled craft status is the NVQ Level 3. JTL therefore endorses the contents of this second of two reference books in so far as it supports candidates to achievement of full electrical craft status. This material has been developed from JTL resources with kind permission.

Acknowledgements
The authors and publisher would like to thank the following individuals and organisations for permission to reproduce photographs:
Alamy Images – pages 63, 67
Alamy Images/Arthur Parker – page 61
Alamy Images/David Templeton – page 167
Arco – pages 50, 51 [goggles/helmet]
Art Directors and Trip – pages 46, 51 [mask], 96, 117, 149, 150 [opto-coupler]
Art Directors and Trip/Andrew Lambert – page 97 [all]
Corbis/Reuters/Jeam-Paul Pelissier – page 29
David J. Green – page 182 [Strap saddle]
Dreamstime.com/Christina Richards – page 41
Dreamstime.com/Edite Artmann – page 51 [powered breathing]
Dreamstime/Mark Yull – page 227 [High voltage]
Dreamstime.com/Photomo – page 1
Pearson Education Ltd/Jules Selmes – page 49
iStockPhoto.com/Tony Tremblay – page 22
Photographers Direct/Daniel Walsh Ltd – page 52
Photographers Direct/Malcolm Furrow – page 124
Photographers Direct/Maxim Sivyi Photography – page 150 [transistor]
Photographers Direct/Peter Trower – page 53
Photographers Direct/Sid Frisby – page 43
Science Photo Library – page 81
Shutterstock/Dainis Derics – page 227 [distribution]
Shutterstock/IRC – page 13 [respirator]
Shutterstock/JoLin – page 55
Shutterstock/Kamil Fazrin Rauf – page 13 [mask]
Shutterstock/Robert Asento – page 226
All other images © Pearson Education Ltd/Gareth Boden

Contents

Introduction iv

Chapter 1 **Statutory regulations and organisational requirements** 1

Chapter 2 **Safe working practice and emergency procedures** 41

Chapter 3 **Effective working practices** 67

Chapter 4 **Electrical systems and components** 81

Chapter 5 **Supply systems, protection and earthing** 167

Chapter 6 **Electrical machines and motors** 231

Chapter 7 **Safe, effective and efficient completion of installations** 261

Chapter 8 **Testing and commission** 323

Chapter 9 **Fault diagnosis and repair** 395

Chapter 10 **Restoring systems to working order** 435

Index 488

Introduction

What is this book?

This book is designed to specifically support the syllabus of the level 3 of the C&G 2330 certificate. It builds upon the material covered in the level 2 book, which supports the level 2 of the same qualification.

This book also covers the material needed for the underpinning knowledge elements of the C&G 2356 NVQ in Electro-technical Services – Electrical Installation (Buildings and Structures). Visit our website **www.heinemann.co.uk <http://www.heinemann. co.uk/>** to download specific mapping to the NVQ syllabus.

How this book can help

This book includes several features designed to help you make progress and reinforce learning:

- **Illustrations and photographs** – clear drawings and photographs showing essential information in a clear and concise way.
- **Margin notes** – short hints to aid good practice and reinforce information.
- **End of chapter activities** – these allow you to put some of the skills you have learnt into practice.
- **Knowledge checks** – use these to test yourself on the information you have learnt.

17th Edition IEE Wiring Regulations

This book has now been updated in accordance with the BS7671:2008 (17th Edition, IEE Regulations) standards. A fundamental change from the 16th to the 17th edition is that phase conductors are now known as line conductors. It is important to note that this terminology refers to the name of the conductor itself. Particularly when dealing with a.c. theory, the terms 'phase' and 'line' are still used to avoid confusion when dealing with specific concepts and values of voltage and current.

Remember whenever carrying out any electrical work please consult the current edition of the Regulations.

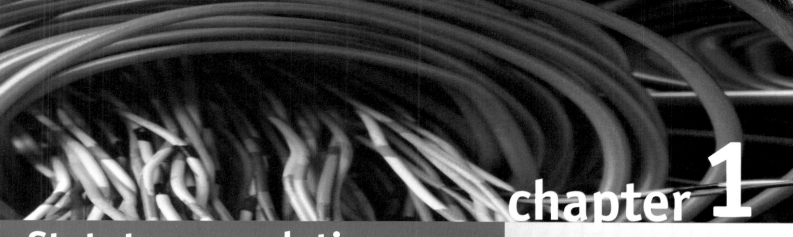

Statutory regulations and organisational requirements

chapter 1

Unit 1 outcome 1

Within the construction industry, there is a great deal of legislation and regulations that govern safe working. All employers have duties under the Health and Safety at Work Act to ensure the health and safety of themselves and others, such as their employees. Employers therefore have an obligation to provide a safe working environment.

In the event of an accident it is the responsibility of the employer to take a report of the incident. These work alongside risk assessments to help avoid a repeat of the incident in the workplace.

This chapter builds on concepts first explored at Level 2.

On completion of this chapter the candidate will be able to:

- select appropriate Personal Protective Equipment (PPE) and safety equipment to complete a designated task

- recognise a series of warning, prohibition and mandatory safety signs

- complete an accident report

- carry out and prepare a risk assessment

- list factors within the electro-technical sector causing change and the importance of good customer relations.

Legislation

At the end of this topic area the candidate will be able to identify basic safety regulations and their application to electro-technical operations, as well as stating employers' responsibilities to maintain safety.

Regulatory requirements, statutory and non-statutory

As we saw at Level 2 there are a number of statutory and non-statutory laws and regulations that detail the procedures that need to be followed to work safely. Many of these were covered in the Level 2 book, including the following.

Statutory regulations:

- Health and Safety at Work Act 1974
- Electricity Safety, Quality and Continuity Regulations 2002
- Electricity at Work Regulations 1989
- Management of Health and Safety at Work Regulations 1999
- Control of Substances Hazardous to Health Regulations 2002
- Provision and Use of Work Equipment Regulations 1998

Non-statutory regulations:

- BS 7671 Requirements for Electrical Installations

These are explored further here together with additional regulations.

Health and Safety at Work Act 1974 (HASAWA)

This Act contains a number of regulations. These are also used by government authorities to control the standard of working conditions throughout industry. All employers are covered by the HASAWA. The Act places certain specific duties on both employers and employees which must be complied with by law.

Each employee, and this includes *you*, is also required by law to assist and co-operate with their employer and others in making sure that safe working environments are maintained, that all safety equipment is fully and correctly used, and that all safety procedures are followed.

There are many sections to the Act. The main ones that will affect you are as follows.

Section 1   Gives the general purposes of Part 1 of the Act, which are to maintain or improve standards of health and safety at work, to protect other people against risks arising from work activities, to control the storage and use of dangerous substances and to control certain emissions into the air.

Section 2   Contains the duties placed upon employers with regard to their employees.

Section 3	Places duties on employers and the self-employed to ensure their activities do not endanger anybody (with the self-employed, that includes themselves) and to provide information, in certain circumstances, to the public about any potential hazards.
Section 4	Places a duty on those in control of premises, which are non-domestic and used as a place of work, to ensure they do not endanger those who work within them. This extends to plant and substances, means of access and egress as well as to the premises themselves.
Section 6	Places duties on manufacturers, suppliers, designers, importers etc. in relation to articles and substances used at work. Basically they have to research and test them, and supply information to users.
Section 7	Places duties upon employees.
Section 8	Places a duty on everyone not to intentionally or recklessly interfere with or misuse anything provided in the interests of health, safety and welfare.
Section 9	States that employers may not charge their employees for anything done for, or equipment provided for, health and safety purposes under a relevant statutory provision.

The duties of both employers (Section 2) and employees (Section 7) were covered fully at Level 2. Their responsibilities include providing:

- a safe place of work, with safe and risk-free access to and exit from the place of work
- (and maintaining) safe plant and equipment
- safe systems of work
- a safe working environment
- safe methods for handling, storing and transporting goods and materials
- reports of accidents
- information, instruction, training and supervision of employees
- a safety policy.

Electricity Safety, Quality and Continuity Regulations 2002

These regulations specify power quality and supply continuity requirements to ensure an efficient and economic supply service for customers.

The duty holders identified in the Regulations are generators, distributors, suppliers and meter operators, including licensed and non-licensed duty holders. It should be noted that contractors and agents of duty holders, parties constructing networks, persons installing connections, persons operating small-scale embedded generators and consumers may also have duties under the Regulations.

The requirements of the Regulations apply to public and private operators and to electricity networks used to supply consumers in England, Wales and Scotland.

The Regulations are mainly concerned with the electricity transmission and distribution systems, the overhead service lines to consumers' premises and the apparatus therein up to the consumers' terminals, as all of these are accessible to the public. However, street furniture falling within the scope of the Regulations would include streetlights, traffic signals, bollards, advertising hoardings, bus shelters, public telephones etc. situated on or adjacent to roads, streets and footpaths.

Electricity at Work Regulations 1989 (EAWR)

Made under the HASAWA 1974, these regulations came into force on 1 April 1990. Every employer and self-employed person has a duty to conform to these regulations. Penalties can be imposed. We looked at how this Act defined employers and employees at Level 2.

The level of responsibility you hold to make sure the regulations are met depends on the amount of control you have over electrical safety in any particular situation. A person may find him or herself responsible for causing danger to arise elsewhere in an electrical system, at a point beyond his or her own installation. This situation may arise, for example, if you energise a circuit while somebody is working in a different room on that circuit. This is obviously a dangerous situation. Because such circumstances are 'within their control', the effect of Regulation 3 is to bring responsibilities for compliance with the rest of the Regulations to that person, thus making him/her a duty holder.

Absolute/reasonably practicable

Duties in some of the regulations are either regarded as 'absolute' meaning they absolutely *have* to be met or, if they have a qualifying term applied to them, 'reasonably practicable'. The meaning of reasonably practicable has been well established in law. The interpretations below are given as a guide only.

Absolute

If the requirement in a regulation is 'absolute', for example if the requirement is not qualified by the words 'so far as is reasonably practicable', the requirement must be met regardless of cost or any other consideration.

Reasonably practicable

Someone who is required to do something 'so far as is reasonably practicable' must think about the amount of risk of a particular work activity or site and, on the other hand, the costs in terms of the physical difficulty, time, trouble and expense which would be involved in taking steps to reduce the risks to health and safety of a particular work process.

For example, in your own home you would expect to find a fireguard in front of a fire to prevent young children from touching the fire and being injured. This is a cheap and effective way of preventing accidents and would be a reasonably practicable situation. If the cost or technical difficulties of taking certain steps to prevent those risks are very high, it might not be reasonably practicable to take those steps. The greater the degree of risk, the less weight that can be given to the cost of measures needed to prevent that risk.

In the context of the Regulations where the risk is very often that of death from electrocution and where the nature of the precautions which can be taken are so often very simple and cheap, e.g. insulation surrounding cables, the level of duty to prevent that danger approaches that of an absolute duty.

Here is a summary of the regulations you are most likely to have to comply with and whether they are regarded as 'absolute' or 'reasonably practicable'.

Regulation 4

Standard of duty: *reasonably practicable*

All electrical systems shall be constructed and maintained to prevent danger. All work activities are to be carried out so as not to give rise to danger.

Regulation 5

Standard of duty: *absolute*

No electrical equipment is to be used where its strength and capability may be exceeded so as to give rise to danger.

Regulation 6

Standard of duty: *reasonably practicable*

Electrical equipment sited in adverse or hazardous environments must be suitable for those conditions.

Regulation 7

Standard of duty: *reasonably practicable*

Permanent safeguarding or suitable positioning of live conductors is required.

Regulation 8

Standard of duty: *absolute*

Equipment must be earthed or other suitable precautions must be taken, e.g. the use of residual current devices, double insulated equipment, reduced voltage equipment etc.

Regulation 9

Standard of duty: *absolute*

Nothing is to be placed in an earthed circuit conductor that might, without suitable precautions, give rise to danger by breaking the electrical continuity or introducing a high impedance.

Regulation 10

Standard of duty: *absolute*

All joints and connections in systems must be mechanically and electrically suitable for use.

Regulation 11

Standard of duty: *absolute*

Suitable protective devices should be installed in each system to ensure all parts of the system and users of the system are safeguarded from the effects of fault conditions.

Regulation 12

Standard of duty: *absolute*

Where necessary to prevent danger, suitable means shall be available for cutting off the electrical supply to any electrical equipment. (Note: drawings of the distribution equipment and methods of identifying circuits should be readily available. Ideally, mains-signed isolation switches should be provided in practical work areas.)

Regulation 13

Standard of duty: *absolute*

Adequate precautions must be taken to prevent electrical equipment, which has been made dead in order to prevent danger, from becoming live while any work is carried out.

Regulation 14

Standard of duty: *absolute*

No work can be carried out on live electrical equipment unless this can be properly justified. This means that risk assessments are required. If such work is to be carried out, suitable precautions must be taken to prevent injury.

Regulation 15

Standard of duty: *absolute*

Adequate working space, adequate means of access and adequate lighting shall be provided at all electrical equipment on which or near which work is being done in circumstances that may give rise to danger.

Regulation 16

Standard of duty: *absolute*

No person shall engage in work that requires technical knowledge or experience to prevent danger or injury unless he or she has that knowledge or experience, or is under appropriate supervision.

Management of Health and Safety at Work Regulations 1999

The main requirement on employers is to carry out a risk assessment. Employers with five or more employees need to record the significant findings of the risk assessment. Risk assessment can be straightforward in a simple workplace, such as a typical office. However, it can become complicated if it deals with serious hazards such as those on a construction site, nuclear power station, chemical plant or an oil rig. We will look more at risk assessment throughout this chapter.

Besides carrying out a risk assessment, employers also need to:

- make arrangements for implementing the health and safety measures identified as necessary by the risk assessment

- appoint competent people (often themselves or company colleagues) to help them implement the arrangements

- set up emergency procedures

- provide clear information and training to employees

- work together with other employers sharing the same workplace.

Control of Substances Hazardous to Health Regulations 2002 (COSHH)

COSHH guards against hazardous substances, such as glues, paints, fumes, dust, bacteria etc. Employers must therefore ensure the following.

- They must work out what hazardous substances are used in your workplace and find out the risks to people's health from using these substances.

- Employers must decide what precautions are needed before starting work with hazardous substances.

- They must prevent people being exposed to hazardous substances and, where this is not reasonably practicable, control the exposure. If it is reasonably practicable, exposure must be prevented by changing the process or activity so that the hazardous substance is not required or generated, or by replacing it with a safer alternative or using it in a safer form, e.g. pellets instead of powder. If prevention is not reasonably practicable, exposure should be adequately controlled by one or more of the measures outlined in the Regulations, e.g. total enclosure of the process. For a **carcinogen** special requirements apply. Only as a last resort should personal protective equipment be provided.

- Employers must make sure that control measures are used and maintained properly and that safety procedures are followed. Engineering controls and respiratory protective equipment have to be examined and, where appropriate, tested at suitable intervals.

Definition

Carcinogen
– any substance which exposes the user to the risk of cancer

- If required, employers must monitor exposure of employees to hazardous substances. Occupational Exposure Limits or OELs (threshold limits for concentrations of hazardous substances in the air) are approved by HSC and have legal status under COSHH.

- Employers are required to carry out health surveillance where their assessment has shown that this is necessary or COSHH makes specific requirements. This might involve examinations by a doctor or trained nurse. In some cases trained supervisors could, for example, check employees' skin for dermatitis or ask questions about breathing difficulties where work involves substances known to cause asthma.

- If required, employers must prepare plans and procedures to deal with accidents, incidents and emergencies.

- They must provide their employees with suitable information, instruction and training about the nature of the substances they work with or are exposed to, the risks created by exposure to those substances and the precautions they should take.

- They must provide employees with sufficient information and instruction on control measures (their purpose and how to use them), how to use personal protective equipment and clothing provided, results of any exposure monitoring and health surveillance (without giving people's names) and emergency procedures.

Provision and Use of Work Equipment Regulations 1998 (PUWER)

These Regulations require that any risks to people's health and safety, from equipment they use at work, be prevented or controlled. In addition to the requirements of PUWER, lifting equipment is also subject to the requirements of the Lifting Operations and Lifting Equipment Regulations 1998.

In general terms, the Regulations require that equipment provided for use at work is:

- suitable for the intended purpose

- safe for use, maintained in a safe condition and, in certain circumstances, inspected to ensure this remains the case

- used only by people who have received adequate information, instruction and training; and accompanied by suitable safety measures, e.g. protective devices, markings, warnings.

Generally, any equipment which is used by an employee at work is covered, for example hammers, knives, ladders, drilling machines, power presses, circular saws, photocopiers, lifting equipment (including lifts), dumper trucks and motor vehicles. If an employer allows employees to provide their own equipment, it will be covered by PUWER and the employer will need to make sure it complies with the Regulations.

The Regulations cover places where the HASAWA applies – these include factories, offshore installations, offices, shops, hospitals, hotels and places of entertainment etc. PUWER also applies in common parts of shared buildings and temporary places of work such as construction sites.

While the Regulations cover equipment used by people working from home, they do not apply to domestic work in a private household.

Portable Appliance Testing (PAT)

There is no specific legislation requiring portable appliance testing to be carried out. However, compliance with existing legislation results in a duty of care being placed on employers to ensure that the workplace and its equipment are safe for use. As we can see from the following examples taken from legislation, it is vital that Portable Appliance Testing be carried out to ensure that employers meet their legal obligations.

The Health and Safety at Work Act 1974 puts a duty of care upon both the employer and the employee to ensure the safety of all persons using the work premises.

The Management of Health and Safety at Work Regulations 1999 requires that every employer shall make suitable and sufficient assessment of the risks to health and safety of:

- employees, from any potential hazards to which they are exposed while at work
- persons not in their employment, arising out of, or in conjunction with, conduct by the employer or the business.

The Provision and Use of Work Equipment Regulations 1998 states every employer shall ensure work equipment is maintained in an efficient state, working order and good repair.

The Electricity at Work Regulations 1989 states the employer shall prevent, so far as is reasonably practicable, the threat of danger in all **systems** and **electrical equipment** and to see that this is maintained.

It is clear that these acts and regulations apply to all electrical equipment used in, or associated with, places of work. This scope extends from distribution systems down to the smallest pieces of electrical equipment. The implication is that there is a requirement to inspect and test all types of electrical equipment in all work situations. The process of checking equipment defects by a visual inspection linked to a testing programme is covered in Chapter 8.

Control of Major Accidents and Hazards Regulations 1999 (COMAH)

COMAH applies mainly to the chemical industry but also to some storage activities, explosives and nuclear sites and other industries where threshold quantities of dangerous substances identified in the Regulations are kept or used.

Definition

Electrical equipment – includes anything used, intended to be used or installed, to generate, provide, transmit, transform, rectify, convert, conduct, distribute, control, store, measure or use electrical energy

System – an electrical system in which all the electrical equipment is, or may be, electrically connected to a common source of electrical energy. This includes both the source and the equipment

The main aim of COMAH is to prevent and reduce the effects of major accidents involving dangerous substances such as chlorine, liquefied petroleum gas, explosives and arsenic pentoxide, which can cause serious damage/harm to people and/or the environment.

The general duty on all operators that underpins all the regulations is to take all measures necessary to prevent major accidents and limit their consequences to people and the environment. This applies to all establishments within scope.

By requiring measures both for prevention and mitigation (making less severe), there is recognition that all risks cannot be completely eliminated. Thus, the phrase 'all measures necessary' is taken to include this principle and a judgement must be made about the measures in place. Where hazards are high, then high standards will be required to ensure risks are acceptably low.

Noise and Statutory Nuisance Act 1993

This Act was designed to control and eliminate the nuisance created by noise in the street. Although it is mainly used to combat unsocial behaviour, Section 3 is concerned with audible intruder alarms.

Any person who installs an audible intruder alarm on or in any premises shall:

- ensure that the alarm complies with any prescribed requirements
- ensure that the local authority is notified within 48 hours of the installation.

A person who (on or after the first appointed day) becomes the occupier of any premises on, or in which, an audible intruder alarm has been installed, shall not permit the alarm to be operated unless:

- the alarm complies with requirements
- the police have been notified in writing of the names, addresses and telephone numbers of the current keyholder
- the local authority has been informed of the address of the police station to which the notification above has been given.

The responsibility for enforcement for this lies with the local authority. Local authorities have the power to enter premises and turn off the alarm or issue fixed penalty fines of between £500 and £5000.

Noise Act 1996

This Act is about ensuring that the public is protected from the nuisance of noise within the hours of darkness – meaning from 11pm to 7am. Once again, the enforcing authority is the local authority. As above, they have the power to enter a property and seize equipment, as well as the power to issue a fixed penalty notice of £100 or a summary conviction and much higher fine. The level of noise when measured from within the complainant's premises should not exceed 35dB(A), or 10dB(A) above the 'underlying level' (background noise).

Remember

A statutory nuisance could arise from the poor state of your premises or any noise, smoke, fumes, gases, dust, steam, smell, effluvia, the keeping of animals, deposits and accumulations of refuse and/or other material

Remember

Where appropriate, use mains-generated electricity in preference to diesel generators. This will help you to reduce noise levels and to reduce the risk of pollution through fuel spillage

Reporting of Injuries, Diseases and Dangerous Occurrences Regulations 1995 (RIDDOR)

The Reporting of Injuries, Diseases and Dangerous Occurrences Regulations 1995 (RIDDOR) place a legal duty on employers, the self-employed and those in control of premises to report some work-related accidents, diseases and dangerous occurrences to the relevant enforcing authority for their work activity. This can be the Health and Safety Executive (HSE) or one of the local authorities (LAs).

The law requires the following work-related incidents to be reported:

- deaths

- injuries where an employee or self-employed person has an accident which is not major but results in the injured person being away from work or unable to do the full range of their normal duties for more than three days (including any days they wouldn't normally be expected to work such as weekends, rest days or holidays) not counting the day of the injury itself

- injuries to members of the public where they are taken to hospital

- work-related diseases

- dangerous occurrences where something happens that does not result in a reportable injury but which could have done.

Incident Contact Centre

Reporting accidents and ill health at work is a legal requirement. The information enables the enforcing authorities to identify where and how risks arise and to investigate serious accidents. The enforcing authorities can then advise on preventive action to reduce injury, ill-health and accidental loss – much of which is uninsurable.

Previous systems required employers to complete an accident report form and notify the relevant authorities. However, sometimes people were unsure to which authority they should be reporting.

To make things easier, and to remove all confusion, in April 2001 the HSE set up the Incident Contact Centre, which provides a central point for employers to report incidents irrespective of whether their business is HSE or local authority enforced.

As employers can now report incidents over the telephone to the Incident Contact Centre, the need for the old F.2508 Accident Report Form has been largely removed (the Incident Contact Centre now completes the forms on an employer's behalf).

The centre is based in Caerphilly (Wales) and allows individuals to make their report to a single point irrespective of their location. The centre will then send the report to the correct enforcing authority. The centre will also send a copy of their final report on the incident to the employer. This meets the statutory obligation for employers to keep records of all reportable incidents for inspection and also allows the employer to correct any errors or omissions.

The quickest way to notify the centre is by telephone. The number is: 0845 300 9923.

Did you know?

If, as a result of your works, levels of noise, vibration, dust or odour are created that cause nuisance to the surrounding community, the Local Authority Environmental Health Department can stop your works

Reporting requirements

In each case, provided that the incident is reported through the Incident Contact Centre, there is no need to complete an F.2508 Accident Report Form or F.2508A Disease Report Form, although the centre will ask for brief details about the business, the affected person or persons and the incident or disease.

A report needs to be made in the following circumstances.

Death or major injury

If there is an accident connected with work and:

- an employee or a self-employed person working on your premises is killed or suffers a major injury (including as a result of physical violence)
- a member of the public is killed or taken to hospital.

The enforcing authority must be informed by telephone via the Incident Contact Centre (see above) without delay (normally within 24 hours). A list of the major injuries covered by the legislation can be found in Chapter 2 of this book.

Over-three-day injury

If there is an accident connected with work (including an act of physical violence) and an employee or self-employed person working on the premises suffers an over-three-day injury, then the enforcing authority must be informed by telephone via the Incident Contact Centre without delay (but no longer than ten days).

Disease

If a doctor notifies an employer that an employee suffers from a reportable work-related disease, then the enforcing authority must be informed by telephone via the Incident Contact Centre as soon as possible after notification (but no longer than ten days).

Dangerous occurrence

If something happens which does not result in a reportable injury, but which clearly could have done, it may be a dangerous occurrence that must be reported.

The enforcing authority should be informed by telephone via the Incident Contact Centre immediately (but no longer than ten days).

Operational and bystander requirements

At the end of this topic the candidate will be able to describe safety requirements in terms of equipment that is required as personal protective equipment (PPE), including respiratory protective equipment, and the need to protect bystanders.

At Level 2 we looked at PPE that needs to be provided under the Personal Protective Equipment (PPE) at Work Regulations 1992. This covers eye, hand, foot and head protection. It is part of an employer's responsibilities to reduce workplace risk by providing such material free of charge. It is essential that it is suitable for protecting the user against the risk for which it is issued. The types of PPE were fully covered at Level 2.

Respiratory protective equipment (RPE)

We all need clean air to live, and we need correctly functioning lungs to allow us to inhale that air. Fumes, dusts, airborne particles, such as asbestos, or just foul smells, such as in sewage treatment plants, can all be features of construction environments.

A range of **respiratory protective equipment** (RPE) is available, from simple dust protection masks to half-face respirators, full-face respirators and powered breathing apparatus. To be effective these must be carefully matched to the hazard involved and correctly fitted. You may also require training in how to use them properly.

A simple dust protection mask

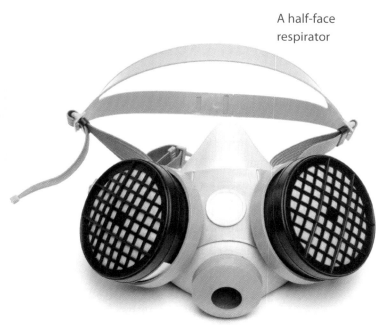

A half-face respirator

Bystander safety requirements

NO Access to
the Public

General
Danger

Warning Tape Signifiying Danger

Figure 1.01 Common
warning signs found in
the workplace

Section 3 of the Health and Safety at Work Act 1974 places a duty on employers to ensure their work activities do not endanger anybody, including members of the public. It also requires, in certain circumstances, for information to be made aware to the public about any potential hazards. Section 7 of the Act places a duty upon you to take reasonable care at work for the health and safety of yourself and others who may be affected by what you do or do not do.

It therefore follows that for any work activity on which you are engaged that can cause harm to others, precautions have to be taken to prevent injury and danger. The use of barriers around the work area would prevent people walking into your work area. Putting up warning notices, advising of your presence and any danger, will also help.

A combination of these two methods is the most effective – after all, if we were just to rely on signs we could not guarantee that people could read or interpret them correctly.

When working on steps or a portable scaffold the effect of putting warning barriers and notices around is twofold. It prevents people walking into your walk area, reducing the risk of them:

● being injured from falling objects

● knocking you off the equipment.

It is important to have provision for safe and easy access and exit from a building or site. The reasons and provisions made for this are covered at Level 2.

Implementing and controlling health and safety

On completion of this topic the candidate will be able to describe the implementation of health and safety legislation within individual organisations and recognise designated warning signs.

As we can see there is an awful lot of legislation to be read and interpreted correctly if the workplace is to be made as safe as possible. Keeping up to date with and understanding the implications of this legislation can be daunting, so how do we manage it?

Strangely, there is no set method for distributing new or amended health and safety requirements to employers. It is up to every employer to ensure that they are meeting legal requirements; lack of knowledge is not regarded as an excuse in a court of law.

The means of keeping up to date vary from company to company and often depend on their size and structure. In smaller companies the owner or manager is likely to have responsibility, but pressure on their time might be very high. Many smaller companies choose to hire the services of an external health and safety consultant.

The consultant will normally visit the employer and carry out an inspection of the main premises and types of site and activities undertaken. The consultant will then produce the necessary paperworks, e.g. risk assessments safety policy, COSHH assessments etc. and provide a manual for the employer to refer to and implement.

Some employers use a consultant on a one-off basis and then choose to maintain the information and systems provided themselves. However, there are risks associated with this as employers must ensure that the manual remains current with all developments. Other employers engage a consultant on an ongoing basis where the consultant monitors all legislation, codes of practice and relevant information and continually provides the employer with updated information, documentation and services.

Large companies often have their own specialist member(s) of staff with responsibility for health and safety.

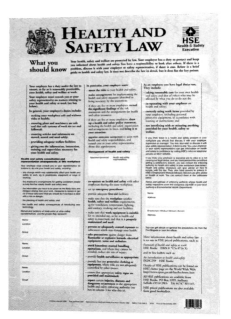

Displaying a health and safety notice in the workplace is compulsory

Health and safety responsibility

Responsibility begins with senior management. Strong leadership is vital in the implementation of effective health and safety risk control. Everyone should know and believe that management is committed to continuous improvement in health and safety performance. Consequently, management should explain its expectations, and how the organisation and procedures will deliver them. Although health and safety functions can (and should) be delegated, legal responsibility for health and safety (including where it involves members of the public) rests with the employer.

These general duties are expanded and explained in the Management of Health and Safety at Work Regulations 1999. Employers must:

- assess work-related risks for both employees and those not in their employ

- have effective arrangements in place for planning, organising, controlling, monitoring and reviewing preventive and protective measures

- appoint one or more competent persons to help in undertaking the measures needed to comply with health and safety law

- provide employees with comprehensible and relevant information on the risks they face and the preventive and protective measures that control those risks.

Find out

What is your employer's approach to health and safety? Can you see any ways in which it could be improved?

Did you know?

Employers are covered by Sections 2 and 3 of the Health and Safety at Work Act 1974. Employers need to prepare – and make sure that workers know about – a written statement of the company health and safety policy and the arrangements in place to put it into effect

Health and safety policy statement

By law (HASAWA 1974, Section 2(Part 3)), anyone employing five or more people must have a written health and safety policy, which includes arrangements for putting that policy into practice. This should be central to achieving acceptable standards, reducing accidents and cases of work-related ill-health, and it shows employees that their employer cares for their health and safety. A health and safety policy statement will include reference to:

- the company's objectives to meet its health and safety requirements
- the company responsibilities under the Health and Safety at Work Act
- employees' responsibilities to meet their requirements under the Health and Safety at Work Act
- who is responsible for what within the organisation, with regards to safety
- joint consultations, or arrangements to consult with staff, over issues of promoting safety and health at work
- a review, or a commitment to review and update the document and policies, in order to keep abreast of changes and/or the introduction of new regulations and codes of practice.

Approved Codes of Practice (ACOPs)

ACOPs give practical advice on how to comply with the law and are approved by the HSE, with the consent of the Secretary of State.

Failure by an employer to follow an ACOP doesn't make them liable to any civil or criminal proceedings. However, in any criminal proceedings where an employer is said to have committed an offence by contravening any requirement or prohibition imposed for which there was an Approved Code of Practice at the time of the alleged contravention, the court will use the ACOP to demonstrate the employer's non-compliance. In such circumstances the employer's only defence would be to prove that it was using another appropriate method/system to a higher standard.

Safe working practices and procedures

Every year thousands of people have accidents at work. In 2004 in the UK there were 235 deaths at work, 28,000 major accidents and approximately 130,000 accidents which resulted in the victims being off work for more than three days. This resulted in hundreds of thousands of hours lost to industry that year. Of course these were just the recorded accidents – we can only speculate how many hundreds of thousands of accidents occurred that didn't warrant reporting.

Through the implementation of the Health and Safety at Work Act, and the Management of Health and Safety at Work Regulations, it is intended to bring about safe working practice and procedures, designed to keep to a minimum the risks associated with work and reduce the number of accidents.

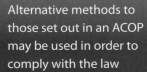

Remember

Writing a health and safety policy statement should be more than just a legal requirement – it should show the commitment of an employer to planning and managing health and safety

Remember

What is written in the policy has to be put into practice. The true test of a health and safety policy is the actual conditions in the workplace, not how well the statement has been written

Did you know?

Alternative methods to those set out in an ACOP may be used in order to comply with the law

The safety policy puts in place a structure to manage health and safety issues within an organisation, with those people named in the document responsible for various elements of that safety policy. The document also lays out the procedures to be adopted. This includes the assessment of work-related risk, effective planning, organisation and control of preventative and protective measures, the provision of comprehensible and relevant information on the risks to be faced by the workforce and the necessary protective measures to be taken.

Safety signs

These were covered at Level 2. The main examples are shown below. Most signs are recognised by shape and colour. Some contain only symbols; others information such as height and distance.

Prohibition signs

Shape: Circular

Colour: Red borders and cross bar. Black symbol on white background

Meaning: Shows what must NOT be done

Example: No smoking

Warning signs

Shape: Triangular

Colour: Yellow background with black border and symbol

Meaning: Warns of hazard or danger

Example: Danger: electric shock risk

Mandatory signs

Circular

White symbol on blue background

Shows what must be done

Wear eye protection

Information or safe condition signs

Square or rectangular

White symbol and/or wording on green background

Indicates or gives information on safety provision

First aid post

Figure 1.02 Common safety signs found in the workplace

The roles, responsibilities and powers

On completion of this topic the candidate will be able to outline the roles, responsibilities and powers of the safety officer, safety representative, HSE inspectors and environmental health officers.

The safety officer

The safety officer is a suitably qualified member of staff with delegated responsibility for all things related to health and safety, and is answerable to the company managers. Their position within the company is to look after the company interests, protecting the workforce and thereby protecting the organisation from legal action. Their role and responsibilities are likely to include:

- arranging internal and external training for employees on safety issues
- monitoring and implementing codes of practice and regulations
- the update and display of information
- liaison with external agencies such as the HSE
- carrying out and recording regular health and safety inspections and risk assessments
- advising on selection of, training in, and use and maintenance of PPE
- maintenance of accident reports and records.

Should an accident occur, the safety officer would lead the investigations, identify the causes and advise on any improvements in safety standards that need to be made.

The safety representative

The Safety Representatives and Safety Committees Regulations 1977 were originally made under Section 15 of the Health and Safety at Work Act 1974, although they have since been amended. The Regulations and associated Codes of Practice provide a legal framework for employers and trade unions to reach agreement on arrangements for safety representatives and safety committees to operate in their workplace. The safety representative is often, but not necessarily, a trade union member and their role is similar to that of the safety officer. However, their focus is different as their role is to protect the workforce. Their duties would include:

- making representations to the employer on behalf of members on any health, safety and welfare matter
- representing members in consultation with the HSE inspectors or other enforcing authorities
- inspecting designated workplace areas at least every three months
- investigating any potential hazards, complaints by members and causes of accidents, dangerous occurrences and diseases
- requesting facilities and support from the employer to carry out inspections and receive legal and technical information
- paid time off to carry out the role and to undergo either TUC or union-approved training.

The Health and Safety Commission (HSC) has issued guidance for employers who do not recognise independent trade unions. For all but the smallest companies, they recommend setting up a safety committee of members drawn from both management and employees. This should help employers comply with their legal duties, particularly under Section 2 (Part 4) of the Health and Safety at Work Act 1974 and the Health and Safety (Consultation with Employees) Regulations 1996.

Under the Health and Safety (Consultation with Employees) Regulations, any employees not in groups covered by trade union safety representatives must be consulted by their employers; the employer can choose to consult with them directly or through the elected representatives.

Health and Safety Commission and Health and Safety Executive

The Health and Safety Commission (HSC) and the Health and Safety Executive (HSE) are responsible for the regulation of almost all the risks to health and safety arising from work activity in the UK. Their mission is to protect people's health and safety, by ensuring risks in the changing workplace are properly controlled. Among other things, they look after health and safety in nuclear installations and mines, factories, farms, hospitals and schools, offshore gas and oil installations, the gas grid, railways and the movement of dangerous goods and substances.

HSC members are appointed by the appropriate Secretary of State, and the HSC in turn appoints members of the HSE. The role of the HSC is to review the effectiveness of current health and safety policy and, if changes are required, they formulate these changes into proposals for government to debate in Parliament and create new laws or amend old laws.

Did you know?

Local authorities are responsible to the HSC for enforcement in offices, shops and other parts of the services sector

HSE inspectors

To help enforce the law, HSE inspectors have a range of statutory powers. Consequently they can, and sometimes do, visit and enter premises without warning. If there is a problem the inspector can do the following.

Did you know?

The HSE has about 4000 employees throughout the UK including advisers, scientific and medical experts and inspectors

- Issue an **improvement notice**. This notice will say what needs to be done, why and by when. The time period within which to take the remedial action will be at least 21 days to allow the duty holder time to appeal to an employment tribunal if they so wish. The notice also contains a statement that in the opinion of an inspector an offence has been committed.

- Issue a **prohibition notice**. Where an activity involves, or will involve, a risk of serious personal injury, the inspector can serve a prohibition notice stopping the activity immediately, or after a specified time period, and not allowing it to be resumed until remedial action has been taken. The notice will explain why the action is necessary.

- Take legal action. In certain circumstances, the inspector may also wish to start legal proceedings. Improvement and prohibition notices, and written advice, may be used in court proceedings.

Before deciding on a course of action, HSE inspectors apply a concept known as 'proportionality'. Effectively, this means that the degree of enforcement action to be taken will be in proportion to the degree of risks discovered.

Environmental health officers

Environmental health officers are employed by local authorities and they inspect commercial businesses such as warehouses, offices, shops, pubs and restaurants within a borough area.

They have the right to enter any workplace without giving notice, though they may in practice give notice. Normally, the officer looks at the workplace, work activities and management of health and safety, and checks that the business is complying with health and safety law.

Environmental health officers may offer guidance or advice to help businesses. They may also talk to employees and their representatives, take photographs and samples, serve improvement notices and take action if there is a risk to health and safety that needs to be dealt with immediately.

Penalties for health and safety offences

(Penalties are current at the time of writing (August 2007). These penalties can change from time to time.)

The Health and Safety at Work Act 1974, Section 33 (as amended), sets out the offences and maximum penalties under health and safety legislation.

Failing to comply with an improvement or prohibition notice, or a court remedy order (issued under the HASAWA, Sections 21, 22 and 42 respectively) will incur the following penalty:

- Lower court maximum: £20,000 and/or six months' imprisonment
- Higher court maximum: unlimited fine and/or two years' imprisonment.

Breach of Sections 2–6 of the HASAWA, which set out the general duties of employers, self-employed persons, manufacturers and suppliers to safeguard the health and safety of workers and members of the public who may be affected by work activities, will incur the following penalty:

- Lower court maximum: £20,000
- Higher court maximum: unlimited fine.

There can be other breaches of the HASAWA and breaches of 'relevant statutory provisions' under the Act, which include all health and safety regulations. The legislation imposes both general and more specific requirements, such as requirements to carry out a suitable and sufficient risk assessment or to provide suitable personal protective equipment. Breaches incur the following penalties:

- Lower court maximum: £5000
- Higher court maximum: unlimited fine.

Remember

Everyone can be prosecuted by HSE if they are employed. Prosecution would only happen as a result of continuously failing to implement safe working practices, where such failure has resulted in injury or death

Remember

The more severe the offence, the more severe the punishment

Did you know?

Information on health and safety issues can be obtained from local authorities and emergency services as well as the HSE

Sources of information

On completion of this topic the candidate will be able to explain where information on health and safety can be obtained from.

Apart from via the HSE itself, information can also be sought from local authorities and emergency services. Health and safety literature and information and advice can be obtained from a number of sources:

- the Health and Safety Executive (www.hse.gov.uk)
- Her Majesty's Stationary Office (HMSO)
- libraries and the Internet in general
- environmental health officers at the local council
- specialist health and safety consultants
- emergency services.

Accident prevention

On completion of this topic, the candidate will be able to describe the conditions leading to accidents in the workplace and the means of controlling them.

How do we prevent accidents? At Level 2 we looked at the legal requirements, material problems and causes of accidents, from 'environmental', such as defective machinery or tools, to 'human' causes, such as carelessness or lack of training.

Health and safety in the workplace is something we each need to take personal responsibility for and, by thinking about what we are doing or are about to do, we can avoid most potentially dangerous situations. We can also do this by reporting potentially dangerous situations to the correct people.

Equally, we can demonstrate a responsible approach by reporting potentially dangerous situations, hazards or activity to the correct people. Even if the hazard is something you can easily fix yourself – for example, you might move a cable that was causing a trip hazard – you must still report it to your supervisor. The fact that the cable was there at all might be indicative of someone not doing their job properly and it could happen again.

You may also have a role to play in other circumstances. For example, at present premises are inspected by local fire authorities. If you are aware of something unsafe, such as a fire exit that has been blocked, then tell them about it.

As with the risk assessment process, the secret is to be aware of all possible danger in the workplace and have a positive attitude towards health and safety. Follow safe and approved procedures where they exist and always act in a responsible way to protect yourself and others, and the construction site can be a happy and safe working environment for many years.

Remember

Prevention is better than cure!

Remember

The first person to alert about a health and safety issue should be your site supervisor or safety officer

Safety tip

If the hazard is something that can be easily fixed, and as long as it is not dangerous to yourself, then you should fix it

Using the right equipment can reduce the danger of a hazard

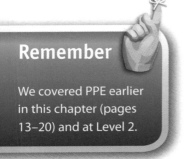

Remember

We covered PPE earlier in this chapter (pages 13–20) and at Level 2.

Of course, it follows that action can be taken to reduce the number of accidents. The following are some examples of how it is possible to reduce the risk.

Eliminate the hazard

This is the easiest way of dealing with a problem. For example, if a cable was to be run at high enough level that access equipment was required to use it, this would clearly have some danger associated with it. If we were to re-route the cable, so it is now run at a lower level, then we will have eliminated the hazard.

Replace the hazard with something less dangerous

In the example above, if we were not able to route the cable any other way, we would then have to look at our choice of access equipment. The use of an extension ladder would be possible but too dangerous; therefore, we would use a mobile tower. However, it may be considered a safer option to use a hydraulic scissor lift or platform while wearing a safety harness.

Guarding the hazard

If the potential hazard is from a moving part of a machine then it makes sense to make sure that there is a guard fitted that prevents any operator coming into contact with the moving part. With the use of electrical control circuits it can be made impossible to use the machine without the guard in place.

Personal protection

Many hazards can be reduced by the wearing of personal protective equipment. This should only be seen as a last resort, however, as this protective measure should be used in conjunction with the measures previously mentioned.

Safety education and publicity

All of the above protective measures will fail if the operator/installer has no idea of what they're doing. Health and safety training is essential if people are to be made aware of the hazards and the precautions to be taken to reduce those hazards to acceptable levels. Unless you are able to recognise the potential for danger, it is impossible to take action to avoid it.

The accident prevention measures discussed above form the basis of risk assessment.

Accident reporting

On completing this topic the candidate will be able to describe the procedures for reporting accidents.

Employers legally (under RIDDOR; see page 11) have to keep a record of any reportable injury, disease or dangerous occurrence. As previously mentioned on page 11, the Health and Safety Executive has set up an Incident Contact Centre, giving employers the facility to report incidents over the telephone. This eliminates the need for the employer to complete Accident Report Form F.2508. The Incident Contact Centre forwards a copy of every report form it completes to the respective employer for them to check and hold as a record. The importance of checking this and any other document cannot be over emphasised.

Small scale, non-reportable incidents and injuries

To deal with small scale, non-reportable incidents and injuries, most employers hold an accident book on site. Should an employer wish to continue with the paper-based system, the HSE has launched a new Accident Report Book which will help organisations comply with the Data Protection Act 1998. Approved by the Information Commissioner, it has been revised as most existing accident books store personal details and information, which can then be seen by anyone reading or making an entry.

The new book allows for accidents to be recorded but individuals' details can be held separately in a secure location. It contains information on first aid and how to manage health and safety information to help prevent accidents from happening in the first place.

The previous version, produced by the Department for Work and Pensions (DWP) and other similar books are not compliant with the Data Protection Act, and the DWP has given responsibility for the publication to HSE. The Information Commissioner ruled that businesses had to change their accident book to comply with the Data Protection Act by 31 December 2003.

New-look accident book, which doesn't contain individuals' personal details

Any report of an accident must contain the following information.

- The name and occupation of the casualty.
- The date on which the accident occurred.
- Information on the injuries sustained or any loss.
- A summary of what has happened.
- A description of events leading up to the accident.
- Details of any witnesses present, names and addresses, and what they saw.
- Conclusions about how the accident occurred and the key points that resulted in the accident.
- Recommendations of how the accident may be avoided in future, for example the fitting of guards or training that may be required.
- Supporting evidence. This is an essential part of the report that could in future support claims for industrial injury compensation. If possible use photographs, video or diagrams of the area to help explain the circumstances.
- Signature of the person/persons making the report. This could include the casualty or not depending upon injuries sustained.
- Date the report was made, as this could be different from the day of the accident.

Risk assessment

On completion of this topic the candidate will know how to carry out and prepare a report identifying potential health hazards.

The basics of how to carry out and prepare a risk assessment are covered at Level 2.

Reporting hazards

On completion of this topic the candidate will be able to describe the procedures used to notify/report hazards to appropriate people.

The communication process

All the ideas we have looked at so far in this chapter involve communication. Communication is about much more than just speaking and writing. You communicate an enormous amount by how you look, the gestures and facial expressions you use, and the way you behave. Even something as simple as a smile can make a big difference to your communication.

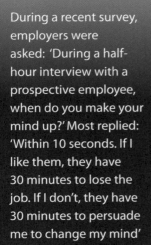

Remember

A risk assessment is nothing more than a careful examination of what, during work activities, could cause harm to people, so that an employer can weigh up whether it has taken enough precautions to prevent harm

Did you know?

During a recent survey, employers were asked: 'During a half-hour interview with a prospective employee, when do you make your mind up?' Most replied: 'Within 10 seconds. If I like them, they have 30 minutes to lose the job. If I don't, they have 30 minutes to persuade me to change my mind'

The benefits of good communication are huge. Good attitude, appearance and behaviour will immediately put customers at their ease and earn respect from others. Clearly conveyed thoughts and ideas, when backed up with good working practices and procedures, will invariably improve productivity, increase profitability and produce satisfied customers.

It is often the case that reports and letters will need to be generated to people within the organisation, to communicate either the findings of investigations into incidents or the results of a risk assessment in the workplace. Sometimes this information will also have to be sent to people outside of the organisation.

Here, we can only talk in generic terms as individual companies have their own procedures and paperwork in place. Remember though that the people you are trying to communicate with will also determine the method by which you contact them. For example, if as a result of an incident in the workplace you needed to summon the **Works Rescue Team**, you would probably do so by using a mobile phone, shortwave radio or simply by sending someone to fetch them.

Communication with a **Fire Officer** may first be by telephone. This initial contact will then need some form of written communication to confirm what has been said – in this case it is likely to be in the form of a letter.

The writing process

Writing isn't too difficult if you break it down into steps and then focus on one step at a time. You can tackle almost any writing project if you follow this method. The main steps are as follows.

Step 1 Have a clear idea about the overall purpose of your letter or report. Think about exactly what you want to say, and why. Most people get stuck because they have not thought carefully about what they are trying to achieve.

Step 2 Gather all the information you will need.

Step 3 Plan the logical order for presenting your ideas. Write a list of headings and check that each one follows on from the previous one. If it is a report, the first paragraph should be a summary of the rest of the document and, if you want your reader to take some action, say so in your last paragraph. Do not start writing in detail until you have completed this step.

Step 4 Now that you know what each section or paragraph will contain, go ahead and write each one. Then read it through again and, if necessary, edit what you have written. Finally, present it as a properly formatted and typed document.

Remember

Try to use simple language. Don't try to impress the reader with your huge vocabulary – it will only frustrate or annoy them. Don't use jargon or abbreviations unless you are sure your reader is familiar with them

Check your work before sending it. We all make mistakes. Checking your work (known as proofreading) is a very important aspect of writing. Pay attention to grammar, spelling and punctuation. If you are using word-processing software to write your letter or report, it will often indicate when something might be wrong. Get into the habit of finding out what the problem is and correcting it. However, always remember that the software cannot detect every mistake. The following sentence has several errors, but they won't show up on the computer screen:

'Two many electricians were note iced to be erecting the too extract fans in 2 the to staff toilets.'

A dictionary can help: if you aren't sure how to spell a word, look it up. If possible, don't proofread something immediately after you have written it. You are more likely to spot any mistakes and find better ways of saying things if you go back to it later.

Environmental Management Systems (EMS)

On completing this topic the candidate will be able to explain Environmental Management Systems (BSEN ISO 14001).

Concern for the environment and awareness of the need to improve management of resources is on the increase. The issue has found its way into the political arena and our everyday lives, and is creating pressures on businesses to demonstrate a commitment to minimising their impact on the environment.

An effective EMS certified to ISO 14001 or registered to EMAS (the European Eco-Management and Audit Scheme) can help an organisation to operate in a more cost-efficient and environmentally responsible manner by managing its activities while also complying with relevant environmental legislation and its own environmental policy.

There are a number of key benefits associated with the implementation of a certified EMS:

- demonstrating conformance
- management confidence
- improved management of environmental risk
- independent assessment
- compliance with legislative and other requirements
- continual improvement
- reduction in costs
- reduction in supply chain pressures.

Companies seeking certification from the accreditation body will need to prove they have considered the impact upon the environment of the design, installation and the disposal of waste, and show, where necessary, that alternative methods and materials have been used. In the area of waste disposal, materials must be disposed of in the appropriate registered waste disposal sites.

Compliance with BSEN ISO 14001, taking into account all of the existing environmental legislation, will be covered fully in the next topic.

Environmental legislation

On completion of this topic the candidate will be able to describe (briefly) environmental legislation as it applies to electro-technical industries.

Environmental Protection Act 1990

As well as introducing controls on pollution to the air, this Act also places a duty of care on all those involved in the management of waste, whether the collection, disposal or treatment of controlled waste.

Pollution Prevention and Control Act 1999

In terms of environmental law this is a major Act, as it recognises the pollutant released into the environment through one medium (air) which leads to an increase in pollution in other mediums such as water or land. This Act sets out to create an integrated approach. The main sections of the Act are as follows.

- **Part 1** deals with integrated pollution control by her Majesty's Inspectorate of Pollution, and air pollution control by local authorities.

- **Part 2** deals with wastes on land.

- **Part 3** deals with any statutory nuisance.

- **Part 4** deals with litter.

- **Part 5** builds on the Radioactive Substances Act 1960.

- **Part 6** deals with genetically modified organisms.

- **Part 7** protects nature conservation.

- **Part 8** handles miscellaneous issues, including contaminated land.

Those processes subject to an integrated approach are:

- air emissions

- processes which give rise to significant quantities of special waste that is toxic or has a high flammability

- processes giving rise to emissions to sewers. This includes substances such as mercury and cadmium, and many pesticides subject to approval of the Environmental Agency.

Pollution Prevention and Control Regulations 2000

Within the UK, the new Pollution Prevention and Control (PPC) regime implements the EU's Integrated Pollution Prevention and Control (IPPC) Directive. The implementing legislation for IPPC (collectively referred to as the PPC Regulations) is the Pollution Prevention and Control (England and Wales) (Amendment) Regulations 2002, with equivalent Regulations governing Scotland and Northern Ireland.

PPC Part A (A(1) and A(2) in England and Wales) includes new issues not previously covered by IPPC such as:

- energy efficiency
- waste minimisation
- vibration
- noise.

The new regime also requires an effective system of management to be implemented to ensure that all appropriate pollution prevention and control measures are taken. Special emphasis is placed on the application of Best Available Techniques (BAT) to reduce the environmental impact of the process.

The Dangerous Substances and Preparations (Safety) and Chemicals (Hazard Information and Packaging for Supply) Regulations 2002 concern the marketing and use of substances and preparations that contain carcinogens, **mutagens** and substances toxic to reproduction (collectively known as CMRs), and the subsequent restriction of their supply to the general public.

Clean Air Act 1993

The Clean Air Act applies to all small businesses that burn fuels in furnaces or boilers, or that burn material in the open, including farms, building sites and demolition sites. It requires them to prevent the emission of dark smoke.

Radioactive Substances Act 1993 (RSA93)

This Act applies to any organisation that is involved in the handling, storing and transportation of radioactive materials. A requirement of the Act is for qualified experts to be employed. They will need to have relevant expertise, training and knowledge to ensure all requirements are met under RSA93. Responsibility for enforcement is with the **Environment Agency**.

Radioactive materials are used in the course of exploration, exploitation and construction programmes. Specialist contractors use a variety of sealed and unsealed radioactive sources. Unsealed sources may be used during grouting or cementing operations, or for reservoir or installation equipment tracing studies. Sealed radioactive sources may be used for radiography, well logging or in liquid level and density gauges.

Companies involved in the use of radioactive materials must give regular checks to their staff for contamination. All operatives working with radioactive materials are required to wear a dosimeter pocket card to register their exposure to the radioactive material.

The dosimeter pocket card warns other workers of the danger

Controlled Waste Regulations 1998

Under the Controlled Waste Regulations 1998, the duty of care requires that you ensure all waste is handled, recovered or disposed of responsibly, that it is handled, recovered or disposed of only by individuals or businesses that are authorised to do so, and that a record is kept of all wastes received or transferred through a system of signed Waste Transfer Notes.

Special waste

If the material you are handling has hazardous properties, it may need to be dealt with as **special waste**.

Dangerous Substances and Preparations and Chemicals Regulations 2000

These Regulations were introduced to prohibit the supply of substances that are classified as carcinogenic, mutagenic or toxic. These substances constitute a risk to the general public as they cause cancer, genetic disorders and birth defects. The Regulations place a duty of care on organisations involved in the design and installation of products within buildings to ensure that materials used are not dangerous and harmful to those that will live in, work in or pass through such installations. The enforcement agency operates through the local authority trading standards department.

Did you know?

If you transport your own building or demolition waste, you will need to be registered as a waste carrier with your Environmental Regulator

Changes in demand in the electro-technical sector

On completion of this topic area the candidate will be able to list factors within the electro-technical sector that bring about a change in demand and describe their influence on working patterns and training needs.

Changing demand

Demand can normally do one of two things: increase or decrease. An increase in demand may bring changes to a workplace as a company tries to handle this extra demand. Some companies handle this easily by the recruitment of extra staff and/or the training of both new and existing staff. Extra demand can also sometimes allow individuals within an organisation to progress their career and perhaps move from craft to supervisory duties.

A decrease in demand, irrespective of the cause, invariably means that companies look to save money and, sadly, this can result in cutting back on services and staffing levels.

The reasons for changes in demand are many and varied. A significant factor is the economy of the country. A healthy buoyant economy will result in growth in all areas, as people and organisations will have more money to spend and will look for improvements in property and décor. This in turn feeds itself back as an increased demand for electrical installation work. The opposite is true for a downturn in the economy: when money is tight, people and organisations are not looking to spend, which feeds through into a reduction of work for the electrical contractor.

Another example of a downturn is increased competition. With an increase in the amount of electrical contractors in an area this will almost certainly result in an increased level of competition as they all try to stay in business. It is not unusual for contractors to quote very low prices, with almost no profit margins, to try to make sure that they get the work and secure more customers.

During periods of downturn, productivity becomes more important. When you look at two contractors competing for the same work, if they are paying the same hourly rate and they are buying materials from the same wholesaler, what is the difference in their contract price? It can only be the rate at which the work is installed. If contractors are to win more work, then they must complete projects on time and to the standards required by the customer and still make a profit. Contractors failing to meet the standards and time constraints will go out of business.

Remember

If poorly handled, extra demand can be very stressful and can cause damage to an organisation's reputation

A decrease in demand can prove equally stressful as organisations try to ensure their own existence

As competition increases, so does the marketing of contractor's skills. Some organisations carry out little or no marketing but exist happily on a reputation built upon their ability to regularly carry out high quality work. Other companies create work by 'aggressive' marketing with sales staff, leaflets, letters and advertising. Market forces will often play a role, as some clients are happy to pay for a high quality product and good service, where other clients seem happy 'as long as the job works'.

Technological changes can also affect demand. For example, there are now sections of the industry that make their living by installing the specialist cabling, cat5 and fibre optic cables needed for CCTV, computers and computerised systems. Obviously future technical change can bring new opportunities to the market that contractors can exploit.

Changes to the law can also bring about an increase in demand. For example, a combination of the Health and Safety at Work Act and the Provision of Work Equipment Regulations, to name two, has resulted in the need to have all portable electrical appliances in the workplace tested on an annual basis. There has subsequently been a huge growth within the industry to meet this demand.

Did you know?

Changes in legislation can bring about work as installations are brought up to new required levels and standards

Changes in demand can bring about great change for individuals. Most apprentice electricians will undertake a structured apprenticeship, regarded as broad-based training, that gives a range of experience in most aspects of electrical installation work and consequently most will become qualified to the National Standards. Completion of the **Advanced Apprenticeship** includes key skills in the Application of Number, Numeracy and IT. These are seen as transferable skills that will enable you to move from one type of employment to another.

Other transferable skills that you may acquire over time are those specialisms that you may have worked with, such as intruder alarms, computer network systems and CCTV etc. However, the effects of change are felt differently from company to company and the way that organisations deal with change can vary widely.

If you move between companies you will find that you may need additional training in certain aspects of the job. Or you may find, even if you stay with the same company, that the job changes and you need to learn new skills. Some people see any form of additional training as being a burden they should not have to bear. Others, perhaps more rightly, see this as being the acquisition of new skills that ensure their usefulness to the employer, as well as giving them new skills that can help them remain a viable force in the workplace for many years.

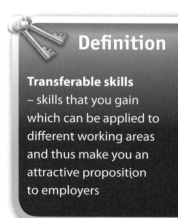

Definition

Transferable skills – skills that you gain which can be applied to different working areas and thus make you an attractive proposition to employers

These new skills also give individuals the opportunity of transferring their skills to other employers. As an example, if you have been trained to cut trunking correctly, then the amount of time it should take to teach you to cut conduit should be small as you already know how to use a hacksaw and measure. In other words, experience of measuring and cutting are the **transferable skills** you bring to the task of cutting conduit.

On the job: Leisure opportunities

Sanjay runs a successful small electrical maintenance business. He employs four permanent electricians and Matt, an apprentice. His clients are mainly other small businesses but he is delighted when his company wins the maintenance contract for a new private leisure centre. However, neither he nor his staff have ever worked on saunas or gym equipment before, and he realises he needs to train up one of his staff. Sanjay is keen that Matt should be trained, but Matt says he doesn't want to do it. Matt's apprenticeship is almost up and Sanjay has to decide whether to give Matt a job at the end of it.

1. Why do you think Matt was not keen to do the extra training? What would your reaction be if you were asked to train for some specialist work?

2. How might Sanjay persuade Matt to take on the training?

3. If you were Sanjay, what would you do now?

Maintenance work can often create a more technical job content since you have to work with all sorts of control systems, from heating systems through to the control of production machinery. Much of this work covers electronics and computerised control. For those that embrace this change and undergo additional training, it is likely that an increase in responsibility and change of working activities may result in an increase in salary or status within the company.

Maintenance work can also bring about the need for a flexible approach to working hours. After all, no one can predict when the sauna in a leisure centre is going to stop working, but we do know that the client will want it repaired as soon as possible.

Flexible working hours is not just the domain of the maintenance electrician. Installation electricians are frequently being required to complete installation out of hours, such as shop refurbishments and shutdowns in places like oil refineries, as well as having to respond to emergency call outs where, for example, the supply has been lost to domestic premises in the middle of the night.

As organisations grow in response to an increasing demand it is essential that everyone in the company knows exactly what their role and responsibilities are. Good communications will ensure that the company is operating as a cohesive unit, with everyone pulling in the same direction for the sake of the organisation. Failure to establish good communications within the organisation will eventually lead to a breakdown of the company structure and the company's fortunes are likely to suffer.

Hopefully, the examples on page 33 demonstrate that you should expect to encounter changes in your working situation and that some of these may result in a change to your original career. It should also show that there will always be a need for continued retraining and updating of individual skills. Each revised edition of the *IEE Wiring Regulations* BS 7671 is another typical example that will create a retraining need.

Changing career

Some people are happy to remain in their original job role until they retire. Others have ambitions to become engineers, supervisors or even manage their own company. If you do have ambitions, then it is definitely an advantage if you can recognise them as early as possible, so that you can plan any relevant education. It is always easier to learn while you are still in education rather than having to relearn a subject in later years.

Career development

We spend a large part of our lives in work. It therefore makes sense to be as happy at work as possible.

For many people this will mean pursuing ambitions and developing their career. For an electrician, there are several options available. The list below is only meant as a guide to typical arrangements; it isn't intended to be definitive or to be followed in any particular order.

Remember

A happy worker is a good worker

Occupation	Qualifications needed
JIB recognised Electrician	NVQ Level 3 + Technical Certificate
JIB recognised Approved Electrician	As above + C & G 2391 + two years' experience
Electrical Supervisor	Approved Electrician + additional experience
Contract/Site Engineer	Supervised + BTEC HNC in Building Services or Foundation degree
Contract Manager	Contract Engineer + BTEC HNC in Building Services or Foundation degree
Consultant	Contract Manager + Degree + Experience

Table 1.01 Industry progression route

Foundation degrees

Traditional degree qualifications are often seen as mainly 'theory' courses. Upon completion the successful students often lack the ability to apply the theoretical knowledge they have gained within real workplace situations.

Foundation degrees are new work-related higher education qualifications, designed to equip young people with the higher-level skills that employers need. The flexibility of this new qualification opens up opportunities for students already in the workplace who thought higher education wasn't for them. Higher education institutions have developed this new qualification directly with employers, to ensure the skills they need are met. It is awarded by both colleges and universities.

The course can be studied over two years or on a pro-rata basis part-time. There are more than 70 foundation degrees on offer throughout the country, and courses include aircraft engineering, classroom assistance, construction, e-commerce, hospitality, multimedia design, police studies and textiles.

Did you know?

Once you have completed the foundation degree you can progress to a full honours degree with just a further 15 months of study

Customer relationships and customer protection

On completion of this topic area the candidate will be able to describe why it is important for continued trading to maintain good customer relations. The reader will also be able to outline measures designed for customer protection.

Customer relations

The success of any company depends upon its ability to win work through the competitive tendering process and through the referral of satisfied customers. For the vast majority of organisations working in the electro-technical sector, the majority of the work comes from referrals made by previous customers and inclusion on preferred contractor's lists.

For a customer to be able to refer a contractor to someone else, they will need to have had a good experience when dealing with them. They should have come to the conclusion that the organisation as a whole, from the management through to the installation electrician, are **'fit for purpose'**.

Factors that would encourage such a view would be based on the following

1. *Performance*

This would be reflected in the way that everybody who came into contact with the customer conducted themselves and the quality of the work produced.

2. *Cost*

The price charged for the job has to be reflected in the work produced. Having decided to part with their hard earned cash, customers have to feel that they are getting value for money.

Completion dates and reliability

Additional factors a customer would consider are completion dates and reliability – these go hand-in-hand.

When working in someone's home or office you should do the following.

- Take care to protect their property. Use dust covers, and ask them to remove objects that might get broken or damaged.

- If there are pets or small children around, ask for them to be kept well away from the working area.

- Make sure you inform the customer before using hazardous substances. Take the correct precautions and respect any instructions you get from the customer.

- Before starting work, make sure that you understand exactly what the client expects.

- If you have recommendations for improvements or alterations, take time to discuss these with the client and give clear technical information if it is needed.

Before starting any job, a programme of work has to be agreed with the customer and other contractors. On large installations the programme will be very complex and involve many trades and tasks, and will be spread over weeks, months and, for some jobs, years. On small jobs, the programme will be less complex, involving fewer trades and work activities. However, both will have a **developed timeline**, with a start date, interim dates for activities to be achieved and a **completion date**.

Failure to meet any of these dates – not starting the job on time, not being clear of any area preventing the next trade from starting their work and not completing the job on time – will have an adverse effect on how the customer perceives the organisation.

Failure to meet these agreed targets leaves the impression of an organisation that is **unreliable**. Once this sort of reputation is established, work will start to dry up and contracts will become much harder to win. A reputation for unreliability is hard to shake off.

Definition

Unreliable – a failure to deliver promised commitments

Know your customer

Who is your **customer**? At first, you'll probably think of the person who wants the work done and is paying for it. But does this mean you can ignore everyone else you will meet in your job? A wider definition of a customer is: 'Anyone who has a need or expectation of you.'

Using this definition, almost everyone you work with becomes your customer, and you will, in turn, be theirs. These people may be architects, consultants, clients, other tradespeople or a member of the public. They may ask you to do a complex task, or simply ask you a question. In either case, they will expect a good, polite response. If you treat everyone as a valuable customer and always try to give them your best service, it will bring you many benefits. Dealing with people is an important part of your job. It is never wise to upset people, if only because it may cause problems later for you or your firm.

Remember

Be polite to everyone you come into contact with during your work projects, not just the main 'customer'

DO:	DON'T:
• be honest • be neat and tidy in your personal appearance, and look after your personal hygiene • learn how to put people at ease, and be pleasant and cheerful • show enthusiasm for the job • try to maintain friendly relationships with customers, but don't get over-familiar • know your job and do it well – good knowledge of the installation and keeping to relevant standards gives the customer confidence in you and your company • explain what you are going to do, and how long it will take • if you are not sure about something – ask!	• 'bad mouth' your employer • use company property and materials to do favours for others • speak for your employer when you have no authority to do so • use bad language • smoke on customer premises • gossip about the customer or anyone else • tell lies – the customer will find out eventually if he or she is being misled or ripped off • assume that you know what your employer wants without bothering to ask.

Table 1.02 Somes dos and don'ts to improve customer relations

Always try to provide answers to any questions about the work. You might be asked these questions.

- Is this the right product for the job?
- Will it cost a lot to buy and install?
- Will it do what I need?
- How reliable is it?
- Will I be able to use it?
- How easy is it to repair?
- How long does the guarantee last?
- Will you be finished on time?

Before answering the question, try to understand why the customer is asking it. What do they really want to know? Are they worried that they cannot afford the installation? Have they booked a holiday that starts just after you are scheduled to finish? If you don't know the answer, don't guess. Promise to find out – and then do so! When the job is completed, make sure the client fully understands how to use the installation, and leave behind any manufacturers' user guides or installation manuals. Invite the customer to contact you if there are any problems in the future.

If you follow this advice you will have excellent relationships with your customers. However, in the unlikely event that a dispute arises, you may both need to seek the help of an independent mediator to settle your differences.

Remember

Remember that you are representing your company. People will judge the company by the way you behave. If you do well, your company could get more work from the client. On the other hand, if you don't, contracts may be lost and you may be out of a job

On the job: Cleaning up mess

Ellie is an apprentice electrician who always cleans up any mess she has made. She is working with Joe, the senior electrician. While installing an extra socket outlet in an old people's home, Joe has left a mess in the kitchen; he has also burnt the kitchen work surface with a cigarette.

1. How could this situation have been avoided?

2. What steps could Ellie take to explain the situation to the home's manager?

3. What should happen to Joe when he gets back to the company premises?

Legal obligations of a contract

Pivotal in the relationship between the customer and the contractor is the legal contract that exists between them. A contract is a formal document which sets out the terms of agreement between the two parties. In its simplest form it informs the client of what is to be received in return for the fee paid and informs the contractor of what must be done to receive the fee.

On signing a contract, both parties have agreed to be bound by the terms and conditions of the contract. The client is agreeing to pay the price stated for the work on completion of the installation, while the contractor is agreeing to carry out the work to the standards required (e.g. BS 7671), producing an installation that is fit for purpose, within the agreed time-scale and for an agreed fee (which is provided at a 'reasonable charge').

Should a dispute arise between the contractor and the client, then the contract would form the basis of any settlement, whether derived through personal negotiation or through the involvement of the courts. A decision would be made as to whether the terms and conditions of the contract had been met or whether there had been a breach of contract by either party.

A contract can be a verbal agreement but the pitfalls of this type of contract are that it can become difficult to establish exactly what the contract was in the event of a dispute. Standard forms of contract exist, issued by the Joint Contracts Tribunal (JCT), and are used commonly throughout the industry.

Penalty clauses

Penalty clauses are often inserted into a contract for large projects. These penalties reflect the potential loss to the client should the project not be completed on time.

For example, each year car production plants around the country shut down for a two-week period, as the workforce has a holiday and the company takes the opportunity to refurbish, repair and update its technology.

During this period, many contractors will be working within the car plant. All will have a deadline to meet and all will have penalty clauses inserted in their contracts with the car manufacturer. These will specify amounts of money that will have to be paid should the contract not be completed on time.

If the car plant is unable to open and start production on time, then the plant will lose millions of pounds each day in lost production, wages, and heating and lighting bills etc. It therefore follows that the penalty clause attempts to claw back some of this loss. The contractors may well find themselves being penalised by tens of thousands of pounds per hour if they are found to be the cause of the delay.

FAQ

Q Why do I need to know about health and safety legislation?

A To avoid prosecution and because following it is likely to make the workplace safer for everyone.

Q Do I need to do what my boss says?

A Basically yes, provided this doesn't amount to 'an illegal instruction'. Your boss cannot force you to break the law.

Q Do I have to work on circuits while they are live?

A Normally no. Work cannot be carried out on live circuits unless it is absolutely unavoidable and proper precautions have been taken to prevent injury.

Q Do all portable appliances have to be tested annually?

A No. The frequency of inspection and testing will depend on the likely risk; for example, construction site tools will need testing more frequently than a computer in an office.

Q Do I have to tell someone if I have an accident?

A Yes, you should always tell your employer if you have an accident in the workplace. Make sure it is recorded in the accident report book, as you might need evidence of it later on.

Q Do I have to wear the pair of goggles I've just been given?

A Yes. If your employer provides you with an item of PPE, then you have to wear it whenever you are told to by signs or instructions.

Activity

1. Obtain a copy of your company's health and safety policy and read it. For most organisations this should be a written document. If not, ask your boss to tell you what the company's policy is.

2. Find out if your company has any policy documents or procedures relating specifically to electrical work.

Knowledge check

1. State the name of the main piece of general health and safety legislation in the UK.

2. Which piece of legislation specifically covers electrical matters?

3. Define the terms 'absolute' and 'reasonably practicable'.

4. List the main types of incidents that are 'reportable incidents' under RIDDOR.

5. Describe the four different types of safety signs you may see on a construction site and explain the colour code of these signs.

6. State what three actions an HSE inspector could take against a company with a health and safety problem.

7. State four actions you could take to reduce the risk associated with a hazard.

8. Which statutory regulation specifically deals with pollution caused by litter?

9. State the qualifications needed to become a JIB Approved Electrician.

10. Explain the advantages of having 'legal contracts' between a contractor and a client.

chapter 2

Safe working practices and emergency procedures

Unit 1 outcome 2

As we have seen in Chapter 1, the aim of legislation and regulations is to promote safety in the workplace, both for the employer and the employee. To fufill this, employees have a duty to apply safety issues in their day-to-day job. It is therefore essential that all employees constantly look to maintain and improve their working practices.

In the event of an accident, an employer and employee must know what procedures are in place within the organisation to tackle emergencies. These procedures can save lives so it is vital that you know about them.

This chapter builds on concepts first explored at Level 2.

On completion of this chapter the candidate will be able to:

- recognise potential safety hazards within the workplace
- identify the location of first aid facilities within the workplace
- describe situations where it is necessary to work in isolation
- describe safety procedures to prevent injury or discomfort
- describe precautions against electric shock and the causes of asphyxiation
- outline emergency evacuation procedures at the workplace
- identify possible sources of fire within the workplace
- identify types of fire extinguisher and match them with various types of fires.

Safe systems of working

On completion of this topic the candidate will be able to outline the necessity of a permit to work scheme.

Permit to work

A **permit to work** is a document that specifies the work to be done, the persons involved, when it is going to be done, the hazards involved, the precautions to be taken and the action to be taken in the event of an incident. The permit is only 'active' for a set period, and if you haven't returned by the allocated time you will be investigated. Permission for you or others to return to the work area will be refused until a new permit to work has been issued.

As the permit to work has to be authorised, it is essential that the person issuing the document is competent and fully able to understand the work to be done, any hazards and the proposed system of work and precautions needed. A sound knowledge of current safety legislation is necessary, such as the Confined Spaces Regulations 1997 and COSHH, among others. This person is usually employed by the company as part of its safety team and would report directly to the company's safety officer.

The permit to work form aids communication between everyone involved, and employers must train staff in its use. It should ideally be designed by the company issuing the permit, taking into account individual site conditions and requirements. On certain sites, separate permit forms may be required for different tasks, such as hot work and entry into confined spaces, so that sufficient emphasis can be given to the particular hazards present and precautions required.

A permit to work system is usually found in industrial environments where there is an inherent danger in the work activities which are being carried out. Although this is mainly the case, it does not stop the principles being applied to other areas of work such as construction and maintenance.

Examples of where such a system may be used to protect people include working:

- near or with corrosive or toxic substances (battery rooms etc.)
- in hazardous atmospheres (the petro-chemical industry)
- near live equipment (supply sub-stations)
- underground or in deep trenches
- on or near overhead power lines
- in confined spaces (within tanks and cylinders)
- on the railways (trackside)
- on equipment in production areas such as assembly lines.

All of the above are extremely dangerous environments to be working in, and unless some form of control is used people will be killed or injured.

Remember

COSHH regulations guard against hazardous substances

Lock off procedures

On completion of this topic the candidate will be able to outline the need for lock off procedures.

In this chapter you will explore the need and methods for safe isolation. Part of the safe isolation procedure is the requirement to lock off equipment. One of the requirements of a permit to work system is the use of locking devices to ensure that the supply cannot be restored while other people may be working on, in or around electrical equipment.

In a situation where a permit to work scheme is in place, the same person who issues the permit will also issue the locking devices and keys.

Multi-lock system

Here three padlocks will need to be removed before the locking device can be opened and removed from the system it is locking off. Some multi-lock systems have up to ten padlocks attached to them.

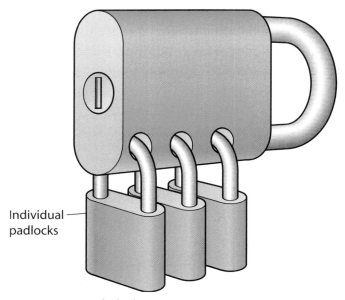

Individual padlocks

Figure 2.01 A multi-lock system

The multi-lock system illustrated above could be used where both a mechanical fitter and electrician need to be working on the same machine. The person issuing the permit to work and the multi-lock device will give the mechanical fitter and electrician a padlock with its own key. The third lock will belong to the person issuing the permit to work.

The electrician and fitter use the multi-lock device and their padlocks to lock off the supply to the equipment. When they have finished their work, they can only remove their particular padlock. This means the supply will not be able to be restored until the person who issues the permit to work has removed their padlock. The permit issuer will not do this until all the other padlocks have been removed. In this way, control is maintained, preventing anyone from being unwittingly electrocuted or injured by the machine.

Working alone

On completion of this topic the candidate will be able to describe situations in which it is inadvisable or unsafe to work in isolation.

A place of work is always a dangerous environment. The completion of an appropriate risk assessment should be sufficient to ensure that all risks have been minimised. However, there will be times when the risk assessment identifies that the actual activity or location are a source of danger that cannot be minimised. In these circumstances it is essential never to work alone.

Confined spaces

Working in confined spaces is controlled by the requirements of the **Confined Spaces Regulations 1997**. In the UK a number of people are killed or seriously injured in confined spaces each year. Those killed include not only the people working in the confined space but also those who tried to rescue them.

What is a confined space?

A confined space can be any space of an enclosed nature where there is a risk of death or serious injury from hazardous substances or dangerous conditions. Some confined spaces are easy to identify – for example, enclosures with a limited opening such as:

- storage tanks
- silos
- reaction vessels
- enclosed drains
- sewers.

Less obvious risks are:

- open top chambers
- vats
- combustion chambers in furnaces
- ductwork
- unventilated or poorly ventilated rooms.

What are the dangers?

Danger can arise in confined spaces because of:

- a lack of oxygen, poisonous gas, fumes or vapour
- liquids and solids which can suddenly fill a space when released, such as grain
- residues left in tanks/vessels

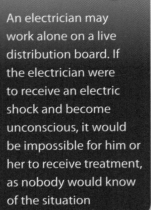

Safety tip

An electrician may work alone on a live distribution board. If the electrician were to receive an electric shock and become unconscious, it would be impossible for him or her to receive treatment, as nobody would know of the situation

- dust in flour mills and cement works

- hot conditions which may lead to a dangerous increase in body temperature.

What does the law require?

For working in confined spaces a risk assessment should identify the hazards and the precautions to be taken. In most cases the assessment will include consideration of:

- the task

- the working environment

- working materials and tools

- the suitability of those carrying out the task

- arrangements for emergency rescue.

If the risk assessment identifies a serious risk of injury from working in the confined space, then:

- avoid entry to the confined space – do the work from the outside

- if entry to a confined space is unavoidable, follow a safe system of work and put in place adequate emergency arrangements before the work starts.

What confined spaces are electricians likely to encounter? From the information above, unventilated or poorly ventilated rooms are identified as confined spaces, therefore working in lofts above properties would fall into this category. The dangers faced could be the hot temperature within the loft area and the possibility of falling through the floor. Electricians frequently work on farms where they may be required to work on blower motors in grain silos. This also falls under the category of a confined space, as well as working in flour mills and cement factories where there would be high concentrations of dust.

Above ground or in trenches

In 2003/2004 there were 67 fatal accidents at work and nearly 4000 major injuries occurred when people were working at height. The **Work at Height Regulations 2005** attempts to address this situation. What is working at height? The Regulations indicate that a place is at 'height' if a person could be injured falling from it, even if he or she is at or below ground level.

Regulation 4 and 6 (1, 2) subsection 17 requires that you must:

- ensure that no work is done at height if it is safe and reasonably practicable to do it at other than height

- ensure that the work is properly planned, appropriately supervised and carried out in a safe way, as is reasonably practicable

- plan for emergencies and rescue

- take account of the risk assessment carried out under Regulation 3 of the Management of Health and Safety at Work Regulations

Remember

Always plan your work, reduce the risks and make sure you have somebody with you in case of emergency

- where the risk of fall cannot be eliminated, use work equipment or other measures to minimise the distance and consequence of a fall, should one occur

- avoid work at height where possible.

Regulation 4 (3, 4) subsection 18 requires you to ensure that the work is postponed while weather conditions endanger health or safety.

Regulations 5 and 6 (5) (b) require employers to ensure that everyone involved in the work is competent (or, if being trained, is supervised by a competent person). This includes involvement in organisation, planning, supervision, and the supply and maintenance of equipment.

Working in excavations

Every year people are killed or seriously injured when working in excavations. Excavation work has to be properly planned, managed, supervised and carried out to prevent accidents.

Planning

Before digging any excavations, it is important to plan against the following:

- collapse of the sides

- materials falling on to people working in the excavation

- people and vehicles falling into the excavation

- people being struck by plant

- undermining nearby structures

- contact with underground services

- hazardous entry and exit points

- fumes

- accidents to members of the public.

Make sure the necessary equipment needed such as trench sheets, props, baulks etc. is available on site before work starts.

Site excavations

Precautions during excavation work

Once work starts:

- wear a hard hat when working in excavations

- prevent the sides and the ends from collapsing by battering them to a safe angle or supporting them with timber, sheeting or proprietary support systems

- never go into an unsupported excavation

- never work ahead of the supports

- never store spoil or other materials close to the sides of excavations – the spoil may fall in and the extra loading will make the sides more prone to collapse

- always make sure the edges of the excavation are protected against falling materials – provide toe boards where necessary

- take steps to prevent people falling into excavations; if the excavation is 2 metres or more deep, provide substantial barriers, e.g. guard rails and toe boards

- keep vehicles away from excavations wherever possible; use barriers if necessary

- where vehicles have to tip materials into excavations, use stop blocks to prevent them over-running; remember that the sides of the excavation may need extra support

- keep workers separate from moving plant such as excavators; where this is not possible, use safe systems of work to prevent people being struck

- make sure that any plant operators are competent

- make sure excavations do not affect the footings of scaffolds or the foundations of nearby structures; walls may have very shallow foundations and these can be undermined by even small trenches

- look around for obvious signs of underground services, e.g. valve covers or patching of a road surface

- use locators to trace any services; mark the ground accordingly

- make sure that the person supervising the excavation work has service plans and knows how to use them

- everyone carrying out the work should know about safe digging practices and emergency procedures

- provide good ladder access or other safe ways of getting into and out of the excavation

- fence off all excavations in public places to prevent people and vehicles falling in

- take precautions (e.g. securely covering excavations) where children might get on to a site out of hours, to reduce the chance of them being injured

- make sure that a competent person supervises the installation at all times.

Remember

Even shallow trenches can be dangerous. You may need to provide supports even if the work only involves bending or kneeling in the trench. You should never work in a trench alone without supervision

Did you know?

An estimated 80% of industrial head injuries are sustained by people who are not wearing any protective equipment

Unguarded machinery

When working near any moving or rotating machinery there is always a risk of injury. As in all situations, a risk assessment should be carried out prior to starting work. This risk assessment requires that the hazards presented by the moving machinery should be assessed and precautions taken. If it is impossible, for production reasons, to stop the machines, then suitable precautions must be taken. These should include:

- putting in place a temporary guard or screen
- ensuring the machine operator knows that you are present
- planning for emergencies and rescue
- having someone work with you
- taking account of the risk assessment carried out under Regulation 3 of the Management of Health and Safety at Work Regulations
- identifying the location of the emergency stop buttons.

Where fire risk exists

There may be times when you work in locations where flammable materials are stored or manufactured. You might also work in oil refineries or sewage treatment works where flammable gases are given off. In these situations you have to recognise the obvious dangers and take precautions to minimise the risk to yourself. These will include:

- using intrinsically safe equipment
- taking account of the risk assessment carried out under Regulation 3 of the Management of Health and Safety at Work Regulations
- planning for emergencies and rescue
- having someone working with you
- using a safe system of work, as prescribed during the safety induction on the site
- identifying the emergency exits
- locating the fire extinguishers
- locating the nearest manual fire alarm point.

Dealing with the dangers from chemical hazards, toxic agents and corrosives was covered at Level 2.

Remember! No matter what the situation, there are three clear rules.

1. A risk assessment is required and followed under Regulation 3 of the Management of Health and Safety at Work Regulations.
2. Have someone working with you.
3. Have a safe system of work and plan for emergencies.

Provision of first aid treatment

On completion of this topic area the candidate will be able to describe the need for, and the provision of, first aid treatment.

What is first aid at work?

While at work people can be injured or become ill. It does not matter whether the injury or the illness is work related or not. What is important is that these people receive attention and that an ambulance is called in serious cases. First aid at work is about the arrangements to ensure this happens. It can save lives and prevent minor injuries becoming major ones.

Health and Safety (First Aid) Regulations 1981

The Health and Safety (First Aid) Regulations 1981 require employers to provide adequate and appropriate equipment, facilities and personnel to enable first aid to be given to employees if they are injured or become ill at work. These Regulations apply to all workplaces, including those with five or fewer employees and to the self-employed.

What is adequate will depend on the circumstances in the workplace. This includes whether trained first aiders are needed, what should be included in a first aid box and if a first aid room is needed. Employers should carry out an assessment of first aid needs to determine this.

That said, the minimum first aid provision on any work site is a suitably stocked first aid box and an appointed person to take charge of first aid arrangements. As a guide, and where there is no special risk in the workplace, a minimum stock of first aid items would be:

- a leaflet giving general guidance on first aid, e.g. HSE leaflet *Basic advice on first aid at work*
- 20 individually wrapped sterile adhesive dressings (assorted sizes)
- two sterile eye pads
- four individually wrapped triangular bandages (preferably sterile)
- six safety pins
- six medium-sized (approximately 12cm x 12cm) individually wrapped sterile unmedicated wound dressings
- two large (approximately 18cm x 18cm) sterile individually wrapped unmedicated wound dressings
- one pair of disposable gloves.

You should not keep tablets or medicines in the first aid box.

A typical first aid kit, including the things mentioned in the list opposite

The list on page 49 is a suggested contents list only; equivalent but different items will be considered acceptable.

Each employer must also inform employees of the first aid arrangements. This is achieved by putting up notices around the premises, on notice boards in the rest rooms, canteen, staff rooms and work areas, telling staff who and where the first aiders or **appointed persons** are and how to summon their assistance. Details like mobile phone numbers or internal telephone numbers should be clearly marked. The notice should also give clear indication as to where the nearest first aid box is located.

Safety procedures

On completion of this topic the candidate will be able to describe the safety procedures required to prevent injury or discomfort to their own or colleagues' skin, eyes, hands and limbs.

Personal hygiene

The working environment can be a dirty and greasy one. There may even be traces of chemicals that could make you unwell. It is foolhardy to go straight from the workplace to the rest area and eat, as all the dirt and grease will transfer to your food and this may make you ill. Personal protective equipment was covered in chapter 1 (page 13) and in greater depth at Level 2. This section will focus on a few specialised pieces of equipment you may not be familiar with.

Hand protection

Use barrier cream before starting work. This fills the pores of the skin with a water-soluble antiseptic cream, so that when you wash your hands the dirt and germs are removed with the cream. Always wash at the end of the work period, before and after using the toilet, and before handling food. Re-apply barrier cream after washing. Do not use solvents to clean your hands. They remove protective oils from the skin and can cause serious problems such as **dermatitis**.

Hands can also be protected by glove. There are different designs of gloves for different tastes.

Gauntlet gloves

Rigger gloves

Eye protection

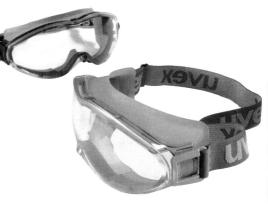

Safety goggles

Helmet with visor screen

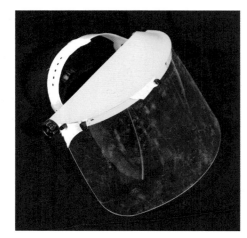

Half-face mask

There are many types of eye protection equipment available – for example safety, spectacles, box goggles, cup goggles, face shields and welding goggles. To be safe you have to use the right type of equipment for the specific hazard you face. Should anything enter your eye, then it needs to be flushed out with a sterile saline solution. Bottles containing such a solution are usually found either in the first aid box or clearly displayed on the wall by the first aid box. Never use a bottle that has previously been opened, as you cannot be sure that it is not contaminated.

Respiratory protective equipment

Breathing in hazardous substances, such as dusts, fumes, vapours, gases or even micro-organisms, can cause damage to health and sometimes lead to death. If direct prevention or control of exposure is not possible, then respiratory protective equipment (RPE) may be needed; this should always be seen as a measure of last resort in the hierarchy of control measures.

Remember

Your eyes are two of your most precious possessions. They are among the most vulnerable parts of your body to injury at work

Powered breathing apparatus

RPE includes a very wide range of devices from simple respirators offering basic protection against low levels of harmful dusts etc. to full-face respirators and powered breathing apparatus. To be effective these must be carefully matched to the hazard involved and correctly fitted. You may also require training in how to use them properly.

Legal requirements for RPE

RPE must be either CE-marked or HSE approved. Before CE-marking became compulsory in 1995, any RPE which was not already marked had to be approved by the HSE. This equipment can still be used as long as it is suitable, maintained and performs correctly.

The use of safety guards, screens and fences

Safety guards

Example of a safety guard

It is totally irresponsible to use any machinery either without a safety guard or with the safety guard disabled. Safety guards have been fitted to equipment and machines for safety. Therefore to disable them or remove them makes the equipment or machine dangerous to use.

The Health and Safety at Work Act 1974, Section 8, places a duty on everyone not to intentionally or recklessly interfere with or misuse anything provided in the interests of health and safety. Simply put, it is therefore a criminal offence to disable or remove a safety guard from a machine.

Safety screens

Occasionally you may find yourself having to carry out work in the back of a live control panel. This may be for operational reasons. In such cases it may not have been possible to safely isolate the supply. Risk assessment of the situation should be undertaken to note the hazards and dangers. Precautions should be identified to minimise those hazards.

Insulated safety screens can be placed between yourself and those live parts. This will considerably reduce the risk to you from working in this area. Insulated safety screens come in various sizes, ranging from only 500mm × 200mm to 2m high × 500mm wide.

Fences

Security on site is of paramount importance. It is essential to keep non-work personnel, particularly children and animals, out of the work site. The provision of fences is usually the responsibility of the main contractor. On some occasions it is possible that the electrical contractor takes on that role. In these cases they will have to provide and maintain security fencing around the site.

When working on the installation or refurbishment of areas within a working factory it is necessary to also provide fencing around the work area.

The type of fence needed is determined by the planned location of the fence. Modern galvanised steel fencing panels that link together can be erected and secured in place by concrete mounting blocks. These are easy to erect and are fairly portable.

Security fencing acts as a barrier to non-work personnel

Dealing with electric shock

On completion of this topic area the candidate will be able to describe the appropriate emergency action to be taken on discovery of someone receiving an electric shock.

We covered basic first aid procedures for those suffering electric shock at Level 2, so we will briefly recap the essentials here.

Electric shock occurs when a person becomes part of an electric circuit. The severity of shock will depend on the level of current, size of voltage and length of time in contact with the body. 50 mA is usually a lethal level.

On finding someone receiving an electric shock you must:

- check for your own safety
- isolate the electric supply
- get help and then begin carrying out basic resuscitation procedures, as covered at Level 2.

Remember

Basic resuscitation procedures are best remembered as the ABC of resuscitation: Airways, Breathing and Circulation

Remember

Be prepared to treat for shock at any time as there are a number of factors, other than electric shock, that may cause people to stop breathing

Precautions against electric shock

On completion of this topic area the candidate will be able to describe the precautions to be taken to reduce the risk of an electric shock while at work.

On average, 20 people are killed a year through receiving an electric shock at work, with a further 30 people receiving injuries, some of which are severe or permanent. Shocks from faulty equipment may also lead to falls from ladders, scaffolds or other work platforms. Those using electricity may not be the only ones at risk, as poor electrical installations and faulty electrical appliances can lead to fires. Yet most of these accidents can be avoided with careful planning and straightforward precautions.

Hazards

Main hazards include:

- contact with live parts causing shock and burns (mains voltage at 230 volts a.c. can kill)
- faults which could cause fires
- fire or explosion where electricity could be the source of ignition in a potentially flammable or explosive atmosphere, e.g. in a spray-paint booth.

Remember that the danger of electric shock can increase depending on the conditions where the installation is used. The harsher the conditions, the greater the risk of electric shock becomes. The dangers from harsh conditions and the precautions that can be taken to reduce the danger are covered at Level 2.

Reduce the voltage

On construction sites where the working conditions are dirty, damp and wet, people using electrically operated power tools are particularly at risk. To reduce the risk it makes sense to reduce the voltage to a level that cannot cause harm.

Research has shown that the body can withstand unlimited contact, without sustaining damage, up to 50 volts. Therefore it would make sense to make all power tools work at 50 volts. However, technology is not quite there yet and, for example, to get an electric drill to give the level of continued performance required, voltage higher than 50 volts is required.

A compromise position has been reached, whereby 110 volt equipment is supplied from a safety isolating transformer with the secondary side centre tapped to earth. In the event of an earth fault this would result in a voltage to earth of 55 volts, protecting the user from electric shock.

Did you know?

Each year about 1000 accidents at work involving electric shock or burns are reported to the Health and Safety Executive (HSE). Around 20 of these are fatal. Most of these fatalities arise from contact with overhead or underground power cables

However, the site offices, cloakrooms and canteens can still be supplied at 230 volts. Temporary lighting can be run at lower voltages, e.g. 12, 25, 50 or 110 volts, and should there be no provision for a 110 volt supply it is now possible to use battery-operated portable equipment.

BS EN 60309-2 plugs and sockets are colour-coded to represent the voltage

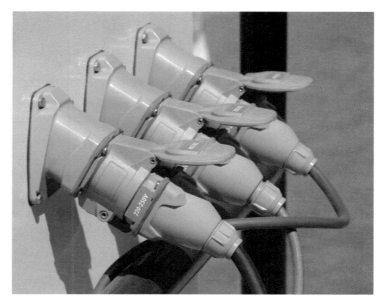

BS EN 60309-2 Plugs and sockets

they are connected to as follows:

- **Red** – 400 V
- **Blue** – 230 V
- **Yellow** – 110 V
- **White** – 50 V
- **Purple** – 25 V

The voltage can also be protected against through the use of a safety device, such as a **RCD** (*residual current device*). More information about RCDs can be found at Level 2.

Safety training

The Electricity at Work Regulations 1989, Regulation 16, requires that no person shall engage in any work activity that requires technical knowledge or experience to prevent danger of injury, unless he or she has that knowledge or experience, or is under appropriate supervision.

It therefore follows that for anybody working in the electro-technical industry they must have undertaken a programme of training. Any such training programme would not only cover the technical aspects but must also include safety training. This safety training should cover the wider aspects of working within the industry, such as manual handling, working heights and basic first aid, while also concentrating on the specialisms within the industry such as working with electricity, safe isolation, and the need to plan and follow safe systems of work.

Warning signs and notices

When working as an electrician it is essential to be able to recognise safety signs that are used in the working environment. Below are examples of warning signs that can be commonly found in the electro-technical industry. Other examples of warning signs can be found on page 14.

DANGER OF DEATH

CAUTION Disconnect electrical supply before working on this equipment

CAUTION 240 volts

CAUTION Burried cables

WARNING Uninterruptible power supply

CAUTION Isolator

DANGER Isolate before removing cover

DANGER Men working on equipment

Do not use

Figure 2.02 Warning signs found in the workplace

Isolation procedures

Safe isolation of electric supplies are covered in depth at Level 2. The safe isolation procedure drawn up by the Joint Industry Board for the Electrical Contracting Industry is as follows:

- identify sources of supply
- isolate
- secure the isolation
- test that the equipment/system is dead
- begin work.

Asphyxiation

On completion of this topic the candidate will be able to outline the causes of **asphyxiation** and the appropriate emergency action to be taken.

The possible causes of asphyxiation in the electro-technical industry are numerous. You must be aware that without taking appropriate precautions it is something that can affect you or others working around you, at any time.

Areas of risk

It would be impossible to list here all the different situations and chemicals you could be working with or near that may result in asphyxiation. We will focus on a couple of key situations.

Confined spaces

Earlier in this chapter (pages 44–46) we outlined the requirements of the Confined Space Regulations and listed the dangers.

Dangers of asphyxiation in confined spaces include working in areas such as the loft above a premises. Such an environment is usually poorly ventilated (if at all), which in itself could be sufficient to cause asphyxiation. However, if this is combined with the hot summer sun beating down on the roof, producing a substantial increase in body temperature of the person working in the space, the chances of asphyxiation increase substantially. If the person working in the loft was also working with PVC adhesive, as used for jointing PVC conduit, then the risk of asphyxiation would be multiplied.

Other work situations where asphyxiation could be a problem are spray booths in paint shops and when working in the petro-chemical industries. In these environments, toxic fumes and gases can be given off as part of the industrial process when several substances are mixed together. In a normal working environment these fumes and gases are controlled and are generally not a problem. However, in the event of an accident, gases can escape and cause asphyxiation. If working in these environments it is always advisable to:

- be aware of the processes
- be aware of the gases that could be given off
- be aware of the consequences
- be aware of the safety procedures to be adopted should an alarm be raised.

Of course, under the Management of Health and Safety at Work Regulations, risk assessment of the work and the work area should have been undertaken, the hazards identified and precautions taken to minimise or remove those hazards. It may be

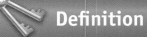

Definition

Asphyxia – definitions vary, from unconsciousness or death resulting from lack of oxygen to suffocation, a condition caused by insufficient intake of oxygen. Causes include choking, toxic gases, exhaust gases (principally carbon monoxide), electric shock, drugs, crushing injuries of the chest, diminished environmental oxygen and drowning

Remember

A confined space is a place which is substantially enclosed

that in the environments described on page 57 it is not possible to shut down the processing plant. If this is the case, then suitable precautions would need to be taken such as using the appropriate PPE including respiratory protective equipment.

Action

If you find someone you believe is suffering from the effects of fume inhalation or asphyxiation, you should do the following.

- Immediately check the area for your own safety – do not put yourself at risk. Remember, you will not be able to help anybody if you are unconscious.

- Open windows and doors if possible.

- Raise the alarm and get help.

- If it is safe to do so, check the casualty.

- Take the casualty out into the fresh air.

- Loosen clothing around the casualty's neck and chest so as not to impede breathing. If unconscious, proceed with resuscitation. Try not to inhale any of the breath that is exhaled by the casualty, as it would be possible for you to be overcome by the fumes in their lungs.

Dangerous occurrences and hazardous malfunctions

On completion of this topic area the candidate will be able to define what is meant by the terms 'dangerous occurrence' and 'hazardous malfunction'.

RIDDOR

RIDDOR stands for the **R**eporting of **I**njuries, **D**iseases and **D**angerous **O**ccurrences **R**egulations 1995. These Regulations came into force on 1 April 1996.

RIDDOR places a duty on employers and the self-employed to report some work-related accidents, diseases and dangerous occurrences. It applies to all work activities.

Dangerous occurrence and hazardous malfunction

If something happens which has not resulted in a reportable injury but which clearly could have done, it may be considered a **dangerous occurrence** and must be reported immediately. If an item of equipment was to fail in its function, and as a result have the potential to cause harm, then this would be defined as a **hazardous malfunction**.

Any such hazardous malfunction or dangerous occurrence must be reported immediately to the enforcing authority and followed up, within ten days, with a completed Accident Report Form (F.2508).

Did you know?

A number of people each year are killed or seriously injured in circumstances similar to those described here

Find out

What equipment and materials, if any, that are common to your working environment could cause you harm or bring about asphyxiation

Reportable dangerous occurrences are:

- collapse, overturning or failure of load-bearing parts of lifts and lifting equipment
- explosion, collapse or bursting of any closed vessel or associated pipework
- failure of any freight container in any of its load-bearing parts
- plant or equipment coming into contact with overhead power lines
- electrical short circuit or overload causing fire or explosion
- any unintentional explosion, misfire or failure of demolition
- an unintended collapse with projection of material beyond a site boundary
- accidental release of a biological agent likely to cause severe human illness
- failure of industrial radiography or irradiation equipment to de-energise or return to its safe position after the intended exposure period
- malfunction of breathing apparatus while in use or during testing immediately before use
- failure or endangering of diving equipment, the trapping of a diver, an explosion near a diver or an uncontrolled ascent
- collapse or partial collapse of a scaffold more than 5 metres high or erected near water where there could be a risk of drowning after a fall
- unintended collision of a train with any vehicle
- dangerous occurrence at a well (other than a water well)
- dangerous occurrence at a pipeline
- failure of any load-bearing fairground equipment, or derailment or unintended collision of cars or trains
- a road tanker carrying a dangerous substance overturning, suffering serious damage, catching fire or the substance being released.

The following dangerous occurrences are all reportable except in relation to offshore workplaces:

- unintended collapse of:
  - any building or structure under construction alteration or demolition where more than five tonnes of material falls;
  - a wall or floor in a place of work
  - any false-work
- explosion or fire causing suspension of normal work for over 24 hours
- sudden, uncontrolled release in a building of:
  - 100kg or more of a flammable liquid
  - 10kg or more of a flammable liquid above its boiling point
  - 10kg or more of a flammable gas
  - 500kg of these substances if the release is in the open air
- accidental release of any substance which may damage health.

Did you know?

The incident can be reported through the **Incident Contact Centre**, Caerphilly Business Park, Caerphilly CF83 3GG, telephone 0845 300 9923 (8.30am-5.00pm) or email: riddor@natbrit.com or visit the website (www.riddor.gov.uk)

Find out

Additional categories of dangerous occurrences apply to mines, quarries, relevant transport systems (railways etc.) and offshore workplaces. Find out what these are

Alarm procedures

At the end of this topic the candidate should be able to describe the procedures to be taken in the event of the sounding of an emergency alarm.

Wherever you may find yourself working, whether on a construction site or within an existing building, it is essential that you familiarise yourself with the evacuation procedures. After all, when the alarm sounds there is no point asking someone 'What do I do?' The person who knows will probably have already gone!

Evacuation system

The evacuation procedure may vary from place to place and may be determined by the physical size of the site to be evacuated. For example, in a relatively small building there may be one alarm whereas on a large site with several buildings the alarm may be zoned. Initially the alarm may be sounded in just one of the buildings or in a part of the building.

Escape routes

All escape routes should be clearly identified by the provision of adequate signage and emergency lighting along the route. Additional illuminated emergency exit signs should be placed over the doors and along the emergency route.

In places open to large numbers of people, where the construction of the building forms corridors and stairwells, fire doors should be installed. If these are electrically controlled then they are designed normally to be held open. However, in the event of a fire they will automatically close. This is not to hinder the movement of people but to prevent the spread of fire.

Figure 2.03 Typical exit direction signs

Within an area there may be more than one route indicated as the escape route. In order to discover the quickest route out of the building you should take time to familiarise yourself with the building and escape routes before you start work.

Assembly points

As with all the issues regarding fire safety, before starting work on any site or in any building, familiarising yourself with the evacuation procedures and assembly points can only serve to preserve your life during an emergency.

In the event of an evacuation it is necessary for it to be orderly and managed to avoid people getting hurt. There is little point in doing repeated fire drills if, after getting out of the building, people are just left milling around and getting in the way of the emergency services. It therefore follows that there should be designated areas where people can assemble. This would also allow for the checking of persons to find out if anyone is missing who might be trapped in the building. The number and positions of assembly points will be determined by the size of the building and the number of staff employed within the building. On large sites there may be several designated assembly points around the site; smaller buildings may have just one. They could be in the car park at the front or rear of the building or at some other convenient space away from the building.

Make sure you know your assembly point in the event of a fire

'Reporting in'

In order to prepare for a possible emergency, fire evacuation procedures need to be managed. Fire marshals need to be appointed, as they will have several functions to perform. These include:

- taking a roll call to ensure everyone is out of the building
- ensuring no one re-enters the building until the all clear is given
- making sure people do not smoke during the evacuation, as the emergency might be a gas leak, not a fire
- liaising with the emergency services
- sounding the all clear so people can return safely into the building.

Safety tip

Failure to report in in the event of an evacuation could result in a fire officer risking his or her life by entering the building to search for you while you are outside at the assembly point

Large sites may have several fire marshals and therefore a reporting hierarchy would need to be established, so that the emergency services do not receive conflicting reports on what happened. It is important to make sure that when you have evacuated the building you report to the fire marshal, either directly or indirectly, by reporting to your supervisor. They in turn will report to the fire marshal that you have safely exited the building.

Stay out

Once you have left the building under no circumstances re-enter until you have been given the all clear by the fire officer or fire marshal. To do so may **risk your life and that of others** who may come to search for you.

Fire safety

On completion of this topic the candidate will be able to describe methods of fire prevention and controlling fires.

The basics of fire and fire fighting-equipment are covered at Level 2. This section will concentrate on legislation affecting fire safety and how to work to prevent fire.

Fire legislation

The current statutes, which fire authorities are given direct responsibility to enforce, are covered under the Regulatory Reform (Fire Safety) Order 2005.

Legislation covering fire safety in the UK went through a dramatic change in October 2006. The Fire Precautions Act 1971 and the Fire Precautions (Workplace) Regulations 1997 were repealed, along with many other fire safety regulations embedded in other statutes. Fire certificates were abolished and employers will become solely responsible for fire safety within their workplaces.

The Regulatory Reform (Fire Safety) Order 2005 is largely based on the Fire Precautions (Workplace) Regulations 1997 and the Dangerous Substances and Explosive Atmosphere Regulations 2002 (DSEAR). The Order applies to all workplaces/premises, and to the self-employed, with only a few minor exceptions. The fire risk assessment element of the Fire Precautions (Workplace) Regulations remains and additional duties include:

- the duty to prevent fire spreading
- the duty to maintain building regulations standards for the use and protection of the fire service
- the duty to appoint one or more employees to assist in ensuring compliance with the Regulations (such as a fire marshal).

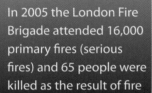

Remember

On smaller sites the role of fire marshal will probably be taken by one person, whatever the situation. Make sure you know who this is

Did you know?

In 2005 the London Fire Brigade attended 16,000 primary fires (serious fires) and 65 people were killed as the result of fire

Did you know?

The new fire regulations state that fire-fighting equipment in the workplace is there only to enable a small fire to be attacked to prevent it from spreading

The purpose of fire-fighting equipment is also clarified, and the order states there should be a named responsible person (normally the employer) for fire safety. As with the Health and Safety at Work Act 1974, it is up to the responsible person to demonstrate that he or she did everything 'reasonably practicable' to prevent injury, and there will be a civil liability if a breach of duty causes harm.

The fire service's role is now that of enforcement, similar to the role of the HSE for general health and safety matters.

Fire prevention

Fires can spread rapidly. Once established, even a small fire can generate sufficient heat energy to spread and accelerate the fire to surrounding combustible materials. Fire prevention is largely a matter of commonsense and good housekeeping. For example, keep the workplace clean and tidy. If you smoke, don't throw lit cigarettes on to the ground, and don't leave flammable materials lying around or near sources of heat or sparks.

From an electrical perspective, make sure that all leads are in good condition, that fuses are of the correct rating and that circuits are not overloaded. Any alterations or repairs to electrical installations must be carried out only by qualified personnel and must be to the standards laid down in the IEE Regulations (BS 7671).

Fire fighting

A fire safety officer once said that the only use people should think of for a portable fire extinguisher 'was to break the window so that you could escape from the building'. The point he was making is that it is dangerous to try to fight a fire, and the use of fire extinguishers should only be considered as a first-response measure, e.g. where the fire is very small or where it is blocking your only means of exit. Fire fighting is a job for the professional emergency service.

Did you know?

In parts of the United States where sprinklers have become compulsory, almost no one dies from fire at home. Sprinklers are fitted in as many rooms as required, fed by small pipes and run off the mains water

Sprinkler systems

To reduce, as far as you possibly can, the risk of dying in a fire, you should have fire sprinklers installed.

Sprinkler system

Sprinklers can be individually heat-activated, so the whole system doesn't go off at once. They rarely get set off accidentally as they need high temperatures to trigger them. They operate automatically whether you are in the building or not. Sprinklers should sound the alarm when they go off – so they alert you and also tackle the fire.

Smoke alarms

Smoke alarms will alert you to slow-burning, smoke-generating fires that may not create enough heat to trigger a sprinkler.

Fire drills

Regular fire drills must be held, and all personnel must be familiar with normal and alternative escape routes. Fire routes should be clearly marked and emergency lighting signs fitted above each exit, where applicable. Make sure you know where your assembly point is. Remember:

- if you discover a fire, raise the alarm immediately
- leave by the nearest exit
- call the fire service to deal with the fire
- if in doubt stay out
- close but don't lock the windows to help starve the fire of oxygen.

Some of the above duties may be allocated to a particular person.

FAQ

Q Do I have to fill in a permit to work form for every job?

A Not usually. A permit to work form is only used if there are dangers in the workplace which must be controlled before or during the work.

Q Do I have to 'lock off' a supply before working on it?

A Yes. It is very much in your interest to ensure that any equipment you are working on cannot inadvertently become live again.

Q My boss insists on keeping a spare key to my padlock 'just in case'. Is this a good idea?

A No. A padlock used for locking off supplies should have only one key.

Q An engineer always locks the supplies off for me and gives me a permit to work. Should I check it?

A Yes. Never take someone's word for it that the correct circuit or installation has been effectively isolated. Check it yourself – it's your life!

Q The safety guard on my machine gets in the way. Can I remove it?

A No. All safety guards should be fitted and properly adjusted whenever the machine is working. If you need to remove the guard to work on the machine you must ensure the machine is switched off and cannot be inadvertently re-started.

Q Is a voltage of less than 50 V safe?

A No. An electric shock of less than 50 V will not cause you any permanent harm but any electrical supply may cause fire, burns, arcing, explosions or mechanical movement, which may cause injury or even death.

Activity

1. Under supervision, carry out a safe isolation procedure on a circuit or installation.

Knowledge check

1. Give three examples of the type of activity which may require the issue of a permit to work.

2. List the dangers of working in a confined space.

3. Why is it important to use barrier cream when working with greasy substances?

4. State four of the hazards associated with electricity.

5. Describe what is meant by a 'competent person'.

6. List the five main steps of a safe isolation procedure, according to the Joint Industry Board.

chapter 3

Effective working practices

Unit 1 outcome 3

Within all careers it is essential that workers use the most effective working practices. This is no different in the electro-technical sector. As we have seen in earlier chapters, by working in an organised manner we can improve both safety and profitability for the business. Good teamworking and thorough technical knowledge makes better electricians.

However, working practices are also there to be improved, and we should always look to improve the skills we have. There are many opportunities for further training and self-improvement within the industry. Never become complacent with the skills you have – always be assessing and looking to improve!

This chapter builds on concepts first explored in chapters 1 and 2 and at Level 2.

On completion of this chapter the candidate will be able to:

- identify the relation between learning and personal performance, and outline the career patterns and training opportunities in the electro-technical sector

- explain the features of teamworking

- explain employment legislation in the electro-technical sector in terms of rights and responsibilities of the individual

- explain best working practice for carrying out electro-technical tasks safely and efficiently

- explain the standards for assessing working practices

- outline the benefits of approving working practices.

Learning and performance

On completion of this topic area the candidate will be able to identify the relationship between personal performance and learning, and outline the education and training opportunities for entry, promotion and transfer within the electro-technical sector.

The electro-technical sector

The electro-technical sector is no different from any other sector of industry. Competitiveness and performance are key to the success of any organisation. The same can be said at a personal level. If an individual is to be successful, he or she will need to be both good performers and competitive. Both of these go hand in hand with a desire for lifelong learning. Studies have shown that, in the vast majority of cases, people who embrace the idea of continued learning are those who go on to be successful workers and some of the best earners.

Like many craft-based industries, the electro-technical sector is one in which the person at the top of the organisation frequently often started at the bottom. For most small- and medium-sized businesses this is almost certainly always the case. Larger organisations often use graduate employees in some of the higher positions.

We looked at career progression in the industry at Level 2. Other qualifications that are appropriate to the electro-technical industry are:

- City and Guilds 2392: Certificate in Fundamentals of Inspection Testing and Initial Verification
- City and Guilds 2391: Inspection and Testing of Installations, and the Design and Verification of Installations
- City and Guilds 2382: Certificate in the Requirements for Electrical Installations.

City and Guilds 2330 is designed to support and help career progression within the electro-technical industry, and can be applied to six occupational pathways:

- Level 3 certificate in electro-technical technology – installation (buildings and structures)
- Level 3 certificate in electro-technical technology – electrical maintenance
- Level 3 certificate in electro-technical technology – instrumentation and associated equipment
- Level 3 certificate in electro-technical technology – high wire electrical systems and associated equipment
- Level 3 certificate in electro-technical technology – electrotechnical panel building
- Level 3 certificate in electro-technical technology – electrical machines, rewind and repair.

In theory an electro-technical operative who has the full suite of 2330 qualifications would be able to work anywhere in the electro-technical industry!

Teamwork

On completion of this topic area the candidate will be able to explain the features of teamworking.

Promoting good relationships with fellow workers

A site where everyone works together to finish the job is much happier and more productive than one where people are at loggerheads. Below is a checklist of the things you can do to help make things run smoothly.

- Co-operate with other trades – it's always better than conflict. ✔
- Be patient and tolerant with others. ✔
- Attend site meetings regularly – this helps liaison with other trades. ✔
- Keep to the agreed work programme. ✔
- Do your work in a professional manner. ✔
- Finish your work on time; don't hold others up if you can help it. ✔
- Respond cheerfully to reasonable requests from colleagues. ✔
- Don't leave the site for long periods of time. ✔
- Don't borrow tools and materials unless it is necessary. If you do borrow items return them promptly and in good condition. ✔
- Tell your employer if you have personal or work difficulties – don't be too proud. ✔
- Take good care of your own and others' property. ✔

- Keep noise down, especially from your radio. ✔
- Show respect for everyone on site – make an effort to learn people's names. ✔
- Make sure everyone, including visitors to the site, has the right PPE. ✔
- Report any breakdown in discipline or disputes between co-contractors promptly to the site supervisor. ✔
- Keep a current edition of the *Wiring Regulations* or the *Amicus* guide book with you on site. ✔
- Always do your best to answer questions from visitors or other tradespeople. ✔
- Never play practical jokes on colleagues (for example, hiding tools, lunch boxes, car keys). This can cause bad feeling and may result in injury or accident. ✔

Teamworking

Successful projects require good teamwork, but who exactly is 'the team'? In one sense, everyone working on the project is part of the team. More commonly, though, contractors will see their own group of people as a team, who will be working with other teams of tradespeople to complete the whole project.

Working well with others is all part of the job

Here is one way to look at the development of a team.

1. Forming.

2. Storming.

3. Norming.

4. Performing.

The model gives useful insights whether we understand a team as being just the squad of electricians or everyone working on site. As a team develops, relationships between team members shift and the team leader changes leadership style. As you read through the four stages of team development described on page 71, try to apply the model to teams you have been involved with: for example, a gang of mates at school or a sports team.

Team development

1. Forming

This is the initial stage, when the team has just come together. The members probably don't know one another very well and individual roles and responsibilities are unclear. There is little agreement between members about what the team is trying to do. Some will feel confused and won't know what they should be doing. At this stage the team relies heavily on the team leader for guidance and direction. The leader must provide lots of answers about the team's purpose and objectives and relationships with groups outside the team.

2. Storming

During this stage, team members jockey for position as they try to find themselves a role in the team. The leader might receive challenges to his or her authority from other team members. The team's purpose becomes clearer, but there remains plenty of uncertainty and decisions are hard to achieve because members argue a lot. Small groups or factions may form and there may be power struggles. The team needs to be focused on its goals to avoid being broken up by relationship and emotional issues. Some compromises will be needed to make any progress. The leader has to become less bossy and more of a coach.

3. Norming

This is a more peaceful stage when team members generally reach agreements easily. Roles and responsibilities are clear and accepted by all, and the team works together. Members develop ways and styles of working by discussion and agreement. Big decisions are made by the whole team or the team leader, but smaller decisions are left to individuals or small groups inside the team. The team may also enjoy fun and social activities together. The leader now acts to guide the team gently and enable it to do its job, and has no need to enforce decisions. The team may share some leadership roles.

4. Performing

In this final stage the team knows clearly what it is doing and why. It has a shared vision and needs little or no input from the leader. If disagreements occur they are tackled positively by the team itself. The team works together towards achieving the goal and copes with relationship, style and process issues along the way. Team members look after each other. The leader's role is to delegate and oversee tasks and projects, and there is no need for instruction or assistance, except for individual personal development.

Employment legislation

On completion of this topic area the candidate will be able to explain employment legislation within the electro-technical industry in terms of the rights and responsibilities of personnel.

Employment legislation within the electro-technical industry

Within England and Wales, the law regarding employment both protects and imposes obligations on employees during their employment and after it ends. The law sets certain minimum rights and an employer cannot give you less than what the law stipulates. The principal rights and obligations imposed on employers and employees arise from three sources:

- common law, which governs any contract of employment between employer and employee and includes the body of law created by historical practice and decisions

- UK legislation

- European legislation and judgements from the European Court of Justice (ECJ).

UK employment law has been heavily influenced by European law, particularly in the areas of equal pay and equal treatment. Consequently, many of our statutory minimum rights began their life in European legislation.

> **Definition**
>
> **Disability** – a person has a disability if he or she has a physical or mental impairment that has a substantial and long-term effect on his or her ability to carry out normal everyday activities

Employment Rights Act 1996

Subject to certain qualifications, employees have a number of statutory minimum rights (e.g. the right to a minimum wage) and the main vehicle for employment legislation is the Employment Rights Act 1996 – Chapter 18. If you did not agree certain matters at the time of commencing employment, your legal rights will apply automatically. The Employment Rights Act 1996, deals with many matters such as: right to statement of employment, right to pay statement, minimum pay, minimum holidays, maximum working hours and right to maternity/paternity leave. A 2002 amendment to this Act made provision for statutory rights to paternity and adoption leave.

Sex Discrimination Act 1975

The Sex Discrimination Act 1975 makes discrimination unlawful on the grounds of sex and marital status and, to a certain degree, gender reassignment. The Act originated out of the Equal Treatment Directive, which made provisions for equality between men and women in terms of access to employment, vocational training, promotion and other terms and conditions of work.

The Equal Opportunities Commission (EOC) has since published a Code of Practice. While this is not a legally binding document, it does gives guidance on best practice in the promotion of equality of opportunity in employment, and failure to follow it may be taken into account by the courts.

Data Protection Act 1998

The Information Commissioner enforces and oversees the Data Protection Act 1998 and the Freedom of Information Act 2000. The Commissioner is an independent supervisory authority reporting directly to the UK Parliament and has an international role as well as a national one.

The Data Protection Act 1998 outlines key principles to make sure that information is handled properly. It states that data must be: fairly and lawfully processed for limited purposes, not kept for longer than necessary, processed in line with individuals' rights, not transferred to other countries without adequate protection, secure, accurate, adequate, relevant and not excessive. All data controllers must keep to these principles.

Human Rights Act 1998

The Human Rights Act 1998 covers many different types of discrimination, including some which are not covered by other discrimination laws. However, it can be used only when one of the other 'articles' (the specific principles) of the Act applies, such as the right to 'respect for private and family life'.

Furthermore, rights under the Act can only be used against a public authority (for example the police, a local council or Jobcentre) and not a private company. However, court decisions on discrimination will generally have to take into account what the Human Rights Act says. The Act guards rights to liberty and security, prohibits discrimination, and protects freedoms of speech, expression and assembly.

Race Relations Act 1976 and Amendment Act 2000

When originally passed, the Race Relations Act 1976 made it unlawful to discriminate on racial grounds in relation to employment, training and education, the provision of goods, facilities and services, and certain other specified activities. The 1976 Act applied to race discrimination by public authorities in these areas but not all functions of public authorities were covered.

The 1976 Act also made employers vicariously (explicitly) liable for acts of race discrimination committed by their employees in the course of their employment, subject to a defence that the employer took all reasonable steps to prevent the employee discriminating. The 2000 Act extended this by outlawing race discrimination in functions not previously covered and placed a duty on public authorities to eliminate unlawful discrimination.

Disability Discrimination Act 1995

The Disability Discrimination Act tackles the discrimination faced by many people with disabilities. This Act gives disabled people rights in the areas of employment, access to goods, facilities and services and buying or renting land or property. The employment rights and first rights of access came into force in December 1996; further rights of access came into force on 1 October 1999; the final rights of access came into force in October 2004.

In addition, this Act allows the government to set minimum standards so that disabled people can use public transport easily.

Types of discrimination and victimisation

Much of this legislation deals with discrimination. Therefore, it is useful to quickly outline the nature of different types of discrimination. There are two main forms of discrimination – direct and indirect.

Direct discrimination

Direct discrimination occurs when someone is treated less favourably because of their sex, race or disability. It is tested by comparing how someone of an alternative sex or racial group or who is not disabled was treated in the same circumstances.

An example of direct discrimination could be a woman of superior qualifications and experience being denied promotion in favour of a less experienced and less qualified man.

Indirect discrimination

Indirect discrimination occurs where the effect of certain requirements, conditions or practices imposed by an employer have an adverse impact disproportionately on one group or other. Courts tend to consider three factors.

1. Whether the number of people from a racial group or of one sex that can meet the job criteria is considerably smaller than the rest of the population.

2. Whether the criteria cannot actually be justified by the employer as being a real requirement of the job, i.e. an applicant who could not meet the criteria could still do the job as well as anyone else.

3. Whether, because the person cannot comply with these criteria, he or she has actually suffered in some way (this may seem obvious but a person cannot complain unless he or she believes that they have lost out in some way).

With cases of indirect discrimination, employers may argue that discrimination is required for the job. For example, one individual claimed indirect discrimination on religious grounds against his employer as he was requested to shave off his beard. The court agreed that discrimination had been applied, but as the employer was a factory involved in food preparation the particular case was rejected on the grounds of hygiene.

Positive discrimination and positive action

Positive discrimination occurs when someone is selected to do a job purely on the basis of their gender or race, not on their ability to do the job. This is illegal under the Sex Discrimination Act and the Race Relations Act, and is generally unlawful other than for **genuine occupational requirements**.

Positive action is activity to increase the numbers of men, women or minority ethnic groups in a workforce where they have been shown through monitoring to be under-represented. This may be in proportion to the total employed by the employer or in relation to the profile of the local population.

Did you know?

An employer cannot argue it was not their intention to discriminate; the law only considers the end effect

Definition

Genuine occupational requirements – where an employer can demonstrate that there is a genuine identified need for someone of specific race, gender etc. to the exclusion of other races, genders etc.

Project management

On completion of this topic area the candidate will be able to explain how to carry out electro-technical tasks safely and efficiently.

Every project is unique but many tasks are common to them all. If a project is to be completed successfully, safely and efficiently, it will require planning. The project manager will, with the help of others, create a **work plan**.

Typically, a work plan will include the following activities:

Definition

Work plan – a strategy that can be applied to a whole project (or part of one) to ensure that the installation is carried out safely and efficiently

- checking the drawings, instructions and specifications; it is a waste of time, money and materials if unrequired work is completed

- checking that the work area and environment are suitable and safe at all times

- creating a logical sequence for all the activities and identifying the tasks to be done

- listing the tools, materials and equipment needed for the project, and making sure they are available on site at the right time

- finding out what skills are required – whether areas of work require specialist knowledge or ability

- co-ordinating with other contractors, preparing programmes of work and managing the installation process

- avoiding clashes with other trades and ensuring key dates are met

- making sure that the site, workforce and installation comply with all appropriate safety legislation and codes of practice

- ensuring the project has been completed satisfactorily to customer's specified requirements and to BS 7671 through the process of commissioning and by inspection, testing and certification.

Depending on the size of the project, one person or several may have responsibility for devising and monitoring the work plan. On a large project, contract managers and engineers, project engineers, safety officers and site supervisors may all be involved. For a small project, many of these tasks (if not all) may fall to the electrician on site. This might include producing drawings and specifications.

Quality management systems

On completion of this topic area the candidate will be able to explain the standards for assessing working practices and procedures.

Quality systems

Although the legislation we have looked at in chapters 1 and 2 places a legal requirement on both employer and employee, none of it reflects the quality of an employer. There are two commonly used standards that help to signify the quality of a company: Investors in People and ISO 9001.

Investors in People

Investors in People (IiP) is a national quality standard which sets a level of good practice for improving the performance of an organisation through its people. Developed in 1990, the standard sets out a level of good practice for training and development of people to achieve business goals.

The IiP standard provides a national framework for improving business performance and competitiveness through a planned approach to setting and communicating business objectives and then developing people to meet these objectives. The aim is to create an environment where what people can do, and are motivated to do, matches what the organisation needs them to do. Because the award of IiP status is time restricted, the process should bring about a culture of continuous improvement.

The IiP standard is based on four key principles:

- commitment to investing in people to achieve business goals
- planning how skills, individuals and teams are to be developed to achieve these goals
- taking action to develop and use necessary skills in a well-defined and continuous programme directly tied to business objectives
- evaluating the outcomes of training and development for an individual's progress towards goals, the value achieved and future needs.

ISO 9001

Customers are becoming better informed and their expectations are growing. For any business, the only way to keep up is to offer a commitment to quality. In fact, any organisation, whatever its size or industry sector, can introduce a quality management system such as **ISO 9001**.

The system should ensure consistency and improvement of working practices, which in turn should provide products and services that meet customers' requirements. ISO 9001 is the most commonly used international standard that provides a framework for an effective quality management system.

The benefits of implementing a quality management system include:

- policies and objectives set by top management
- understanding customers' requirements with a view to achieving customer satisfaction
- improved internal and external communications
- greater understanding of the organisation's processes
- understanding how statutory and regulatory requirements impact on the organisation and its customers
- clear responsibilities and authorities agreed for all staff
- improved use of time and resources
- reduced wastage
- greater consistency of products and services
- improved staff morale and motivation.

Revised in December 2000, the existing three standards within ISO 9000 (ISO 9001, ISO 9002 and ISO 9003) were merged into a single standard, ISO 9001. Consequently, the ISO 9000 suite of standards was restructured to comprise four core standards:

- ISO 9000 Concepts and Terminology
- ISO 9001 Requirements for Quality Assurance
- ISO 9004 Guidelines for Quality Management of Organisations
- ISO 9011 Guidelines for Auditing Quality Management Systems (formerly ISO 10011).

To be registered a quality management system must meet the requirements set out in ISO 9001. All the other standards in the system then exist to help an organisation implement an effective quality management system and ultimately help gain approval.

ISO 9001:2000, the requirement standard, now includes the following main sections:

- Quality Management System
- Management Responsibility
- Resource Management
- Product Realisation
- Measurement Analysis and Improvement.

Benefits of improving working practices and procedures

On completion of this topic area the candidate will be able to outline the benefits of improving working practices and procedures.

A quality management system in accordance with ISO 9001:2000 will provide an organisation with a set of processes that ensure a common sense approach to the management of the organisation.

The system is internally audited in a year-on-year rolling programme. This ensures that the system is working effectively. Once a year an external audit is conducted by an inspector qualified to assess compliance with ISO 9001. The approval of the inspector will enable the organisation to continue to display the quality management ISO 9001 logo. This method of continual assessment ensures consistency and improvement of working practices, which in turn should provide products and services that benefit both the customer and the organisation.

The benefits of improving working practices and procedures include:

- increased customer satisfaction, through greater understanding of requirements and consistency of product
- improved productivity and clearer responsibilities
- more efficient use of resources and communication
- reduced wastage and therefore increased profitability.

FAQ

Q I don't feel part of the team. What can I do?

A Don't worry, it usually takes time before new people are fully accepted as part of the team. Just be your normal, chatty, helpful, kind, honest self and I'm sure things will work out just fine!

Q I think I've been discriminated against. What can I do?

A In the first instance, you need to talk to your employer and explain why you think you've been discriminated against. It may have been completely unintentional and the employer may not have realised that their actions were discriminatory. If the discrimination continues you may need to consult your union representative, if you have one. In extreme cases, you may have to take your case to an employment tribunal.

Q Why do I have to use a work plan? I'm not in it for the paperwork.

A Unfortunately, paperwork is a necessary evil! Constructing a building is a very complex task and people need to know what they are supposed to be doing and when.

Q My company is an 'Investor in People'. What does that mean?

A Most companies rely absolutely on having staff who know what they are doing. Investors in People is an award given to companies that value their staff and provide training to keep them up to date. You should always tell your company if there are training courses that you think would help you to do your job better.

Activity

1. Find out if your company has any written policies of non-discrimination, ethical practices or staff development.

Knowledge check

1. List the three main City and Guilds qualifications likely to be held by a JIB Approved Electrician.

2. State the four phases that a developing team will go through and describe what happens in each.

3. Which piece of legislation makes it illegal to discriminate against a person because of the colour of their skin?

4. Under what circumstances can a company apply 'positive discrimination' to an employee or a prospective employee?

5. List and describe what activities a work plan involves.

6. List the four key principles of the Investors in People scheme.

chapter 4

Electrical systems and components

Unit 1 outcome 4

As an electrician, it is essential that your technical knowledge is as advanced as possible. At Level 2 (pages 91–118) we covered many of the principles of basic electron theory and the concepts behind electro-technical work. Now we will focus on the application of these theories in a practical setting.

This chapter builds on concepts first explored at Level 2.

On completion of this chapter the candidate will be able to:

- describe resistors, by defining resistance and Ohm's law

- describe magnetism and magnetic circuits by defining magnetic fields and magnetic flux

- describe inductance and inductive components

- describe capacitors by defining capacitance

- state the effects of resistance, inductance, capacitance and impedance in a.c. circuits

- describe semiconductor devices in rectifier circuits and transistors

- describe basic electronic circuits and components

- describe luminaire components.

Resistance

At Level 2 we looked at resistors in circuits. On completion of this topic area the candidate will have an understanding of resistance and how it may be calculated.

Electron flow

At Level 2 we considered the amount of electrons flowing in a conductor every second and the force that pushes them along. But does anything ever interfere with this flow?

Well, yes it does. All materials offer a certain amount of resistance to the flow of current for a given size. Think of it as a sort of 'friction' effect in which the electrons moving through the material generate heat – a bit like rubbing your hands together.

Some materials have a very low resistance to the flow of electric current. They can transmit electrical energy around a building or across a large distance without losing a lot to the atmosphere in the form of waste heat. Metals such as copper, aluminium, gold and silver are good conductors of electricity and are used for electrical cables and connectors.

The resistance of most materials is affected by temperature and most metals have what is known as a 'positive temperature coefficient of resistance', which means that as the temperature goes up, the resistance increases. Of course the opposite is also true; as the temperature decreases, the resistance will reduce. At about –237°C (or absolute zero), the theoretical resistance would be zero.

Don't forget there is a place for materials with a high electrical resistance. Because their resistance is high it is very difficult for electric current to flow through them which makes them very good insulators of electricity.

Most cables are a combination of both conductors and insulators. Conductors provide a low resistance path to allow the current to flow freely around the circuit and insulators stop the electricity leaking out of the cable.

There is a scientific law that we can apply to resistance. This is Ohm's law. Ohm's law was introduced at Level 2. However, we will now recap briefly on the principles behind this law.

Did you know?

Conductors with zero resistance would result in significant savings in power losses in overhead transmission lines but, unfortunately, you use more energy cooling the conductors than you save in power losses!

Ohm's law

So far we have established that **current** is the amount of electrons flowing past a point every second in a conductor and that a force known as the electromotive force (e.m.f. or **voltage**) is pushing them. We now also know that even good conductors will try to oppose the current, by offering a **resistance** to the flow of electrons.

Ohm's law describes the link between **current** in a conductor, the force (**e.m.f.** or **voltage**) that pushes electrons around the conductor and the **electrical resistance** to this flow.

In simple language we could write Ohm's Law as follows:

The amount of electrons passing by every second will depend upon how hard we push them, and what obstacles are put in their way.

We can prove this is true, because if we increase the voltage (push harder), then we must increase the number of electrons that we can get out at the other end. Try flicking a coin along the desk. The harder you flick it, the further it travels along the desk. This is what we mean by **directly proportional**. If one thing goes up (voltage), then so will the other thing (current). Now place some obstacles in the way of the coin (this is the resistance). Even though it is flicked with the same strength, it does not go as far. This is **indirectly proportional** – one thing has gone up (resistance) but the other has gone down (current).

Ohm's law can be expressed by the following formula:

$$\text{Current (I)} = \frac{\text{Voltage (V)}}{\text{Resistance (R)}}$$

Remember

Remember Ohm's law,

$$I = \frac{V}{R}$$

Resistivity
Conductor resistance

Electrons find it easier to move along some materials than others, and each material has its own resistance to the electron flow. This individual material resistance is **resistivity**, represented by the Greek symbol rho (ρ) and measured in ohm metres (Ωm).

To summarise: the amount of electrons that can flow along a conductor will be affected by how far they have to travel, what material they have to travel through and how big the object is that they are travelling along.

Think about it: Would you rather run for 100 metres or 25 miles? Which is easier: to walk along a 3 metre high corridor, or to crawl along a 1 metre high pipe on your stomach? The length (or distance) you have to travel and the cross-sectional area over which you travel will affect the speed of your journey.

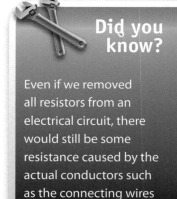

Did you know?

Even if we removed all resistors from an electrical circuit, there would still be some resistance caused by the actual conductors such as the connecting wires

As an electrical formula, this is expressed as follows:

$$\text{Resistance} = \frac{\text{Resistivity} \times \text{Length}}{\text{Cross-sectional area}}$$

or

$$R = \frac{\rho \times L}{A}$$

We find the value of resistivity for each material by first measuring the resistance of 1 metre cube of the material. Then, as cable dimensions are measured in square millimetres (i.e. 2.5mm^2), this figure is divided down to give the value of a 1 millimetre cube.

This resistivity, as we found out earlier, is given in Ωmm, or in other words we will encounter a resistance of so many ohms for metre forward that we travel through the conductor. The accepted value for copper is 17.8 $\mu\Omega$mm; the accepted value for aluminium is 28.5 $\mu\Omega$mm.

Let us now look at some typical questions involving resistivity.

Example 1

Find the resistance of the field coil of a motor where the conductor cross sectional area (csa) is 2mm², the length of wire is 4000m and the material resistivity is 18 $\mu\Omega$mm.

$$R = \frac{\rho \times L}{A}$$

Problem 1   **Problem 2**

$$R = \frac{18}{1,000,000} \times \frac{4,000,000}{2}$$

What has happened here?

Problem 1: The value of ρ is given in millionths of an ohm millimetre. 18 $\mu\Omega$mm, is 18 millionths of an ohm millimetre and we therefore write it as 18 divided by one million, or:

$$\frac{18}{1,000,000}$$

Problem 2: Remember, when doing calculations all units should be the same. The length is in metres but everything else is in millimetres. Therefore, as all units must be the same and there are 1000mm in a metre, 4000m should be expressed as 4,000,000mm.

So back to the calculation:

$$R = \frac{18}{1,000,000} \times \frac{4,000,000}{2}$$

Therefore:

$$R = \frac{72,000,000}{2,000,000} = 36 \ \Omega$$

Another way of doing this calculation, **without the calculator**, would have been to cancel the zeros down (division):

$$R = \frac{18}{1,000,000} \times \frac{4,000,000}{2}$$

Leaving us with:

$$R = \frac{18 \times 4}{2} = \mathbf{36 \ \Omega}$$

Remember

All measurements should be of the same type, i.e. metres or millimetres. In these examples metres have been used, hence the csa is multiplied by 10^3 and the resistivity is multiplied by 10^{-6}. This ensures that all answers will be in the same units

Example 2

A copper conductor has a resistivity of 17.8 $\mu\Omega$mm and a csa of 2.5mm². What will be the resistance of a 30m length of this conductor?

If: Then: So:

$$R = \frac{\rho \times L}{A}$$ $$R = \frac{17.8 \times 30 \times 10^{-9}}{2.5 \times 10^{-6}}$$ $$R = 0.2136\ \Omega$$

Example 3

A copper conductor has a resistivity of 17.8 $\mu\Omega$mm and is 1.785mm in diameter. What will be the resistance of a 75m length of this conductor?

To enable this question to be answered, we must first convert the diameter into the csa. This is carried out by using one of the following formulas, which you may remember from your school days.

(a) $csa = \dfrac{\pi d^2}{4}$ Where d = diameter

Or:

(b) $csa = \pi r^2$ Where r = radius and π = **3.142**

Using the first formula:

(a) $csa = \dfrac{\pi d^2}{4}$

Step 1: Put in the correct values:

$$csa = \frac{3.142 \times 1.785 \times 1.785}{4}$$

Step 2: Multiply the top line:

$$csa = \frac{10.01}{4}$$

Step 3: Divide by 4:

$$csa = \textbf{2.5mm}^2$$

Therefore, using this method, the csa is 2.5mm².

Using the second formula:

$$csa = \pi r^2$$

Step 1: Put in the correct values:

$$csa = 3.142 \times 0.8925 \times 0.8925$$

Step 2: Multiply out:

$$csa = 2.5mm^2$$

Using the second method, the csa is still $2.5mm^2$. We can now proceed with the example:

$$R = \frac{\rho \times L}{A}$$

Step 1: Put in the correct values:

$$R = \frac{17.8 \times 10^{-6} \times 75 \times 10^3}{2.5}$$

Step 2: Calculate out the top line:

$$R = \frac{17.8 \times 10^{-3} \times 75}{2.5}$$

Which is the same as:

$$R = \frac{17.8 \times 75}{2.5 \times 10^3}$$

So:

$$R = \frac{1335}{2500}$$

Therefore: **R = 0.534 Ω**

Magnetism and electromagnetism

On completion of this topic area the candidate will be able to describe magnetism and magnetic circuits.

Magnetism

We covered the basic principles of magnetism and magnetic circuits at Level 2. There we saw that magnetism is a fundamental force (like gravity) and arises due to the movement of electrical charge. Materials attracted to magnets, such as iron, steel, nickel and cobalt, also have the ability to become magnetised and are known as magnetic materials.

A permanent magnet is a material inserted into a strong magnetic field, which then exhibits a magnetic field of its own even when removed from the original field. This allows it to exert a force on other magnetic materials. Only a change in environment (temperature, de-magnetising field etc.) can have an impact on the strength of the magnet. The ability of the magnet to withstand different environments helps define its capabilities and applications.

Magnetic fields are created from two atomic sources: the spin and orbital motions of electrons. The magnetic field looks like a series of closed loops that start at one end (pole) of the magnet, arrive at the other and then pass through the magnet to the original start point.

Each one of these lines is called a line of magnetic flux and has the following properties:

- they will never cross but may become distorted
- they will always try to return to their original shape
- they will always form a closed loop
- outside the magnet they run north to south
- the higher the number of lines of magnetic flux, the stronger the magnet.

Magnetic flux

If we counted the number of lines, we would establish the magnetic flux (measured in webers or Wb). The more lines there are, the stronger the magnet. This number of lines is called the **flux density** (measured in teslas, or T – the number of webers per square metre).

We use the following formula to express this:

$$\text{Flux density B (tesla)} = \frac{\text{magnetic flux (webers)}}{\text{CSA}} = \frac{\Phi}{A} \, \text{Wb/m}^2$$

Of course in reality these lines do not exist – we just use the lines to assist our understanding. How do we know they are there? If a piece of paper is placed over a bar magnet and iron filings are sprinkled on to the paper, the filings will fall into line with the magnetic flux.

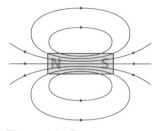

Figure 4.01 Bar magnet

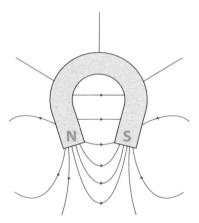

Figure 4.02 Horseshoe magnet

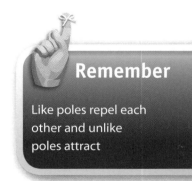

Remember

Like poles repel each other and unlike poles attract

Here is an example to help you understand what is going on. The field pole of a motor has an area of 5cm² and carries a flux of 80 μWb. What will the flux density be? Remembering the formula:

$$\text{Flux density B (tesla)} = \frac{\text{magnetic flux (webers)}}{\text{CSA}} = \frac{\Phi}{A} \text{ Webers/m}^2$$

We need to make allowance for the area being given in cm² and the flux in μWb. Therefore:

$$\text{Flux density B (tesla)} = \frac{80 \times 10^{-6} \text{ Wb}}{5 \times 10^{-4} \text{ m}^2} = 0.16 \text{ T}$$

Electromagnetism

An electromagnet is produced where there is an electric current flowing through a conductor. This magnetic field is proportional to the current being carried since the larger the current the greater the magnetic field. An electromagnet is defined as being a temporary magnet because the magnetic field can only exist while there is a current flowing.

Figure 4.03 Lines of magnetic forces set up around a conductor

The direction of the magnetic field is traditionally determined using the 'screw rule'. Think of a normal right-hand threaded screw, with the top as the direction of the current and the rotation of the screw equal to the direction of rotation of the magnetic field.

Rotation of screw = Rotation of magnetic field

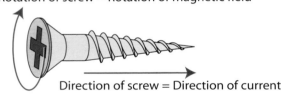

Direction of screw = Direction of current

Figure 4.04 The screw rule

Inductance and inductive components

On completion of this topic area the candidate will be able to describe inductance and inductive components.

We are now going to look more closely at the electromagnet and its effects. When current is passed along a conductor a magnetic field is set up around that conductor, always in a clockwise direction in relation to the current causing the magnetic field.

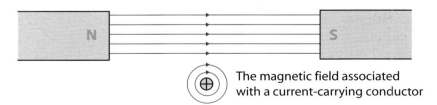

The magnetic field associated with a current-carrying conductor

Figure 4.05 The magnetic field associated with two fixed poles

If the conductor is then formed into a coil of several turns and a current passed through it, the magnetic field created around each segment of the coil combines with the magnetic field of the segments on either side to create an overall magnetic effect similar to that of a bar magnet.

The strength of the magnetic field produced by such a coil can be fairly weak. However, if we place within the coil a soft iron rod the effect is to strengthen the magnetic effect by increasing the magnetic flux density, as the lines of magnetic flux are concentrated through the magnetic material.

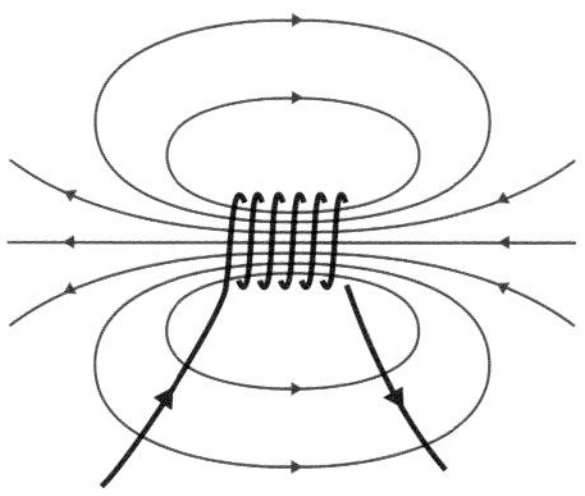

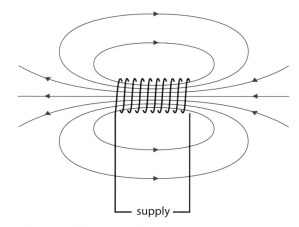

Figure 4.06 A conductor wrapped into a coil

Electromagnetic induction

As we have previously mentioned if a current passes along a conductor then a magnetic field is set up around that conductor. Since the basic effects of an electric current are reversible, if a current produces a magnetic field then a magnetic field must be able to produce a current. This is called electromagnetic induction.

Stated simply: if a conductor is moved through a magnetic field, provided there is a closed circuit then a current will flow through it.

We know that we need a 'force' to drive electrons along a conductor and we can say that an e.m.f. or voltage must be producing the current. In this situation we are causing an e.m.f. and this is known as the induced e.m.f. It will have the same direction as the flowing current.

If we were to pass an electric current through a conductor this would generate a uniform magnetic field around the conductor and at right angles to the conductor. The strength of this magnetic field is directly proportional to the current flowing in the conductor.

The strength of this magnetic field can be further increased by coiling the conductor to form a solenoid.

If the coil were connected to a d.c. supply the only resistance to the current flow would be the resistance of the conductor itself. However, if the coil were connected to an a.c. supply the situation must be looked at differently.

Any change in the magnetic environment of a coil of wire will cause a voltage (e.m.f.) to be 'induced' in the coil. No matter how the change is produced, the voltage will be generated. The change could be produced by, for example, changing the magnetic field strength, moving a magnet toward or away from the coil, moving the coil into or out of the magnetic field or rotating the coil relative to the magnet.

supply

Figure 4.07 A solenoid

The alternating current creates the effect of a continuously changing magnetic field inside the coil; this effect reacts with the flow of current and opposes it.

Inductance is typified by the behaviour of a coil of wire in resisting any change of electric current through the coil. The SI unit of inductance is known as the henry, symbol H. The symbol for inductance is L.

The unit of inductance is given to be the rate of change of current in a circuit of 1 amp per second which produces an induced electromotive force of 1 volt.

Values of inductors range from about 0.1 microhenry, written as 0.1 μH, to 10 henries (H).

The inductance of a coil can be altered by:

- changing the number of turns of wire on the coil
- changing the material composition of the core (air, iron or steel)
- changing the diameter of the coil
- changing the material composition of the coil.

Remember the solenoid which was covered at Level 2? You will realise that an e.m.f. is induced only when we have a changing situation. What could change in the solenoid setup? Well, the following could change:

- the number of turns in the coil (N)
- the rate of change of current flowing in the coil – how quickly the current alternates in the coil
- the rate of change of magnetic flux – how quickly the magnetic flux changes.

In the nineteenth century a rather clever scientist named Michael Faraday spent a lot of time looking at magnetic induction. He came up with a law that tells how much e.m.f. is induced when a conductor is moving in a magnetic field.

Faraday found that, for a conductor, the induced e.m.f. is given by the following equation:

$$\text{e.m.f.} = \frac{\Delta \Phi}{\Delta t}$$

So we can find the induced e.m.f. by knowing the rate of change of flux. This is simply the same as how quickly the conductor cuts the lines of flux.

So, for a coil of N turns:

$$\text{e.m.f.} = -N\frac{\Delta \Phi}{\Delta t}$$

And we can find the induced e.m.f. by knowing the inductance and rate of change of current. These equations are true for both self- and mutual inductance.

$$\text{e.m.f.} = -L\frac{\Delta I}{\Delta t}$$

If the current in a coil of inductance 0.1H rises from 0A to 10A in 0.01 seconds, the induced e.m.f. will be:

$$\text{e.m.f.} = -L\frac{\Delta I}{\Delta t} = -0.1\text{H} \times \frac{10\text{A}}{0.01\text{s}} = 100\text{V}$$

Remember

A solenoid is a long hollow cylinder around which is wound a uniform coil of wire. It produces a similar magnetic field to a bar magnet. A solenoid can be activated with an electric switch

Note: In all these equations there is a negative (–) sign. This is because any induced e.m.f. will always be in opposition to the changes that created it.

When a number of inductors need to be connected together to form an equivalent inductance, they follow the same rules as for resistors:

- to increase inductance, connect inductors in series

- to decrease inductance and increase the current rating, connect inductors in parallel.

To show the effects of inductance we can plot a graph of current against time.

It takes time to build up to maximum current; however, this is important when connected to an a.c. supply because the rate of change of current with time can be calculated and adjusted so that a smoothing effect can be produced in the a.c. If the coil is suddenly switched off, the magnetic field collapses and a high voltage is induced across the circuit. This effect is used for starting fluorescent tube circuits.

Electromagnetic induction forms the basis for the generation of electricity.

Dynamic induction

If a conductor is passed through a magnetic field so that the conductor cuts at right angles to the lines of magnetic flux, an e.m.f. is produced within the conductor. If the conductor was formed to create a circuit, then a current would flow. This movement of the conductor through the magnetic field is known as dynamic induction.

Static induction

If a magnet is passed over a conductor so that the lines of magnetic force cut through the conductor, then an e.m.f. will be induced in the conductor. If the conductor is then formed to create a circuit, then a current would flow. This process by which an e.m.f. is produced as the result of a magnetic field being passed over a stationary conductor is called static induction.

Mutual induction

In the previous examples, we considered permanent magnets and conductors. However, the effect would be the same if we were to pass a conductor through a magnetic field created by an electromagnet. This is called mutual induction and forms the basis of transformers.

In a **double wound transformer**, when current passes through the primary winding a magnetic field is set up around the primary winding. The supply to the winding is a.c. Therefore the magnetic field around the primary winding grows and collapses rapidly, 100 times per second. The lines of magnetic flux also grow and collapse at the same speed. As they do so the lines of magnetic flux cut through the conductors that form the secondary winding inducing an e.m.f. which can then be used to supply equipment.

Definition

A double wound transformer – has separate windings : one for the primary side and one for the secondary side

Self-induction

In an **auto transformer** the principle of self-induction is used for the production of the e.m.f.

Around each part of the coil, a magnetic field is created. On a 50 Hz supply the magnetic field will grow and collapse 100 times each second as the magnetic field around one part of the coil grows and collapses. The lines of magnetic flux cut through those parts of the coil that are adjacent to it, inducing in the coil an e.m.f. in the opposite direction to the current that originally created it.

In the case of the transformer, tap-off leads are attached to the coil, to give the output voltage.

Commonplace inductive equipment

Equipment with inductive components is quite common and very varied. Remember, any conductor formed into a coil and connected to an a. c. supply is an inductor. Examples of their use can be found in:

- electric motors
- fluorescent light fittings
- transformers
- contactors.

- bells
- actuators
- motor starters

Force on a current-carrying conductor in a magnetic field

Nearly all motors work on the basic principle that when a current-carrying conductor is placed in a magnetic field it experiences a force.

We have already seen on page 88 that, if we place a current-carrying conductor into a magnetic field, a magnetic field is set up in a clockwise direction in relation to the direction of the current. This can best be remembered by imagining a corkscrew being twisted into a cork; you need to turn the corkscrew to the right (this symbolises the magnetic field) and the corkscrew is moving away from you, which symbolises the current flowing away from you.

Figure 4.08 The magnetic field when the conductor is placed between the poles

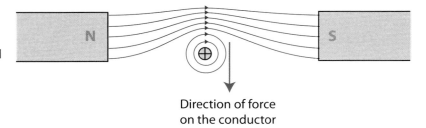

Direction of force on the conductor

If we move the conductor further into the field, as shown in Figure 4.08, we can note the following:

- the main field now becomes distorted
- the field is weaker below the conductor because the two fields are in opposition
- the field is stronger above the conductor because the two fields are in the same direction and aid each other. Consequently the force moves the conductor downward.

If either the current through the conductor or the direction of the magnetic field between the poles is reversed, the force acting on the conductor tends to move it in the opposite direction.

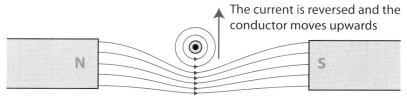

The current is reversed and the conductor moves upwards

Figure 4.09 Reversing the current

The direction in which a current-carrying conductor tends to move when it is placed in a magnetic field can be determined by **Fleming's left-hand (motor) rule**. This rule states that if the first finger, the second finger and the thumb of the left hand are held at right angles to each other as shown in Figure 4.10, then with the first finger pointing in the direction of the Field (N to S) and the second finger pointing in the direction of the current in the conductor, the thumb will indicate the direction in which the conductor tends to move.

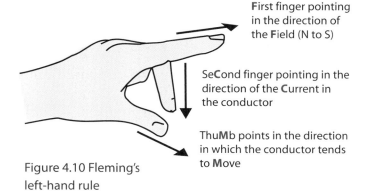

First finger pointing in the direction of the Field (N to S)

SeCond finger pointing in the direction of the Current in the conductor

ThuMb points in the direction in which the conductor tends to Move

Figure 4.10 Fleming's left-hand rule

Calculating the force on a conductor

The force that moves the current-carrying conductor that is placed in a magnetic field depends on the strength of the magnetic flux density (B), the magnitude of the current flowing in the conductor (I), and the length of the conductor in the magnetic field (l).

The following equation expresses this relationship:

Force (F) = B × I × l

In this equation, B is in tesla, l is in metres, I is in amperes, F is in newtons.

Example 1

A conductor some 15 metres in length lies at right angles to a magnetic field of 5 tesla. Calculate the force on the conductor when:

    (a) 15 A flows in the coil

    (b) 25 A flows in the coil

    (c) 50 A flows in the coil.

Answer: Using the formula $F = B \times I \times l$:

    $F = 5 \times 15 \times 15 = 1125$ N

    $F = 5 \times 25 \times 15 = 1875$ N

    $F = 5 \times 50 \times 15 = 3750$ N.

Example 2

A conductor 0.25m long situated in, and at right angles to, a magnetic field experiences a force of 5 N when a current through it is 50 A. Calculate the flux density.

Answer: Transpose the formula $F = B \times I \times l$ for (B):

$$B = \frac{F}{I \times l}$$

Substitute the known values into the equation:

$$B = \frac{5}{0.25 \times 50} = 0.4 \text{ T}$$

Capacitors

On completion of this topic area the candidate will be able to describe capacitors, their construction, function and dangers.

Just as resistors enable us to introduce known amounts of resistance into a circuit to serve our purposes, so we can use components known as capacitors to introduce capacitance into a circuit. Like resistance, capacitance always exists in circuits – though, as you will see when we have discussed the subject in more detail, capacitance exists *between* conductors whereas resistance exists in conductors.

Basic principles

A capacitor is basically two metallic surfaces, usually referred to as plates, separated by an insulator commonly known as the dielectric. The plates are usually, though not necessarily, metal and the dielectric is any insulating material. Air, glass, ceramic, mica, paper, oils and waxes are some of the many materials commonly used. The common symbols used for capacitors are identified in Figure 4.11.

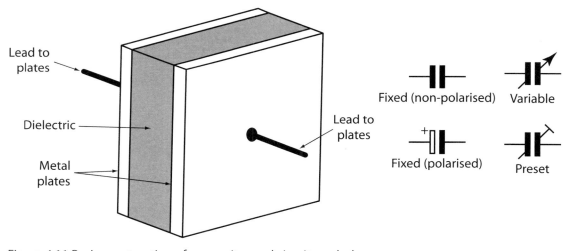

Figure 4.11 Basic construction of a capacitor and circuit symbols

The two plates are not in contact with each other so do not form a circuit in the same way that conductors with resistors do. However, the capacitor stores a small amount of electric charge and, as such, it can be thought of as a small rechargeable battery which can be quickly recharged.

The capacitance of any capacitor depends on three factors.

1. It depends on the working area of the plates, i.e. the area of the conducting surfaces facing each other. We can think of the degree of crowding of excess electrons near the surface of one plate of a capacitor (and the corresponding sparseness of electrons near the surface of the other) as being directly related

Safety tip

Never pick a capacitor up by the terminals as it may still be charged and you will receive a shock. Always ensure the capacitor has been discharged before handling. Some capacitors have a discharge resistor connected in the circuit for this reason

Remember

In any application the capacitor to be used should meet or preferably exceed the capacitor voltage rating. The voltage rating is often called the 'working voltage' and refers to d.c. voltage values. When applied to a.c. circuits the peak voltage value must be used as a comparison to the d.c. working voltage of a capacitor to the **potential difference (p.d.)** applied across the capacitor, for example connecting it directly across a battery. If we increase the area of the plates then more electrons can flow on to one of the plates before the same degree of crowding is reached. The battery voltage determines this level of crowdedness. There is of course a similar increased loss of electrons from the other plate. The working area of the plates is directly proportional to the capacitance. If we double the area of the plates we double the capacitance of the capacitor.

2. It depends on the separation between the plates. The capacitance effect depends on the forces of repulsion or attraction caused by an electron surplus or shortage on the plates. The further apart the plates are, the weaker these factors become. As a result, the degree of crowding of electrons on one plate (and the shortage of electrons on the other) produced by a given p.d. across the capacitor decreases.

3. The capacitance depends on the nature of the **dielectric** or spacing material used.

Capacitor types

There are two major types of capacitor, fixed and variable, both of which are used in a wide range of electronic devices. Fixed capacitors can be further subdivided into electrolytic and non-electrolytic types, and together they represent the majority of the market.

All capacitors possess some resistance and inductance because of the nature of their construction. These undesirable properties result in limitations, which often determine their applications.

Fixed capacitors

Electrolytic capacitors

These capacitors have a much higher capacitance, volume for volume, than any other type. This is achieved by making the plate separation extremely small by using a very thin dielectric (insulator). The dielectric is often mica or paper.

The main disadvantage of an electrolytic capacitor is that it is polarised and must be connected to the correct polarity in a circuit, otherwise a short circuit and destruction of the capacitor will result.

Figure 4.12 illustrates a newer type of electrolytic capacitor using tantalum and tantalum oxide to give a further capacitance/size advantage. It looks like a raindrop with two leads protruding from the bottom. The polarity and values may be marked on the capacitor (see photograph in Polarity section on page 144); or value may be marked by a colour code (see Figure 4.79 on page 148) can be used.

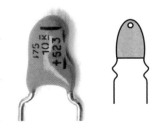

Figure 4.12 Tantalum capacitor

Non-electrolytic capacitors

There are many different types of non-electrolytic capacitor. However, only mica, ceramic and polyester are of any significance. Older types using glass and vitreous enamel are expected to disappear over the next few years and even mica will be replaced by film types.

Mica

Mica is a naturally occurring dielectric and has very high resistance. This gives excellent stability and allows the capacitors to be accurate within a value

Figure 4.13 Mica capacitor

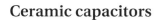

of ±1 per cent of the marked value. Since costs usually increase with increased accuracy, they tend to be more expensive than plastic film capacitors. They are used where high stability is required, for example in tuned circuits and filters required in radio transmission. Figure 4.13 illustrates a typical mica capacitor.

Ceramic capacitors

These consist of small rectangular pieces of ceramic with metal electrodes on opposite surfaces. Figure 4.14 illustrates a typical ceramic capacitor. These capacitors are mainly used in high frequency circuits subjected to wide temperature variations. They have high stability and low loss.

Figure 4.14 Ceramic capacitor

Polyester capacitors

These are an example of a plastic film capacitor. Polypropylene, polycarbonate and polystyrene capacitors are other types of plastic film capacitors. They are widely used in the electronics industry due to their good reliability and relative low cost but are not suitable for high frequency circuits. Figure 4.15 illustrates a typical polyester capacitor; however, they can also be a tubular shape (see Figure 4.16).

Figure 4.15 Polyester capacitor

Figure 4.16 Tubular capacitor

Did you know?

Mica is a common rock-forming mineral. You find it in rocks such as granite and some sandstones and mudstones

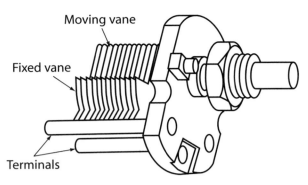

Moving vane

Fixed vane

Terminals

Figure 4.17 A typical variable capacitor used for tuning

Variable capacitors

Variable capacitors generally have air or a vacuum as the dielectric, although ceramics are sometimes used. The two main sub-groups are tuning and trimmer capacitors.

Tuning capacitors

These are so called because they are used in radio tuning circuits and consist of two sets of parallel metal plates. One plate is isolated from the mounting frame by ceramic supports while the other is fixed to a shaft which allows one set to be rotated into or out of the first set. The rows of plates interlock like fingers but do not quite touch each other.

Trimmer capacitors

These are constructed of flat metal leaves separated by a plastic film and can be screwed towards each other. They have a smaller range of variation than tuning capacitors, so are only used where a slight change in value is needed.

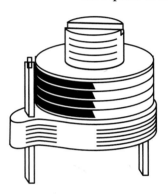

Figure 4.18 A typical capacitor used as a trimmer

Electrostatics and calculations with capacitors

The charge stored on a capacitor is dependent on three main factors: the area of the facing plates; the distance between the plates; and the nature of the dielectric. The charge stored by a capacitor is measured in coulombs (Q) and is related to the value of capacitance and the voltage applied to the capacitor:

Charge (coulombs) = Capacitance (farads) × Voltage (volts)
$Q = C \times V$

The formula for energy stored in a capacitor can be calculated by using the formula:

$W = \dfrac{1}{2} CV^2$

Capacitors in combination

At Level 2, we looked at how resistors may be joined together in various combinations of series or parallel connections. Figures 4.19 and 4.20 illustrate the equivalent capacitance (C_t) of a number of capacitors. C_t can be found by applying similar formulae as for resistors. However, these formulae are the opposite way round to series and parallel resistors.

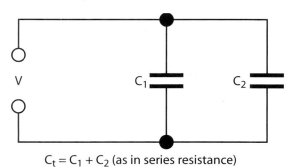

$C_t = C_1 + C_2$ (as in series resistance)

Figure 4.19 Capacitors connected in parallel

$\frac{1}{C_t} = \frac{1}{C_1} + \frac{1}{C_2}$ (as in parallel resistance)

or

$C_t = \frac{C_1 \times C_2}{C_1 + C_2}$ (when there are two capacitors in series)

Figure 4.20 Capacitors connected in series

Example 1

Capacitors of 10 μF and 40 μF are connected in series and then in parallel. Calculate the effective capacitance for each connection.

Series:

$\frac{1}{C} = \frac{1}{C_1} + \frac{1}{C_2}$

$\frac{1}{C_t} = \frac{1}{10\,\mu F} + \frac{1}{40\,\mu F}$

$\frac{1}{C_t} = \frac{4\,\mu F + 1\,\mu F}{40\,\mu F}$

$\frac{1}{C_t} = \frac{5\,\mu F}{40\,\mu F}$

Therefore:

$\frac{C_t}{1\,\mu F} = \frac{40\,\mu F}{5\,\mu F}$

$C_t = 8\,\mu F$

Parallel:

$C_t = C_1 + C_2$

$C_t = 10\,\mu F + 40\,\mu F$

$C_t = 50\,\mu F$

Example 2

Three capacitors of 30 μF, 20 μF and 15 μF are connected in series across a 400 V d.c. supply. Calculate the total capacitance and the charge on each capacitor.

$$\frac{1}{C_t} = \frac{1}{C_1} + \frac{1}{C_2} + \frac{1}{C_3} = \frac{1}{30\ \mu F} + \frac{1}{20\ \mu F} + \frac{1}{15\ \mu F} = \frac{9\ \mu F}{60\ \mu F}$$

Therefore:

$C_t = 6.66\ \mu F$

Q, the charge, is common to each capacitor. Therefore:

$Q = C \times V = 6.66 \times 10^{-6} \times 400$

Therefore:

$Q = 2.664$ millicoulombs

Example 3

Three capacitors of 30 μF, 20 μF and 15 μF are connected in parallel across a 400 V d.c. supply. Calculate the total capacitance, the total charge and the charge on each capacitor.

$C_t = C_1 + C_2 + C_3 = 30 + 20 + 15 = 65\ \mu F$

Total charge $Q = C \times V$

$Q_t = C_1 \times V = 65 \times 10^{-6} \times 400 = 26 \times 10^{-3}$ coulombs

$Q_1 = C_1 \times V = 30 \times 10^{-6} \times 400 = 12 \times 10^{-3}$ coulombs

$Q_2 = C_2 \times V = 20 \times 10^{-6} \times 400 = 8 \times 10^{-3}$ coulombs

$Q_3 = C_3 \times V = 15 \times 10^{-6} \times 400 = 6 \times 10^{-3}$ coulombs

Charging and discharging capacitors

When a capacitor is connected to a battery, positive and negative charges are deposited on the capacitor plates and the capacitor will charge up. These charges build up fast but not instantaneously; they follow a set pattern of charge. A typical charge and discharge circuit is illustrated in Figure 4.21.

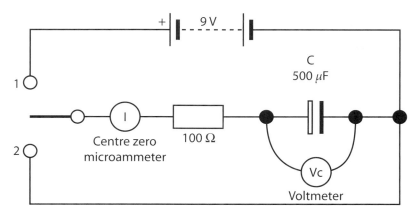

Figure 4.21 A typical charge and discharge circuit

When the switch contact is connected to 1, current flows from the battery to the capacitor plates and the electrolytic capacitor will charge up to the same voltage as the battery.

From the time the current (measured by the microammeter) is switched on a charge curve starts that is related to the charge voltage (measured by the voltmeter) and time. The maximum voltage that the capacitor can take is that of the supply; this then sets the voltage scale from 0 volts to maximum voltage (9 V in this example).

The timescale is divided into sections called time constants. The pattern of charge is always the same. Figure 4.22 illustrates a graph of voltage against time for charging the capacitor.

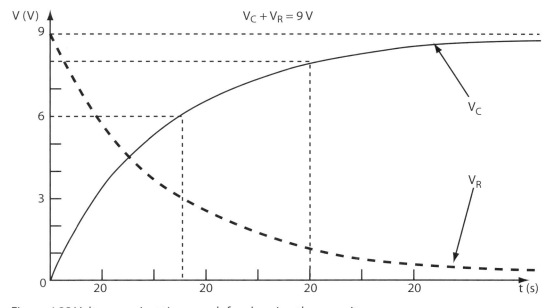

Figure 4.22 Voltage against time graph for charging the capacitor

At 'switch on' the voltage (V_C) across the capacitor is zero and the voltage (V_R) across the resistor is at maximum supply voltage.

As the current flow charges up the capacitor plates, V_C increases with the charging of the capacitor plates. V_R reduces over the charge time because the rate of current flow reduces. This is because the supply voltage and the voltage across the capacitor will be at the same potential at completion of charge, therefore no current will be flowing. The current will follow a similar line to V_R on the graph.

The timescale has been subdivided into equal time periods, with each time period known as a time constant. In the first time period the voltage reaches approximately two-thirds of the maximum volts (6 V); in the next period the charge goes from the finishing point in the first time constant to finish about two-thirds of what is left (2 V).

This pattern continues until the capacitor is fully charged. The rate of charge will be exponential as shown in Figure 4.23 and it will take five time constants for the capacitor to be fully charged. This also applies if the capacitor is discharging through a series resistor; the voltage (V_C) across the capacitor will fall exponentially and it will again take five time constants for the capacitor to be fully discharged.

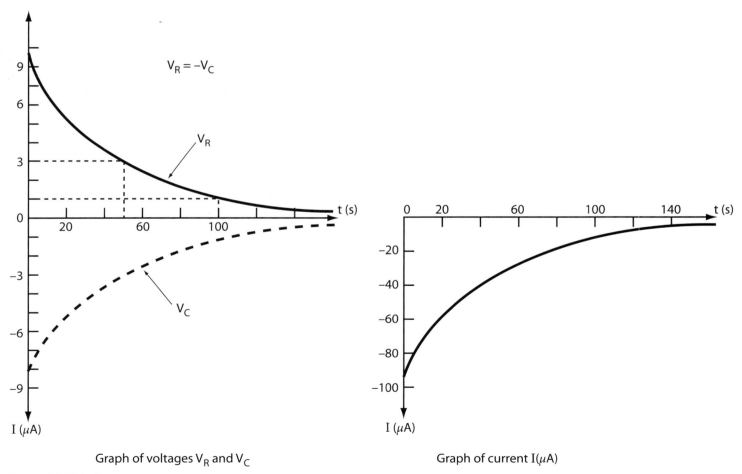

Graph of voltages V_R and V_C Graph of current I(μA)

Figure 4.23 Discharge curve

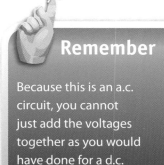

Resistance, inductance, capacitance and impedance

On completion of this topic area the candidate will be able to state the effects of resistance, inductance, capacitance and impedance in a.c. circuits.

The principles of resistance (R), inductance (L), capacitance (C), as well as phasors and impedance (Z) are covered at Level 2. It is strongly advised that you turn back to these pages to refresh your memory on the concepts behind these measurements.

Resistance and inductance in series (RL)

Consider the following diagram.

Here we have a resistor connected in series with an inductor and fed from an a.c.

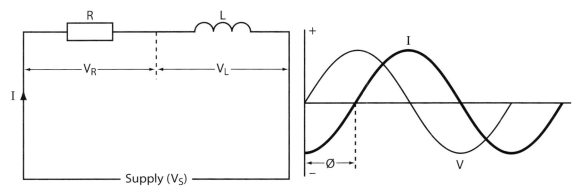

Figure 4.24 Resistor and inductor in series

Remember

Because this is an a.c. circuit, you cannot just add the voltages together as you would have done for a d.c. circuit; you need to construct a phasor diagram to work it out

supply. In a series circuit, the current (I) will be common to both the resistor and the inductor, causing **voltage drop** V_R across the resistor and V_L across the inductor.

The sum of these voltages must equal the supply voltage. Here is how to construct a phasor diagram for this circuit.

In a series circuit we know that current will be common to both the resistor and the inductor. It therefore makes sense to use current as our reference phasor. We also know that voltage and current will be in phase for a resistor. Therefore, the volt drop (**p.d.**) V_R across the resistor must be in phase with the current. Also, in an inductive circuit, the current lags the voltage by 90 degrees.

If the current is lagging voltage, then we must be right in saying that voltage is leading the current.

This means in this case that the volt drop across the inductor (V_L) will lead the current by 90 degrees. When we draw phasors, we always assume that they rotate anti-clockwise and the symbol Ø represents the phase angle.

There are two ways of doing the drawing.

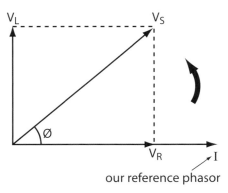

Figure 4.25 Two ways of drawing a phase diagram

In the second example the phasors produce a right-angled triangle. We can therefore use Pythagoras' theorem to give us the formula:

$$V_S^2 = V_R^2 + V_L^2$$

We can then use trigonometry to give us the different formulae, dependent on the values that we have been given:

$$\cos \varnothing = \frac{V_R}{V_S} \qquad \sin \varnothing = \frac{V_L}{V_S} \qquad \tan \varnothing = \frac{V_L}{V_S}$$

Example 1

A coil of 0.15 H is connected in series with a 50 Ω resistor across a 100 V 50 Hz supply. Calculate the following.

(a) The inductive reactance of the coil.
(b) The impedance of the circuit.
(c) The circuit current.

(a) Inductive reactance (X_L)

For inductive reactance, we use the formula:

$$X_L = 2 \pi f L \ (\Omega)$$

Inserting the values, this would give us:

$$X_L = 2 \times 3.142 \times 50 \times 0.15, \text{ therefore } X_L = 47.13 \ \Omega$$

(b) Circuit impedance (Z)

When we have resistance and inductance in series, we calculate the impedance using the following formula:

$$Z^2 = R^2 + X_L^2$$

This becomes:

$$Z = \sqrt{R^2 + X_L^2}$$

In the case of the first formula, isn't this the same as Pythagoras' theorem for a right-angled triangle ($A^2 = B^2 + C^2$)? We therefore sometimes refer to this as the impedance triangle, and it can be drawn for this type of circuit, as shown in Figure 4.26. Here, the angle (Ø) between sides R and Z is the same as the phase angle between current and voltage.

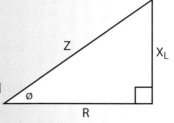

Figure 4.26 Impedance triangle

Using the formula:

$$Z = \sqrt{R^2 + X_L^2}$$

Then: $Z = \sqrt{50^2 + 47.12^2}$ therefore Z = **68.69 Ω**

(c) Circuit current (I)

As we are referring to the total opposition to current, we use the formula:

$$I = \frac{V}{Z} = \frac{100}{68.69} = \mathbf{1.46\ A}$$

Example 2

A coil of 0.159 H is connected in series with a 100 Ω resistor across a 230 V 50 Hz supply. Calculate the following.

(a) The inductive reactance of the coil.
(b) The circuit impedance.
(c) The circuit current.
(d) The p.d. across each component.
(e) The circuit phase angle.

(a) Inductive reactance (X_L)

$X_L = 2\,\pi\,f\,L$ therefore $X_L = 2 \times 3.142 \times 50 \times 0.159 = \mathbf{50\ \Omega}$

(b) Circuit impedance (Z)

$$Z = \sqrt{R^2 + X_L^2} \qquad \text{therefore} \qquad Z = \sqrt{100^2 + 50^2} = \textbf{111.8 } \Omega$$

(c) Circuit current (I)

$$I = \frac{V}{Z} \qquad \text{therefore} \qquad I = \frac{230}{111.8} = \textbf{2.06 A}$$

(d) The p.d. across each component (V)

$$V_R = I \times R \qquad \text{therefore} \qquad V = 2.06 \times 100 = 206 \text{ V}$$

$$V_L = I \times X_L \qquad \text{therefore} \qquad V = 2.06 \times 50 = 103 \text{ V}$$

(e) Circuit phase angle (Ø)

Using a right-angled triangle:

$$\tan \varnothing = \frac{V_L}{V_R} \qquad \text{therefore} \qquad \tan \varnothing = \frac{103}{206} = 0.5$$

If you then enter 0.5 into your calculator and press the INV key, then press TAN, you should get the number 26.6.

Therefore, the current is lagging voltage by **26.6 degrees**

Resistance and capacitance in series (RC)

Consider the following diagram.

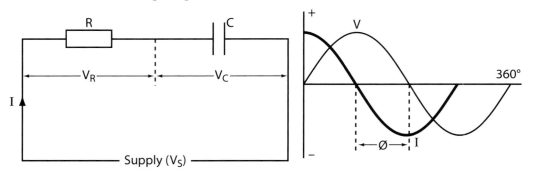

Figure 4.27 A resistor connected in series with a capacitor fed from an a.c. supply

Here a resistor is connected in series with a capacitor and fed from an a.c. supply. Once again, in a series circuit the current (I) will be common to both the resistor and the capacitor, causing voltage to drop (p.d.) V_R across the resistor and V_C across the capacitor.

As with the resistance/inductance (RL) circuit previously, we can take current as the reference phasor. Similarly, the voltage across the resistor will be in phase with that current. Remember also in a capacitive circuit the current leads the voltage by 90 degrees. Therefore we can say that the voltage across the capacitor will be lagging the current. As before, we can now calculate the supply voltage (V_S) by completion of the parallelograms shown in Figure 4.28, as follows:

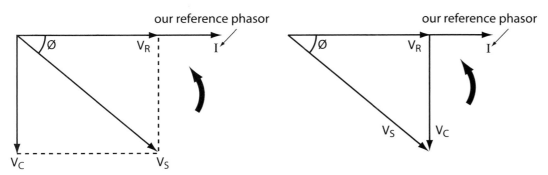

Figure 4.28 Phasor diagrams

As with the inductor, we can apply Pythagoras' theorem and trigonometry to give us the following formulae:

$$V_S^2 = V_R^2 + V_C^2$$

$$\cos \varnothing = \frac{V_R}{V_S} \qquad \sin \varnothing = \frac{V_C}{V_S} \qquad \tan \varnothing = \frac{V_C}{V_R}$$

An example of this is given below.

Example 1

A capacitor of 15.9 μF and a 100 Ω resistor are connected in series across a 230 V 50 Hz supply. Calculate the following.

(a) The circuit impedance.
(b) The circuit current.
(c) The p.d. across each component.
(d) The circuit phase angle.

(a) Circuit impedance (Z)

To be able to find the impedance we must first find the capacitive reactance.

$$X_C = \frac{1}{2 \pi f C}$$

However, as the capacitor value is given in μF, we use $X_C = \dfrac{10^6}{2\pi fC}$

This gives us : $X_C = \dfrac{10^6}{2 \times 3.142 \times 50 \times 15.9} = \dfrac{10^6}{4995.78} = \mathbf{200\ \Omega}$

When we have resistance and capacitance in series, we use the following formula:

$Z^2 = R^2 + X_C^2$ which becomes $Z = \sqrt{R^2 + X_C^2}$

Therefore $\quad Z = \sqrt{100^2 + 200^2} = \sqrt{50\ 000} = \mathbf{224\ \Omega}$

(b) Circuit current (I)

$I = \dfrac{V}{Z}$ $\qquad$ therefore $\qquad I = \dfrac{230}{224} = 1.03\ A$

(c) The p.d. across each component (V)

$V_R = I \times R$ $\qquad$ therefore $\qquad V_R = 1.03 \times 100 = \mathbf{103\ V}$

$V_C = I \times X_L$ $\qquad$ therefore $\qquad V_C = 1.03 \times 200 = \mathbf{206\ V}$

(d) Circuit phase angle (Ø)

Using our right-angled triangle:

$\tan \emptyset = \dfrac{V_C}{V_R}$ $\qquad$ therefore $\quad \tan \emptyset = \dfrac{206}{103} = 2$

If you then enter 2 into your calculator and press the INV key, then press TAN, you should get the number 63.4.

Therefore, the current is leading voltage by **63.4 degrees.**

Resistance, inductance and capacitance in series (RLC)

Consider the following diagram:

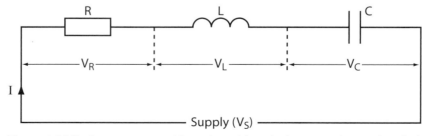

Figure 4.29 Resistor connected in series with an inductor and capacitor, fed from an a.c. supply

Here we have a resistor connected in series with an inductor and a capacitor then fed from an a.c. supply. This is often referred to as an RLC circuit or a **general series** circuit. Again, as we have a series circuit, the current (I) will be common to all three components, causing a voltage drop (p.d.) V_R across the resistor, V_L across the inductor and V_C across the capacitor.

Here V_R will be in phase with the current, V_L will lead the current by 90 degrees (because the current lags the voltage in an inductive circuit) and V_C will lag the current by 90 degrees (because current leads the voltage in a capacitive circuit). Because V_L and V_C are in opposition to each other (one leads and one lags), the actual effect will be the difference between their values, subtracting the smaller from the larger. We can once again calculate V_S by completing a parallelogram as shown in Figure 4.30.

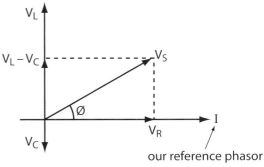

As before, we can now apply Pythagoras' theorem and trigonometry to give us the following formulae depending on whether V_L or V_C is the larger.

$$V_S^2 = V_R^2 + (V_L - V_C)^2 \quad \text{or} \quad V_S^2 = V_R^2 + (V_C - V_L)^2$$

Because I is the same for each component we get:

$$Z = \sqrt{R^2 + (X_L - X_C)^2} \quad \text{or} \quad Z = \sqrt{R^2 + (X_C - X_L)^2}$$

And finally: $\cos \varnothing = \dfrac{V_R}{V_S} \quad \sin \varnothing = \dfrac{V_L - V_C}{V_S} \quad \tan \varnothing = \dfrac{V_L - V_C}{V_R}$

Figure 4.30 Phasor parallelogram

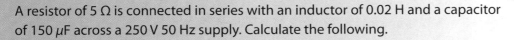

Example

A resistor of 5 Ω is connected in series with an inductor of 0.02 H and a capacitor of 150 μF across a 250 V 50 Hz supply. Calculate the following.

(a) The impedance.
(b) The supply current.
(c) The power factor.

a. Impedance (Z)

In order to find the impedance, we must first find out the relevant values of reactance.

Therefore: $X_L = 2 \pi f L = 2 \times 3.142 \times 50 \times 0.02 = \mathbf{6.28\ \Omega}$

$$X_C = \frac{1}{2 \pi f C} \text{ allowing for microfarads}$$

$$= \frac{10^6}{2 \times 3.142 \times 50 \times 150} = \mathbf{21.2\ \Omega}$$

If you remember, we said that the effect of inductance and capacitance together in series would be the difference between their values. Consequently, this means that the resulting reactance (X) will be found as follows:

$$X = X_L - X_C \quad \text{or, in this case because } X_C \text{ is the larger: } X = X_C - X_L$$

and therefore: $X = 21.2 - 6.28 = \mathbf{14.92\ \Omega}$

We can now use the impedance formula as follows:

$$Z = \sqrt{R^2 + (X_L - X_C)^2} \quad \text{or} \quad Z = \sqrt{R^2 + (X_C - X_L)^2}$$

which gives us

$$Z = \sqrt{5^2 + 14.92^2} = \mathbf{15.74\ \Omega}$$

Note: X_C is greater than X_L. Therefore we subtract X_L from X_C. Had X_L been the higher, then the reverse would be true. Also, as capacitive reactance is highest, the circuit current will lead the voltage. Had the inductive reactance been the higher, then the current would lag the voltage.

b. Supply current (I)

$$I = \frac{V}{Z} = \frac{230}{15.74} = \mathbf{14.6A}$$

c. Power factor (Ø)

$$\cos\varnothing = \frac{V_R}{V_S} = \frac{I \times R}{I \times Z} = \frac{R}{Z} = \frac{5}{15.74} = 0.32$$

Therefore PF = **0.32 leading**

Resistance, inductance and capacitance in parallel

Consider the following diagram

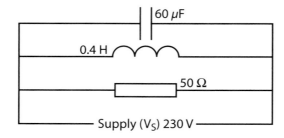

60 μF

0.4 H

50 Ω

Supply (V$_S$) 230 V

Figure 4.31 Resistor, capacitor and inductor in parallel connected to a 50 Hz a.c. supply

There can obviously be any combination of the above components in parallel. However, to demonstrate the principles involved, we will look at all three connected across an a.c. supply.

As we have a parallel circuit, the voltage (V_S) will be common to all branches of the circuit. Consequently, when we draw our parallelogram we will use voltage as the reference phasor.

In this type of circuit the current through the resistor will be in phase with the voltage, the current through the inductor will lag the voltage by 90 degrees and the current through the capacitor will lead the voltage by 90 degrees.

Normally, when we carry out calculations for parallel circuits, it is easier to treat each branch as being a separate series circuit. We then draw to scale each of the respective currents and their relationships to our reference phasor, which is voltage.

As with voltage V_L and V_S in the RLC series circuit, the current through the inductor (I_L) and the current through the capacitor (I_C) are in complete opposition to each other. Therefore, the actual effect will be the difference between their two values. We calculate this value by the completion of our parallelogram. But, bear in mind that the bigger value (I_C or I_L) will determine whether the current ends up leading or lagging.

Example

Using Figure 4.31 as the basis for our example, calculate the circuit current and its phase angle relative to the voltage.

To do this we must first find the current through each branch:

$$I_R = \frac{V}{R} = \frac{230}{50} = \textbf{4.6 A}$$

$$X_L = 2\pi fL = 2 \times 3.142 \times 50 \times 0.4 = \textbf{126 }\Omega$$

Therefore $$I_L = \frac{V}{X_L} = \frac{230}{126} = \textbf{1.8 A}$$

$$X_C = \frac{10^6}{2\pi fC} = \frac{10^6}{2 \times 3.142 \times 50 \times 60} = \textbf{53 }\Omega$$

Therefore $$I_C = \frac{V}{X_C} = \frac{230}{53} = \textbf{4.3 A}$$

The actual effect will be $I_C - I_L$ which gives: $4.3 - 1.8 = 2.5$ A

Now add this to I_R by completing the scale drawing at Figure 4.32. This gives a current I_t of **5.2 A** that is leading voltage by an angle of **28 degrees**.

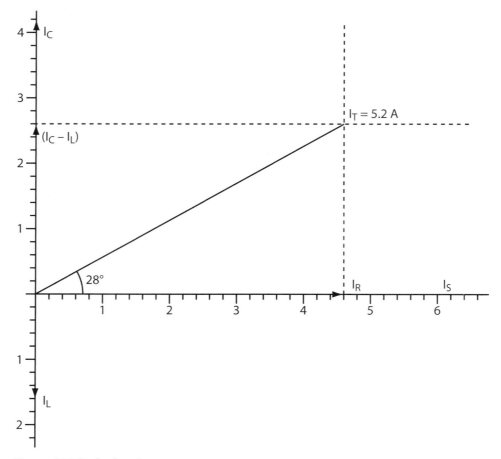

Figure 4.32 Scale drawing

Power in an a.c. circuit

The operation and effect of power in an a.c. circuit was covered at Level 2.

Semiconductor devices

On completion of this topic area the candidate will be able to describe semiconductor devices and identify input and output waveforms for half-wave, full-wave and bridge rectifier circuits.

Semiconductor basics

Think of a semiconductor as being a material with an electrical quality somewhere between a conductor and an insulator, in that it is neither a good conductor nor a good insulator. Typically, we use semiconducting materials, such as silicon or germanium, in which the atoms are arranged in a 'lattice' structure. The lattice has atoms at regular distances from each other, with each atom 'linked' or 'bonded' to the four atoms surrounding it. Each atom then has four **valence electrons**.

Definition

Valence electrons – the electrons in an atom's outermost orbit

However, we have a problem in that, with atoms of pure silicon or germanium, no conduction is possible because we have no free electrons. To allow conduction to take place we therefore add an impurity to the material via a process known as **doping**. When we dope the material we can add two types of impurity:

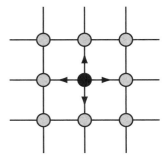

- pentavalent – e.g. arsenic which contains five valence electrons
- trivalent – e.g. aluminium that contains three valence electrons.

Figure 4.33 Lattice structure of semi-conducting material

As we can see by the number of valence electrons in each, adding a pentavalent (five) material introduces an extra electron to the semiconductor and adding a trivalent (three) material to the semiconductor 'removes' an electron (also known as creating a hole).

When there is an extra electron there is a surplus of negative charge; this type of material is called 'n-type'. When we have 'removed' an electron we have a surplus of positive charge; this material is called 'p-type'. It is the use of these two materials that allows us to introduce the component responsible for rectification, called the diode.

The p–n junction

A semiconductor diode is created when we bring together an 'n-type' material and a 'p-type' material to form a p–n junction. The two materials form a barrier where they meet, called the **depletion** layer. In this barrier, the coming together of unlike charges causes a small internal p.d. to exist.

We now need to connect a battery across the ends of the two materials, whereby we call the end of the p-type material the anode and the end of the n-type material the cathode.

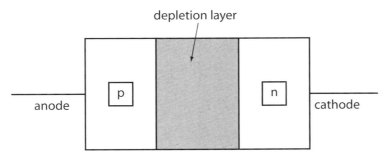

Figure 4.34 p–n diode

If the anode is positive and the battery voltage is big enough, it will overcome the effect of the internal p.d. and push charges (both positive and negative) over the junction. In other words, the junction has a low enough resistance for current to flow. This type of connection is known as being **forward biased**.

Reverse the battery connections so that the anode is now negative and the junction becomes high resistance and no current can flow. This type of connection is known as being **reverse biased**.

When the junction is forward biased, it only takes a small voltage (0.7 V for silicon) to overcome the internal barrier p.d.

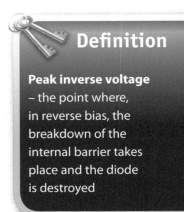

Definition

Peak inverse voltage – the point where, in reverse bias, the breakdown of the internal barrier takes place and the diode is destroyed

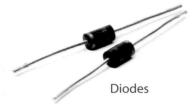

Diodes

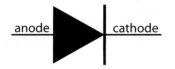

anode   cathode

Figure 4.35 Symbol representing a diode

When reversed biased, it takes a large voltage (1200 V for silicon) to overcome the barrier and thus destroy the diode, effectively allowing current to flow in both directions. As a general summary of its actions, we can therefore say that a diode allows current to flow through it in one direction only.

We normally use the symbol in Figure 4.35 to represent a diode. In this symbol, the direction of the arrow can be taken to represent the direction of current flow.

Zener diode

We have just established that a conventional diode will not allow current flow if reverse biased and below its reverse breakdown voltage. Also, when forward biased (in the direction of the arrow), the diode exhibits a voltage drop of roughly 0.6 V for a typical silicon diode. If we were to exceed the breakdown voltage, it would overcome the internal barrier of the diode, thus allowing current flow in both directions. However, this normally results in the total destruction of the device.

Zener diodes are p–n junction devices that are specifically designed to operate in the reverse breakdown region without completely destroying the device. The breakdown voltage of a zener diode V_z is set by carefully controlling the doping level during manufacture. The breakdown voltage can be controlled quite accurately in the doping process and tolerances to within 0.05 per cent are available, although the most widely used tolerances are 5 per cent and 10 per cent.

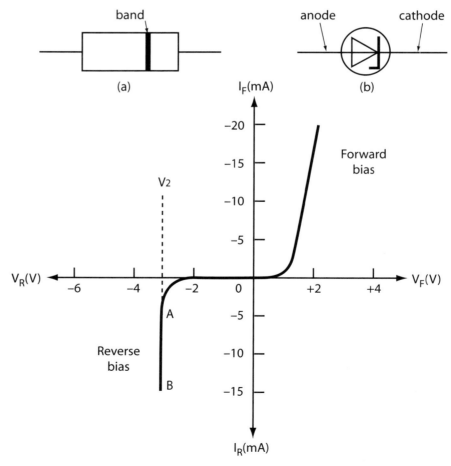

Figure 4.36 Zener diode characteristics

Therefore, a reverse biased zener diode will exhibit a controlled breakdown, allow current to flow and thus keep the voltage across the zener diode at the predetermined zener voltage. Because of this characteristic, the zener diode is commonly used as a form of voltage limiting/regulation device when connected in parallel across a load.

When connected so that it is reverse biased in parallel with a variable voltage source, a zener diode acts as a short circuit when the voltage reaches the diode's reverse breakdown voltage, and therefore limits the voltage to a known value. A zener diode used in this way is known as a shunt voltage regulator (shunt meaning connected in parallel and voltage regulator being a class of circuit that produces a fixed voltage).

For a low current power supply, a simple voltage regulator could be made with a resistor (to limit the operating current) and a reverse biased zener diode as shown in Figure 4.37. Here, V_S is the supply voltage, remembering that V_Z is our zener breakdown voltage.

As a summary, we can therefore say that a zener's properties are as follows.

- When forward biased (although not normally used for a zener) the behaviour is like an ordinary semiconductor diode.

- When reverse biased, at voltages below V_z the device essentially doesn't conduct and behaves just like an ordinary diode.

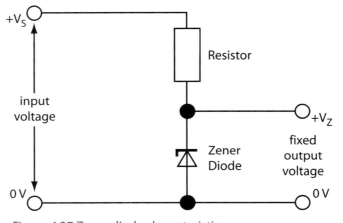

Figure 4.37 Zener diode characteristics

- When reverse biased, any attempt to apply a voltage greater than V_z causes the device to be prepared to conduct a very large current. This has the effect of limiting the voltage we can apply to around V_z.

As with any characteristic curve, the voltage at any given current, or the current at any given voltage, can be found from the curve of a zener diode.

Light emitting diodes (LEDs)

The light emitting diode is a p–n junction especially manufactured from a semiconducting material which emits light when a current of about 10 mA flows through the junction. No light is emitted if the diode is reverse biased. If the voltage exceeds 5 volts then the diode may be damaged; if the voltage exceeds 2 volts then a series connected resistor may be required.

Did you know?

The zener voltage is named after the American physicist Clarence Zener who first discovered the effect

Did you know?

Zener diodes are readily available with power ratings ranging from a few hundred milliwatts to tens of watts. Low power types are usually encapsulated in glass or plastic packages, and heat transfer away from the junction is mainly via conducting along the wires. High power types, like power rectifiers, are packaged in metal cases designed to be fitted to heat sinks so that heat can be dissipated by conduction, convection and radiation

Figure 4.38 illustrates the general appearance of an LED and its symbol.

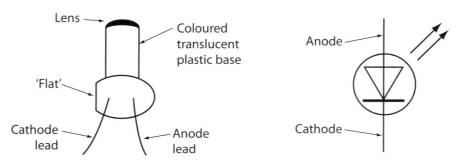

Figure 4.38 Light emitting diode

Diode testing

The p–n junction diode has a low resistance when a voltage is applied in the forward direction and a high resistance when applied in the reverse direction.

Connecting an ohmmeter with the red positive lead to the anode of the junction diode and the black negative lead to the cathode would give a very low reading. Reversing the lead connections would give a high resistance reading in a good component.

Figures 4.39 and 4.40 illustrate these connections.

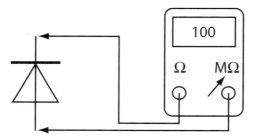

Figure 4.39 High resistance connection

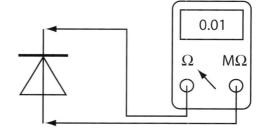

Figure 4.40 Low resistance connection

Infrared source and sensor

An infrared beam of light is projected from an LED which is a semiconductor made from gallium arsenide crystal. The light emitted is not visible light but very close to the white light spectrum. Figure 4.41 shows various housings for both the source output and the sensors within the security alarms industry.

Infrared beams have a receiver which reacts to the beam in differing ways depending upon its use. Infrared sources/receivers are used for alarm detection and as remote control signals for many applications. The passive infrared (PIR) detector is housed in only one enclosure and uses ceramic infrared detectors. The device does not have a projector but detects the infrared heat radiated from the human body.

Thyristors

Sometimes referred to as a 'silicon controlled rectifier' (SCR), the thyristor is a four-layer semiconducting device, with each layer consisting of an alternating n or p type material as shown in Figure 4.42.

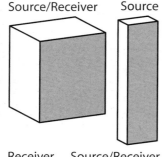

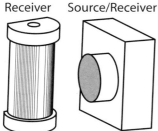

Figure 4.41 Housings for infrared source output and sensors for security alarms

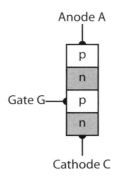

Anode A

p
n
Gate G — p
n

Cathode C

Thyristor

Figure 4.42 Construction of a thyristor

The main terminals (the anode and cathode) are across the full four layers, while the control terminal (the gate) is attached to one of the middle layers. The circuit symbol for a thyristor is shown in Figure 4.43.

Effectively acting as a high speed switch, thyristors are available that can switch large amounts of power (as high as MW). They can therefore be seen in use within **high voltage direct current (HVDC)** systems. These can be used to interconnect two a.c. regions of a power-distribution grid, although the equipment needed to convert between a.c. and d.c. can add considerable cost. That said, above a certain distance (about 35 miles for undersea cables and 500 miles for overhead cables), the lower cost of the HVDC electrical conductors can outweigh the cost of the electronics required.

THY
1A/50

Figure 4.43 Thyristor circuit symbol

Principle of operation

A thyristor acts like a semiconductor version of a mechanical switch, having two states; in other words it is either 'on' or 'off' with nothing in between. This is how they gained their name from the Greek word *thyra* (which means 'door'), the inference being something that is either open or closed.

The thyristor is very similar to a diode, with the exception that it has an extra terminal (the gate) which is used to activate it. Effectively in its normal or 'forward biased' state, the thyristor acts as an open circuit between anode and cathode, thus preventing current flow through the device. This is known as the 'forward blocking' state.

However, the thyristor can allow current to flow through it by the application of a control (gate) current to the gate terminal. It is this concept that allows a small signal at the gate to control the switching of a higher power load. In this respect the thyristor is performing in a similar way to a relay.

Once activated, a thyristor doesn't require a control (gate) current to continue operating and will therefore continue to conduct until either the supply voltage is turned off or reversed, or when a minimum 'holding' current is no longer maintained between the anode and cathode.

These concepts are shown in the Figure 4.44.

Did you know?

The conversion from a.c. to d.c. is known as rectification and from d.c. to a.c. as inversion

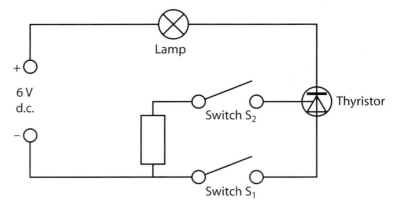

Figure 4.44 Circuit diagram of a thyristor

In Figure 4.44, switch S_1 acts as a master isolator and no supply is present at either the thyristor or at switch S_2. Closing switch S_1 will allow a supply to be present at the thyristor, but there is no signal at the gate terminal as switch S_2 is open and therefore no current will flow to the indicator lamp. However, if we now close switch S_2 the gate will be energised and the thyristor will operate, thus allowing current to flow through it to the indicator lamp. The current at the anode would be large enough in this situation to allow the thyristor to continue operating, even if we opened switch S_2.

The control of a.c. power can also be achieved with the thyristor by allowing current to be supplied to the load during part of each half cycle. If a gate pulse is applied automatically at a certain time during each positive half cycle of the input, then the thyristor will conduct during that period until it falls to zero for the negative half cycle.

You will see from Figure 4.45 that the gate pulse (mA) occurs at the peak of the a.c. input (V). During negative half cycles the thyristor is reverse biased and will not conduct; it will not conduct again until half-way through the next positive half cycle. Current actually flows for only a quarter of the cycle, but by changing the timing of the gate pulses, this can be decreased further or increased. The power supplied to the load can be varied from zero to half wave rectified d.c.

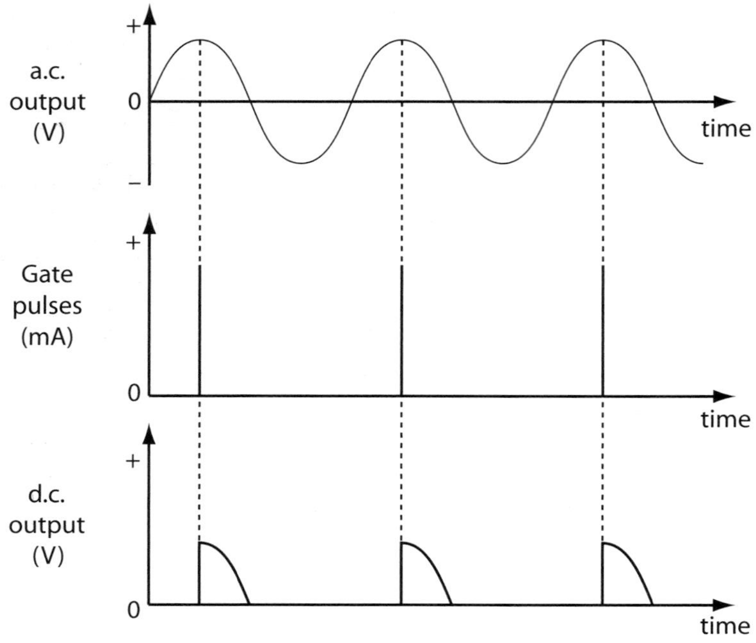

Figure 4.45 Supply of current to load during part of each half cycle

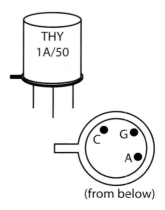

Figure 4.46 Typical small value thyristor

Thyristor testing

To test thyristors a simple circuit needs to be constructed as shown in Figure 4.47. When switch B only is closed the lamp will not light, but when switch A is closed the lamp lights to full brilliance. The lamp will remain illuminated even when switch A is opened. This shows that the thyristor is operating correctly. Once a voltage has been applied to the gate the thyristor becomes forward conducting like a diode and the gate loses control.

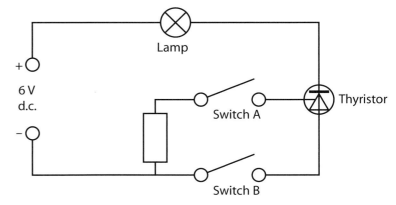

Figure 4.47 Typical small value thyristor

The triac

The triac is a three terminal semiconductor for controlling current in either direction. The schematic symbol for a triac is shown in Figure 4.48.

If we look at Figure 4.48 more closely, we can see that the symbol looks like two thyristors that have been connected in parallel, albeit in opposite facing directions and with only one gate terminal. We refer to this type of arrangement as an inverse parallel connection.

The main power terminals on a triac are designated as MT1 (Main Terminal 1) and MT2. When the voltage on MT2 is positive with regard to MT1, if we were to apply a positive voltage to the gate terminal the left thyristor would conduct. When the voltage is reversed and a negative voltage is applied to the gate, the right thyristor would conduct.

As with the thyristor generally, a minimum holding current must be maintained in order to keep a triac conducting. A triac therefore generally operates in the same way as the thyristor, but operating in both a forward and reverse direction. It is therefore sometimes referred to as a bidirectional thyristor, in that it can conduct electricity in both directions. One disadvantage of this is that triacs can require a fairly high current pulse to turn them on.

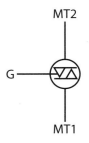

Figure 4.48 Triac symbol

The diac

Before consideration is given to practical triac applications and circuits, it is necessary to examine the diac. This device is often used in triac triggering circuits because it, along with a resistor-capacitor network, produces an ideal pulse-style waveform. It does this without any sophisticated additional circuitry due to its electrical characteristics. Also, it provides a degree of protection against spurious triggering from electrical noise (voltage spikes).

The symbol is shown in Figure 4.49. The device operates like two breakdown (zener) diodes connected in series, back to back. It acts as an open switch until the applied voltage reaches about 32–35 volts, when it will conduct.

Did you know?

Triacs were originally developed for and used extensively in the consumer market. They are used in many low power control applications such as food mixers, electric drills and lamp dimmers etc.

Figure 4.49 Diac symbol

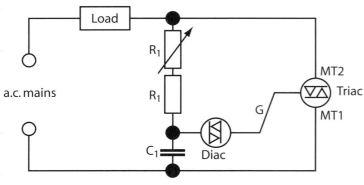

Figure 4.50 GLS lamp dimmer circuit

Lamp dimmer circuit

Figure 4.50 shows a typical GLS lamp dimmer circuit.

The GLS lamp has a tungsten filament, which allows it to operate at about 2500°C, and is wired in series with the triac. The variable resistor is part of a trigger network providing a variable voltage into the gate circuit, which contains a diac connected in series. Increasing the value of the resistor increases the time taken for the capacitor to reach its charge level to pass current into the diac circuit. Reducing the resistance allows the triac to switch on faster in each half cycle. By this adjustment the light output of the lamp can be controlled from zero to full brightness.

The capacitor is connected in series with the variable resistor. This combination is designed to produce a variable phase shift into the gate circuit of the diac. When the p.d. across the capacitor rises, enough current flows into the diac to switch on the triac.

The diac is a triggering device having a relatively high switch on voltage (32–35 volts) and acts as an open switch until the capacitor p.d. reaches the required voltage level.

The triac is a two-directional thyristor, which is triggered on both halves of each cycle. This allows it to conduct current in either direction of the a.c. supply. Its gate is in series with the diac, allowing it to receive positive and negative pulses.

A relatively high resistive value resistor R_2 (100 Ω) is placed in series with a capacitor to reduce false triggering of the triac by mains voltage interference. The capacitor is of a low value (0.1 μF). This combination is known as the **snubber circuit**.

Rectification

Rectification is the conversion of an a.c. supply to a d.c. supply. Despite the common use of a.c. systems in our day-to-day work as an electrician, there are many applications (e.g. electronic circuits and equipment) that require a d.c. supply. The following section looks at the different forms this can take.

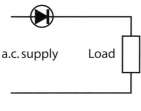

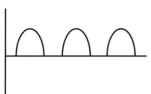

Figure 4.51 Half-wave rectification

Half-wave rectification

We now know that a diode will only allow current to flow in one direction. It does this when the anode is more positive than the cathode. In the case of an a.c. circuit, this means that only the positive half cycles are allowed 'through' the diode and, as a result, we end up with a signal that resembles a series of 'pulses'. This tends to be unsuitable for most applications but can be used in situations such as battery charging. A transformer is also commonly used at the supply side to ensure that the output voltage is to the required level. The waveform for this form of rectification would look as in Figure 4.51.

On the job: On call

You are asked to look at a problem at a small private residential care home. When you arrive the warden explains that there is a problem with the nurse call system. The system has recently been installed but the electrical contractor that fitted it has gone into receivership, so no one is quite sure of its operation. Additionally, the home requires another 'patient call' button to be installed in a further bedroom. The warden has a circuit diagram of the system, which is shown below.

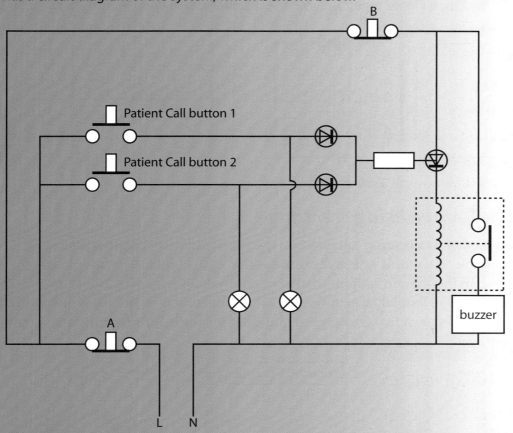

All components, with the exception of the patient call buttons, are located inside the nurse call panel, which is located at the nurses' station. Looking at the diagram, identify the components and prepare a written report for the warden, explaining in writing how the system operates. Now produce a revised circuit diagram to show how an additional patient could be added to the system. (Please assume that the values and ratings of any components for this exercise will be acceptable.) Your new drawing and report will then be held by the warden for future reference.

Full-wave rectification

We have seen that half-wave rectification occurs when one diode allows the positive half cycles to pass through it. However, we can connect two diodes together to give a more even supply. We call this type of circuit **biphase**. In this method we connect the anodes of the diodes to the opposite ends of the secondary winding of a centre-tapped transformer. As the anode voltages will be 180 degrees out of phase with each other, one diode will effectively rectify the positive half cycle and one will rectify the negative half cycle. The output current will still appear to be a series of pulses but they will be much closer together, with the waveform shown in Figure 4.52.

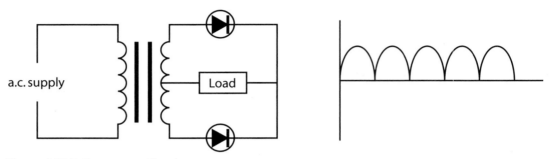

Figure 4.52 Full-wave rectification

The full-wave bridge rectifier

This method of rectification must rely on the use of a centre-tapped transformer but the output waveform will be the same as that of the biphase circuit previously described. In this system are four diodes, connected in such a way that at any instant in time two of the four will be conducting. The connections would be as in Figure 4.53, where we have shown two drawings to represent the route through the network for each half cycle.

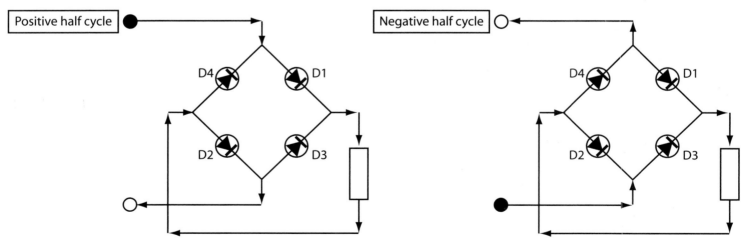

Figure 4.53 Full-wave rectification

The circuits that we have looked at so far convert a.c. into a supply which, although never going negative, is still not a true d.c. supply. This brings us to the next stage of the story – smoothing.

Smoothing

We have seen that the waveform produced by our circuits so far could best be described as having the appearance of a rough sea. The output current is not at a constant value but is constantly changing. This, as we have said before, is acceptable for battery charging but useless for electronic circuits where a smooth supply voltage is required.

To make it useful for electronic circuits, we need to smooth out the waveform by creating a situation that is sometimes referred to as **ripple-free**. In essence there are three ways to achieve this, namely capacitor smoothing, choke smoothing and filter circuits.

Capacitor smoothing

If we connect a capacitor in parallel across the load, then the capacitor will charge up when the rectifier allows a flow of current and discharge when the rectifier voltage is less than the capacitor. However, the most effective smoothing comes under no-load conditions. The heavier the load current, the heavier the ripple. This means that the capacitor is only useful as a smoothing device for small output currents.

Choke smoothing

If we connect an inductor in series with our load, then the changing current through the inductor will induce an e.m.f. in opposition to the current that produced it. Thus the e.m.f. will try to maintain a steady current. Unlike the capacitor, this means that the heavier the ripple (rate of change of current) the more that ripple will be smoothed. Hence the choke is more useful in heavy current circuits.

Filter circuits

This is the name given to a circuit that removes the ripple and is basically a combination of the two previous methods. The most effective of these is the capacitor input filter, shown in Figure 4.54.

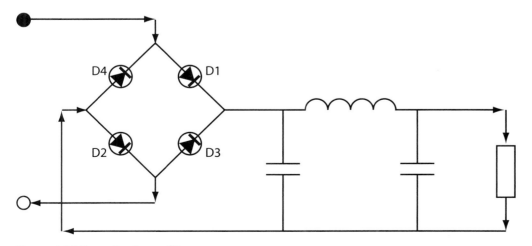

Figure 4.54 Capacitor input filter

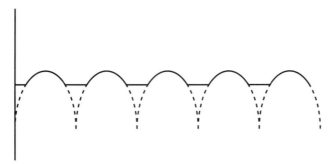

Figure 4.55 Capacitor input filter

The waveform for the filter circuit that we have just spoken about is shown in Figure 4.55, where the dotted line indicates the waveform before smoothing.

Three-phase rectifier circuits

Whereas Figure 4.55 indicates that a reasonably smooth waveform can be obtained from a single-phase system, we can get a much smoother wave from the three-phase supply mains. To do so we use six diodes connected as a three-phase bridge circuit. These types of rectifier are used to provide high-powered d.c. supplies.

Transistors

What is usually referred to as simply a transistor, but is more accurately described as a bipolar transistor, is a semiconductor device. This has two p–n junctions. It is capable of producing current amplification and, with an added load resistor, both a load and voltage power gain can be achieved.

A bipolar transistor consists of three separate regions or areas of doped semiconductor material and, depending on the configuration of these regions, it is possible to manufacture two basic types of device.

When the construction is such that a central n-type region is sandwiched between two p-type outer regions, a pnp transistor is formed, as in Figure 4.56.

Transistor

Did you know?

The term **transistor** is derived from the words 'transfer-resistor'. This is because in a transistor approximately the same current is transferred from a low to a high resistance region. The term **bipolar** means both electrons and holes are involved in the action of this type of transistor

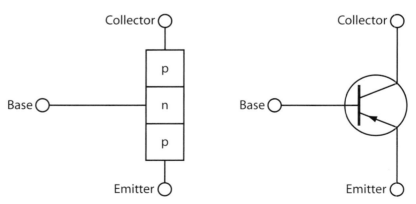

Figure 4.56 pnp transistor and its associated circuit symbols

If the regions are reversed as in Figure 4.57, an npn transistor is formed.

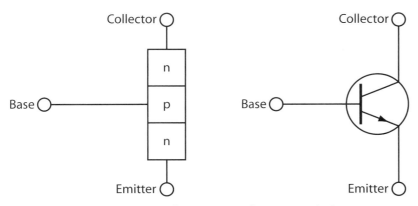

Figure 4.57 npn transistor and its associated circuit symbols

In both cases the outer regions are called the emitter and collector respectively; the central area is called the base.

The arrow in the circuit symbol for the pnp device points towards the base, whereas in the npn device it points away from it. The arrow indicates the direction in which conventional current would normally flow through the device. Electron flow is in the other direction.

Note that in these idealised diagrams the collector and emitter regions are shown to be the same size. This is not so in practice; the collector region is made physically larger since it normally has to dissipate the greater power during operation. Further, the base region is physically very thin, typically only a fraction of a micron (a micron is one-millionth of a metre).

Hard-wire connections are made to the three regions internally; wires are then brought out through the casing to provide an external means of connection to each region. Either silicon or germanium semiconductor materials may be used in the fabrication of the transistor, but silicon is preferred for reasons of temperature stability.

Transistor operation

For transistors to operate three conditions must be met (see Figure 4.58).

1. The base must be very thin.

2. Majority carriers in the base must be very few.

3. The base-emitter junction must be forward biased and the base-collector junction reverse biased.

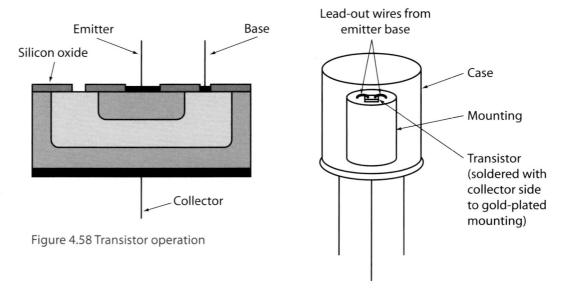

Figure 4.58 Transistor operation

Electrons from the emitter enter the base and diffuse through it. Due to the shape of the base most electrons reach the base-collector junction and are swept into the collector by the strong positive potential. A few electrons stay in the base long enough to meet the indigenous holes present and recombination takes place.

To maintain the forward bias on the base-emitter junction, holes enter the base from the base bias battery. It is this base current which maintains the base-emitter forward bias and therefore controls the size of the emitter current entering the base. The greater the forward bias on the base-emitter junction, the greater the number of emitter current carriers entering the base.

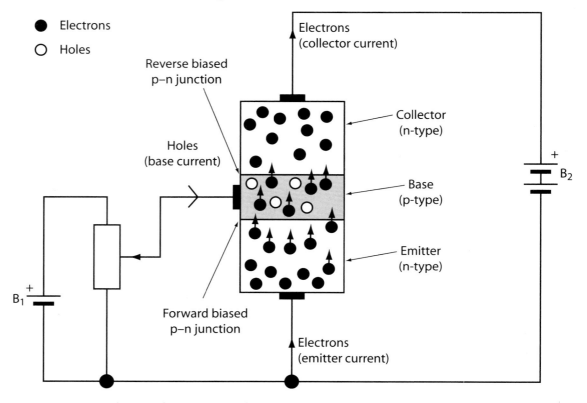

Figure 4.59 Circuit diagram for operation of transistor

The collector current is always a fixed proportion of the emitter current set by the thinness of the base and the amount of doping. Holes from the emitter enter the base and diffuse through it – see Figure 4.59.

Due to the shape of the base most holes reach the base-collector junction and are swept into the collector by the strong negative potential.

Current amplification

Consider, for example, as in Figure 4.60, a base bias of some 630 mV has caused a base current of approximately 0.5 mA to flow, but more importantly has initiated a collector current of around 50 mA.

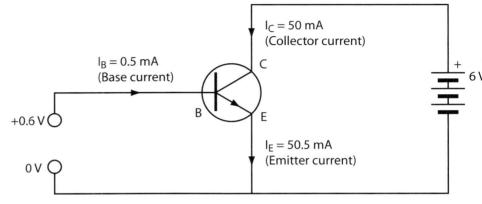

Figure 4.60 Current flow in transistor

Although these currents may seem insignificant, it is the comparison between the base and collector currents which is of interest. This relationship between I_B and I_C is termed the 'static value of the short-circuit forward current transfer'. We normally just call it the gain of the transistor. It is simply a measure of how much amplification we would get. The symbol that we use for this is h_{FE}. This is the ratio between the continuous output current (collector current) and the continuous input current (base current). Thus, when I_B is around 0.5 mA and I_C 50 mA the ratio is:

$$h_{FE} = \frac{I_C}{I_B}$$

$$= \frac{50 \text{ mA}}{0.5 \text{ mA}}$$

(approximately equal to 100.)

Note: There are no units since this is a ratio.

It can, therefore, be said that a small base current initiated by the controlling forward base-bias voltage produces a significantly higher value of collector current to flow, dependent on the value of h_{FE} for the transistor. Thus, current amplification has been achieved.

Voltage amplification

Mention was previously made to the derivation of the word 'transistor', and the device was described as transferring current from a low resistive circuit to approximately the same current in a high resistive circuit. This is an npn transistor so the low resistive reference is the emitter circuit and the high resistive reference is the collector circuit, the current in both being almost identical.

The reason the emitter circuit is classed as having a low resistance is because it contains the forward biased (p–n) base-emitter junction. Conversely, the collector circuit contains the reverse biased (n–p) base-collector junction, which is in the order of tens of thousands of ohms (it varies with I_C). In order to produce a voltage output from the collector, a load resistor (R_B) is added to the collector circuit as indicated in the circuit diagram in Figure 4.61.

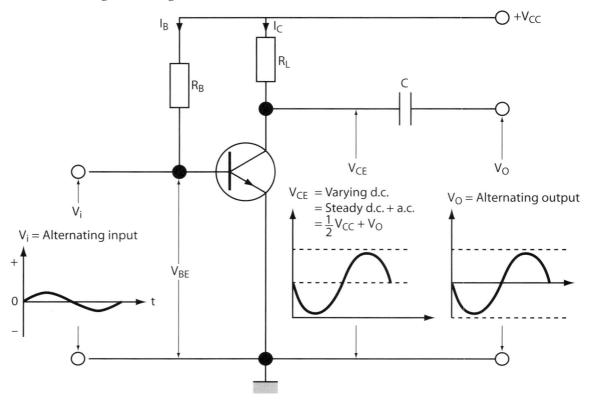

Figure 4.61 Voltage amplification

This shows the simplest circuit for a voltage amplifier. To see how voltage amplification occurs we have to consider that there is no input across V_i, which is called the quiescent (quiet) state. For transistor action to take place the base emitter junction V_{BE} must be forward biased (and has to remain so when V_i goes positive and negative due to the a.c. signal input).

By introducing resistor R_B between collector and base, a small current I_B will flow from V_{CC} through R_B into the base and down to 0 V via the emitter, thus keeping the transistor running (ticking over).

Component resistor values R_B and R_L are chosen so that the steady base current I_B makes the quiescent collector-emitter voltage V_{CE} about half the power supply voltage V_{CC}. This allows V_O to replicate the input signal V at an amplified voltage with a 180 degrees phase shift. When an a.c. signal is applied to the input V_i and goes positive it increases V_{BE} slightly to around 0.61 V. When V_i swings negative, V_{BE} drops slightly to 0.59 V. As a result, a small alternating current is superimposed on the quiescent base current I_B, which in effect is a varying d.c. current.

The collector emitter voltage (V_{CE}) is a varying d.c. voltage or an alternating voltage superimposed on a normal steady d.c. voltage. The capacitor C is there to block the d.c. voltage but allow the alternating voltage to pass on to the next stage. So, in summing up, a bipolar transistor will act as a voltage amplifier if:

- it has a suitable collector load R_L

- it is biased so that the quiescent value V_{CE} is around half the value of V_{CC}, which is known as the class A condition

- the transistor and load together bring about voltage amplification

- the output is 180 degrees out of phase with the input signal as Figure 4.61 indicates

- the emitter is common to the input, output and power supply circuits and is usually taken as the reference point for all voltages, i.e. 0 V. It is called 'common', 'ground' or 'earth' if connected to earth.

Transistor as a switch

We have looked at the transistor as an amplifier of current and voltage. If we connect the transistor as in Figure 4.62, we can operate it as a switch. Compared with other electrically operated switches, transistors have many advantages, whether in discrete or integrated circuit (IC) form. They are small, cheap, reliable, have no moving parts and can switch millions of times per second. The transistor can be thought of as the perfect switch with infinite resistance when 'off', no resistance when 'on' and which changes instantaneously from one state to another, using up no power.

Figure 4.62 shows the basic circuit for an npn common emitter as in previous diagrams with a load resistor R_L connected in series with the supply (V_{CC}) and the collector.

R_B prevents excessive base currents which would seriously damage the transistor when forward biased. With no input across V_I, the transistor is basically turned off. This means that there will be no current (I_C) through R_L; therefore there will be no volt-drop across R_L so the +V_{CC} voltage (6 V) will appear across the output V_{CE}.

If we now connect a supply of between 2–6 V across V_I input, the transistor will switch on, current will flow through the collector load resistor R_L and down to common, making the output V_{CE} around 0 V. From this we can state that:

- when the input $V_i = 0$ V, the output $V_{CE} = 6$ V

- when the input $V_i = 2$ V–6 V, the output $V_{CE} = 0$ V.

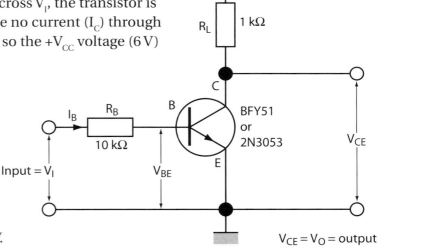

Figure 4.62 Circuit diagram for transistor used as a switch

From this we can see that the transistor is either High (6 V) or Low (0 V), or we can confirm, like a switch, that it is 'On' (6 V) or 'Off' (0 V).

This circuit can be used in alarms and switch relays for all types of processes and is the basic stage for programmable logic control (plc) which uses logic gates with either one or zero to represent what the output is from a possible input.

In Figure 4.63 basic logic gate circuits with their inputs/outputs, 'truth table' and symbols are identified.

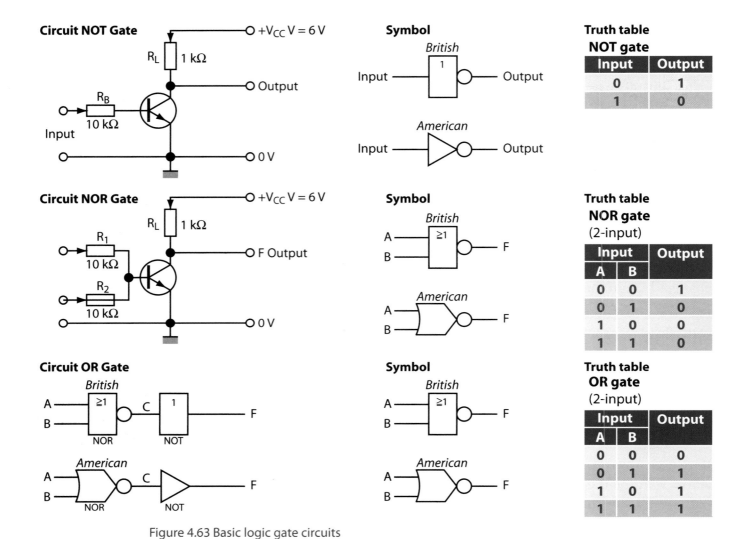

Figure 4.63 Basic logic gate circuits

Circuit NAND Gate

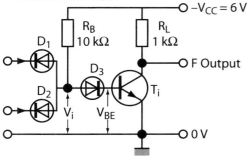

Circuit AND Gate

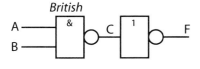

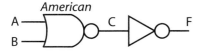

Figure 4.63 Basic logic gate circuits (continued)

Symbol

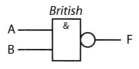

Truth table

NAND gate
(2-input)

A	B	F
0	0	1
0	1	1
1	0	1
1	1	0

Symbol

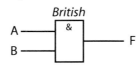

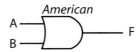

Truth table

AND gate
(2-input)

A	B	C	F
0	0	1	0
0	1	1	0
1	0	1	0
1	1	0	1

Symbol

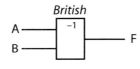

Truth table

Exclusive-OR gate

A	B	F
0	0	0
0	1	1
1	0	1
1	1	0

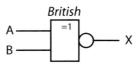

Exclusive NOR gate

Input		Output
A	B	
0	0	1
0	1	0
1	1	0
1	1	1

Testing transistors

As all transistors consist of either an npn or a pnp construction, the testing of them is similar to diodes. Special meters with three terminals for testing transistors are available and many testing instruments have this facility. However, an ohmmeter can be used for testing a transistor to check if it is conducting correctly. The following results should be obtained from a transistor assuming that the red lead of an ohmmeter is positive.

(Note: This is not always the case. With some older analogue meters, the battery connections internally are the opposite way round, so it is always good to check both ways across base and emitter as shown in Figures 4.64 and 4.65.)

A good npn transistor will give the following readings.

- Red to base and black to collector or emitter will give a low resistance reading.
- However, if the connections are reversed it will result in a high resistance reading.
- Connections of any polarity between the collector and emitter will also give a high reading.

A good pnp transistor will give the following readings.

- Black to base and red to collector or emitter will give a low resistance reading.
- However, if the connections are reversed a high resistance reading will be observed.
- Connections of either polarity between the collector and emitter will give a high resistance reading.

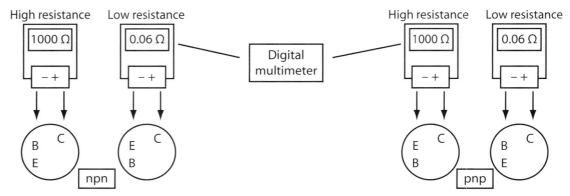

Figure 4.64 Testing transistors with a digital multimeter

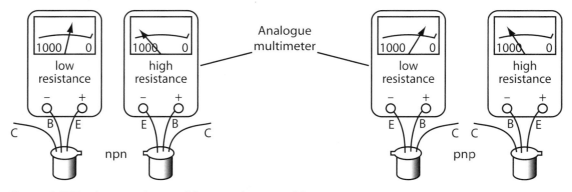

Figure 4.65 Testing transistors with an analogue multimeter

Functions of basic electronic components

On completion of this topic area the candidate will be able to describe the functions of basic electronic components.

Resistors

There are two basic types of resistor: fixed and variable. The resistance value of a fixed resistor cannot be changed by mechanical means (though its normal value can be affected by temperature or other effects). Variable resistors have some means of adjustment (usually a spindle or slider). The method of construction, specifications and features of both fixed and variable resistor types vary, depending on what they are to be used for.

Fixed resistors

Making a resistor simply consists of taking some material of a known resistivity and making the dimensions (csa and length) of a piece of that material such that the resistance between the two points at which leads are attached (for connecting into a circuit) is the value required.

Fixed resistors

Most of the very earliest resistors were made by taking a length of resistance wire (wire made from a metal with a relatively high resistivity, such as brass) and winding this on to a support rod of insulating material. The resistance value of the resulting resistor depended on the length of the wire used and its cross-sectional area.

This method is still used today, though it has been somewhat refined. For example, the resistance wire is usually covered with some form of enamel glazing or ceramic material to protect it from the atmosphere and mechanical damage. The external and internal view of a typical wire-wound resistor is shown in Figure 4.66.

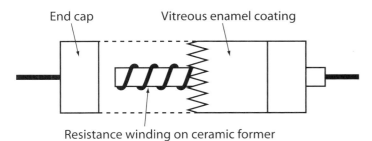

End cap　Vitreous enamel coating

Resistance winding on ceramic former

Figure 4.66 Typical wire wound resistor

Most wire-wound resistors can operate at fairly high temperatures without suffering damage, so they are useful in applications where some power may be dissipated. They are, however, relatively difficult to mass produce, which makes them expensive. Techniques for making resistors from materials other than wire have now been developed for low-power applications.

Resistor manufacture advanced considerably when techniques were developed for coating an insulating rod (usually ceramic or glass) with a thin film of resistive material (see Figure 4.67). The resistive materials in common use today are carbon and metal oxides. Metal end caps fitted with leads are pushed over the ends of the coated rod and the whole assembly is coated with several layers of very tough varnish or similar material to protect the film from the atmosphere and from knocks during handling. These resistors can be mass produced with great precision at very low cost.

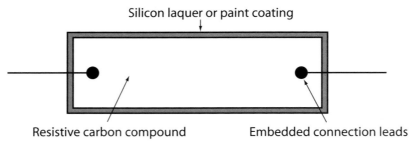

Figure 4.67 Film resistor

Variable resistors

Development of techniques for manufacturing variable resistors closely followed that of fixed resistors, though they required some sort of sliding contact together with a fixed resistor element.

Wire-wound variable resistors are often made by winding resistance wire on to a flat strip of insulating material, which is then wrapped into a nearly complete circle. A sliding contact arm is made to run in contact with the turns of wire as they wrap over the edge of the wire strip, as in Figure 4.68 below. Straight versions are also possible. A straight former is used and the wiper travels in a straight line along it as shown in Figure 4.69.

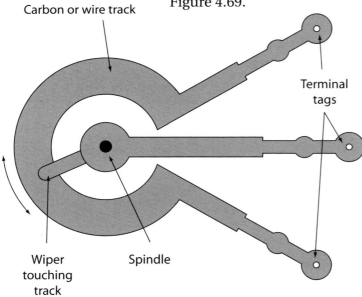

Figure 4.68 Layout of internal track of rotary variable resistor

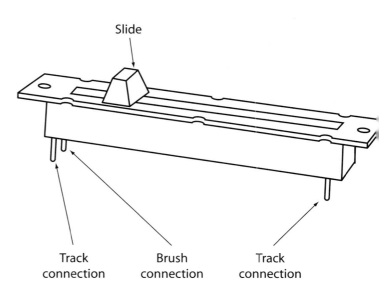

Figure 4.69 Linear variable resistor

While wire-wound resistors are ideal for certain applications, there are many others where their size, cost and other disadvantages make them unattractive. As a consequence, alternative types have been developed.

The early alternative to the wire-wound construction was to make the resistive element (on which the wiper rubs) out of a carbon composition, deposited or moulded as a track and shaped as a nearly completed circle on an insulating support plate. Alternative materials for the track are carbon films or metal alloys of a metal oxide and a ceramic (cermet); again, straight versions are possible.

Preferred values

In theory, it is possible to have resistors in every imaginable resistance value; from zero to tens or hundreds of megohms. In reality, however, such an enormous range would be totally impractical to manufacture and store. From the point of view of the circuit designer, it is not usually necessary either.

So, rather than an overwhelming number of individual resistance values, what manufacturers do is make a limited range of preferred resistance values. In electronics, we use the preferred value closest to the actual value we need.

A resistor with a preferred value of 1000 Ω and a 10 per cent tolerance can have any value between 900 Ω and 1100 Ω. The next largest preferred value, which would give the maximum possible range of resistance values without too much overlap, is 1200 Ω. This can have a value between 1080 Ω and 1320 Ω.

Together, these two preferred value resistors cover all possible resistance values between 900 Ω and 1320 Ω. The next preferred values would be 1460 Ω, 1785 Ω etc.

There is a series of preferred values for each tolerance level as shown in Table 4.01, so that every possible numerical value is covered.

Resistance markings

There is obviously the need for the resistor manufacturer to provide some sort of markings on each resistor so that it can be identified.

The user should be able to tell by looking at the resistor what its nominal resistance value and tolerance are. Various methods of marking this information on to each resistor have been used; sometimes a resistor code will use numbers and letters rather than colours.

E6 series 20% tol	E12 series 10% tol	E24 series 5% tol
10	10	10
		11
	12	12
		13
15	15	15
		16
	18	18
		20
22	22	22
		24
	27	27
		30
33	33	33
		36
	39	39
		43
47	47	47
		51
	56	56
		62
68	68	68
		75
	82	82
		91

Table 4.01 Table of preferred values

Where physical size permits, putting the actual value on the resistor in figures and letters has an obvious advantage in terms of easy interpretation. However, again because of size restrictions, we do not use the actual words and instead use a code system. This code is necessary because, when using small text on a small object, certain symbols and the decimal point become very hard to see. This code system is also commonly used to represent resistance values on circuit diagrams for the same reason.

In reality resistance values are generally given in either Ω, kΩ or MΩ, using numbers from 1–999 as a prefix (e.g. 10 Ω, 567 k Ω etc.). In the code system we replace Ω, kΩ and MΩ and represent them instead by using the following letters:

- Ω = R
- kΩ = K
- MΩ = M.

These letters are now inserted wherever the decimal point would have been in the value. So, for example, a resistor of value 10 Ω resistor would now be shown as 10R, and a resistor of value 567 kΩ resistor would become 567K.

Table 4.02 gives some more examples of this code system. Table 4.03 shows the letters that are then commonly used to represent the tolerance values. These letters are added at the end of the resistor marking so that, for example, a resistor of value 2.7 MΩ with a tolerance of ±10% would be shown as 2M7K.

0.1 Ω	is coded	R10
0.22 Ω	is coded	R22
1.0 Ω	is coded	1R0
3.3 Ω	is coded	3R3
15 Ω	is coded	15R
390 Ω	is coded	390R
1.8 Ω	is coded	1R8
47 Ω	is coded	47R
820 kΩ	is coded	820K
2.7 MΩ	is coded	2M7

Table 4.02 Examples of resistance value codes

F	=	± 1%
G	=	± 2%
J	=	± 5%
K	=	± 10%
M	=	± 20%
N	=	± 30%

Table 4.03 Codes for common tolerance values

Resistor coding

Standard colour code

Many resistors are so small that it is impractical to print their value on them. Instead, they are marked with a code that uses bands of colour. Located at one end of the component, it is these bands that identify the resistor's value and tolerance.

Most general resistors have four bands of colour, but high precision resistors are often marked with a five-colour band system. No matter which system is being used, the value of the colours is the same.

Remember

Before you read a resistor, turn it so that the end with bands is on the left-hand side. Now you read the bands from left to right (as shown in Figures 4.70 and 4.71)

Resistor colour code

Band colour	Value
Black	0
Brown	1
Red	2
Orange	3
Yellow	4
Green	5
Blue	6
Violet	7
Grey	8
White	9
Gold	0.1
Silver	0.01

Tolerance colour code

Band colour	±%
Brown	1
Red	2
Gold	5
Silver	10
None	20

What this means

Band 1 First figure of value
Band 2 Second figure of value
Band 3 Number of zeros/multiplier
Band 4 Tolerance (±%) See below

Note that the bands are closer to one end than the other

Figure 4.70 Resistor and tolerance colour code

Brown	Green	Orange	Gold
1	5	000	5%

Resistor is 15000 Ω or 15K Ω ± 5%

Yellow	Violet	Blue	Silver
4	7	000000	10%

Resistor is 47000000 Ω or 47M Ω ± 10%

Orange	Orange	Brown	Gold
3	3	0	5%

Resistor is 330 Ω ± 5%

Brown	Green	Red	Gold
1	5	00	5%

Resistor is 1500 Ω or 1K5 ± 5%

Figure 4.71 Examples of colour coding

Example 1

A resistor is colour-coded red, yellow, orange, gold. Determine the value of the resistor.

- First band red (First digit) 2
- Second band yellow (Second digit) 4
- Third band orange (No. of zeros) 3
- Fourth band gold (Tolerance) 5%.

The value is 24,000 Ω ±5%.

Example 2

A resistor is colour-coded yellow, yellow, blue, silver. Determine the value of the resistor.

- First band yellow (First digit) 4
- Second band yellow (Second digit) 4
- Third band blue (No. of zeros) 6
- Fourth band silver (Tolerance) 10%.

The value is 44,000,000 Ω ±10%.

Example 3

A resistor is colour-coded violet, orange, brown, gold. Determine the value of the resistor.

- First band violet (First digit) 7
- Second band orange (Second digit) 3
- Third band brown (No. of zeros) 1
- Fourth band gold (Tolerance) 5%.

The value is 730 Ω ± 5%.

Example 4

A resistor is colour coded green, red, yellow, silver. Determine the value of the resistor.

- First band green (First digit) 5
- Second band red (Second digit) 2
- Third band yellow (No. of zeros) 4
- Fourth band silver (Tolerance) 10%.

The value is 520,000 Ω ±10%.

Thermistors

A thermistor is a resistor which is temperature sensitive. The general appearance is shown in the photograph. They can be supplied in various shapes and are used for the measurement and control of temperature up to their maximum useful temperature limit of about 300°C. They are very sensitive and because of their small construction are useful for measuring temperatures in inaccessible places.

Thermistors are used for measuring the temperature of motor windings and sensing overloads. The thermistor can be wired into the control circuit so that it automatically cuts the supply to the motor when the motor windings overheat, thus preventing damage to the windings.

Thermistors can have a temperature coefficient that may be positive (PTC) or negative (NTC). With a PTC thermistor the resistance of the thermistor increases as the surrounding temperature increases. With the NTC thermistor the resistance decreases as the temperature increases.

The rated resistance of a thermistor may be identified by a standard colour code or by a single body colour used only for thermistors. Typical values are shown in Table 4.04.

Colour	Resistance
Red	3,000 Ω
Orange	5,000 Ω
Yellow	10,000 Ω
Green	30,000 Ω
Violet	100,000 Ω

Table 4.04 Colour coding for rated resistance of thermistor

Remember

To help you remember the resistor colour, learn this rhyme:

Barbara	Black	0
Brown	Brown	1
Runs	Red	2
Over	Orange	3
Your	Yellow	4
Garden	Green	5
But	Blue	6
Violet	Violet	7
Grey	Grey	8
Won't	White	9

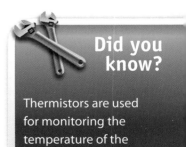

General appearance of a thermistor

Did you know?

Thermistors are used for monitoring the temperature of the water in a motor car

Light dependent resistors

Light-dependent resistors

These resistors are sensitive to light. They consist of a clear window with a cadmium sulphide film under it. When light shines on to the film its resistance varies, with the resistance reducing as the light increases.

These resistors are commonly found in street lighting. You may sometimes observe street lights switching on during a thunderstorm in the daytime. This is because the sunlight is obscured by the dark thunderclouds, thus increasing the resistance, which in turn controls the light 'on' circuit.

Capacitors
Capacitor coding

To identify a capacitor the following details must be known: the capacitance, working voltage, type of construction and polarity (if any). The identification of capacitors is not easy because of the wide variation in shapes and sizes. In the majority of cases the capacitance will be printed on the body of the capacitor, which often gives a positive identification that the component is a capacitor.

The capacitance value is the farad (symbol F); this was named after the English scientist Michael Faraday. However, for practical purposes the farad is much too large, and in electrical installation work and electronics we use fractions of a farad, as follows.

- 1 microfarad = 1 μF = 1×10^{-6} F
- 1 nanofarad = 1 nF = 1×10^{-9} F
- 1 picofarad = 1 pF = 1×10^{-12} F

The power factor correction capacitor found in fluorescent luminaries would have a value typically of 8 μF at a working voltage of 400 V. One microfarad is one million times greater than one picofarad.

The working voltage of a capacitor is the maximum voltage that can be applied between the plates of the capacitor without breaking down the dielectric insulating material.

It was quite common for capacitors to be marked with colour codes but today relatively few capacitors are colour-coded. At one time nearly all plastic foil type capacitors were colour-coded, as in Figure 4.72 (page 141), but this method of marking is rarely encountered. However, it is a useful skill to know and be able to use the colour-coding method as shown in Table 4.05.

This method is based on the standard four-band resistor colour coding. The first three bands indicate the value in normal resistor fashion, but the value is in picofarads. To convert this into a value in nanofarads it is merely necessary to divide by 1000. Divide the marked value by 1,000,000 if a value in microfarads is required. The fourth band indicates the tolerance, but the colour coding is different to the resistor equivalent. The fifth band shows the maximum working voltage of the component. Details of this colour coding are shown in Figure 4.72 and Table 4.05.

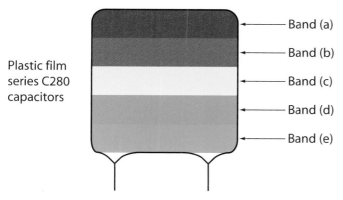

Plastic film series C280 capacitors

Band (a)
Band (b)
Band (c)
Band (d)
Band (e)

Figure 4.72 Capacitor colour band

Standard capacitor colour coding

Colour	1st digit	2nd digit	3rd digit	Tol. band	Max. voltage
Black		0	None	20%	
Brown	1	1	1		100 V
Red	2	2	2		250 V
Orange	3	3	3		
Yellow	4	4	4		400 V
Green	5	5	5	5%	
Blue	6	6	6		630 V
Violet	7	7	7		
Grey	8	8	8		
White	9	9	9	10%	

Table 4.05 Standard capacitor colour coding

Bands are then read from top to bottom. Digit 1 gives the first number of the component value; the second digit gives the second number. The third band gives the number of zeros to be added after the first two numbers and the fourth band indicates the capacitor tolerance, which is normally black 20%, white 10% and green 5%.

Example 1

A plastic film capacitor is colour-coded from top to bottom brown, red, yellow, black, red. Determine the value of the capacitor, its tolerance and working voltage.

- Band (a) – brown = 1

- Band (b) – red = 2

- Band (c) – yellow = 4 multiply by 10,000

- Band (d) – black = 20% tolerance

- Band (e) – red = 250 volts

The capacitor has a value of 120,000 pF or 0.12 μF with a tolerance of 20 per cent and a maximum working voltage of 250 volts.

Example 2

A plastic film capacitor is colour-coded from top to bottom orange, orange, yellow, green, yellow. Determine the value of the capacitor, its tolerance and working voltage.

- Band (a) – orange = 3

- Band (b) – orange = 3

- Band (c) – yellow = 4 multiply by 10,000

- Band (d) – green = 5% tolerance

- Band (e) – yellow = 400 volts

The capacitor has a value of 330,000 pF or 0.33 μF with a tolerance of 5 per cent and a maximum working voltage of 400 volts.

Example 3

A plastic film capacitor is colour-coded from top to bottom violet, blue, orange, black, and brown. Determine the value of the capacitor, its tolerance and working voltage.

- Band (a) – violet = 7

- Band (b) – blue = 6

- Band (c) – orange = 3 multiply by 1000

- Band (d) – black = 20% tolerance

- Band (e) – brown = 100 volts

The capacitor has a value of 76,000 pF or 0.076 μF with a tolerance of 20 per cent and a maximum working voltage of 100 volts.

Often the value of a capacitor is simply written on its body, possibly together with the tolerance and/or its maximum operating voltage. The tolerance rating may be omitted and it is generally higher for capacitors than resistors. Most modern resistors have tolerances of 5 per cent or better, but for capacitors the tolerance rating is generally 10 per cent or 20 per cent. The tolerance figure is more likely to be marked on a close tolerance capacitor than a normal 10 per cent or 20 per cent type.

The most popular form of value marking on modern capacitors is for the value to be written on the components in some slightly cryptic form. Small ceramic capacitors generally have the value marked in much the same way as the value is written on a circuit diagram.

Where the value includes a decimal point, it is standard practice to use the prefix for the multiplication factor in place of the decimal point. This is the same practice as was used for resistors.

The abbreviation μ means microfarad; n means nanofarad; p means picofarad. Therefore:

- a 3.5 pF capacitor would be abbreviated to 3p5

- a 12 pF capacitor would be abbreviated to 12p

- a 300 pF capacitor would be abbreviated to 300p or n30

- a 4500 pF capacitor would be abbreviated to 4n5.

Remember

1,000 pF = 1 nF = 0.001 μF

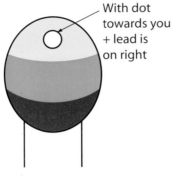

With dot towards you + lead is on right

Figure 4.73 Tantalum capacitor

Polarity

Once the size, type and d.c. voltage rating of a capacitor have been determined it remains to ensure that its polarity is known. Some capacitors are constructed in such a way that if the component is operated with the wrong polarity its properties as a capacitor will be destroyed, especially electrolytic capacitors. Polarity may be indicated by + or – as appropriate. Electrolytic capacitors that are contained within metal cans will have the can casing as the negative connection. If there are no markings a slight indentation in the case will indicate the positive end. Tantalum capacitors have a spot on one side as shown in Figure 4.73. When this spot is facing you the right-hand lead will indicate the positive connection.

Transistors

There are two types of standard transistors, npn and pnp. Each have different circuit symbols. The letters refer to the layers of semiconductor material used to make the transistor. Most transistors used today are npn because this is the easiest type to make from silicon.

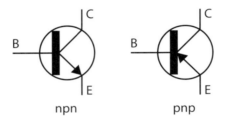

npn   pnp

The leads are labelled base (B), collector (C) and emitter (E). These terms refer to the internal operation of a transistor.

Figure 4.74 npn and pnp transistors

Connecting

Transistors have three leads which must be connected the correct way round. This is important, as a wrongly connected transistor may be instantly damaged when you switch the circuit on. If in doubt, refer to a supplier's catalogue to identify the leads.

Figure 4.75 shows the leads for some of the most common case styles. Note that this diagram shows the view from below the transistor with the leads facing towards you.

Figure 4.75 Common case styles for transistors

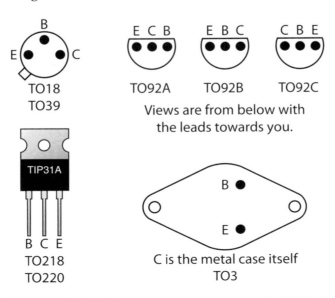

TO18
TO39

E C B
TO92A

E B C
TO92B

C B E
TO92C

Views are from below with the leads towards you.

TIP31A

B C E
TO218
TO220

B ●
E ●
C is the metal case itself
TO3

Transistor codes

There are three main series of transistor codes used in the UK.

Codes beginning with B (or A)

The first letter (B) is for silicon. Sometimes it is A, which is for germanium (rarely used these days). The second letter indicates the type; C means low power audio frequency; D means high power audio frequency; F means low power high frequency. The rest of the code identifies the particular transistor and is determined by the manufacturer. Examples of this type include BC108, BC478.

Codes beginning with TIP

TIP refers to the manufacturer: Texas Instruments Power transistor. The letter at the end identifies versions with different voltage ratings. An example of this type is TIP31A.

Codes beginning with 2N

The initial 2N identifies the part as a transistor and the rest of the code identifies the particular transistor. An example of this type is 2N3053.

Choosing a transistor

Most projects will specify a particular transistor, but if necessary you can usually substitute an equivalent transistor from the wide range available. The most important properties to look for are the maximum collector current I_C and the current gain h_{FE}. To make selection easier most suppliers group their transistors in categories. These are determined either by their typical use or maximum power rating.

To make a final choice you will need to consult the tables of technical data which are normally provided in catalogues. They contain a great deal of useful information but they can be difficult to understand if you are not familiar with the abbreviations used.

The key abbreviations used are as follows:

- I_C max – the maximum collector current
- V_{CE} max – the maximum voltage across the collector-emitter junction
- h_{FE} – the d.c. current gain – 100@20mA means a gain of 100 at 20 mA
- P_{tot} max – the maximum power which can be developed in the transistor.

Tables 4.06 and 4.07 show some examples.

Code	Case style	I_C max	V_{CE} max	h_{FE} min	V_{CE} max	Typical use
BC108	TO18	100 mA	20 V	110	300 mW	Low power
TIP31C	TO220	3 A	100 V	10	40 W	High power

Table 4.06 Technical data for NPN transistors

Code	Case style	I_C max	V_{CE} max	h_{FE} min	V_{CE} max	Typical use
BC108	TO18	200 mA	25 V	120	600 mW	Low power
TIP31C	TO220	3 A	60 V	25	40 W	High power

Table 4.07 Technical data for PNP transistors

Field effect transistors (FETs)

Field effect transistor devices first appeared as separate (or discrete) transistors, but now the field effect concept is employed in the fabrication of large-scale integration arrays such as semiconductor memories, microprocessors, calculators and digital watches.

There are two types of field effect transistor: the junction gate field effect transistor, which is usually abbreviated to JUGFET, JFET or FET; and the metal oxide semiconductor field effect transistor known as the MOSFET. They differ significantly from the bipolar transistor in their characteristics, operation and construction.

The main advantages of an FET over a bipolar transistor are as follows.

- Its operation depends on the flow of majority current carriers only. It is, therefore, often described as a unipolar transistor.

- It is simpler to fabricate and occupies less space in integrated form.

- Its input resistance is extremely high, typically above 10 MΩ especially for MOSFET devices. In practice, this is why voltage measuring devices such as oscilloscopes and digital voltmeters employ the FET in their input circuitry, so that the voltage being measured is not altered by the connection of the instrument.

- Electrical noise is the production of random minute voltages caused by the movement of current carriers through the transistor structure. Since the FET does not employ minority carriers, it therefore has the advantage of producing much lower noise levels compared with the bipolar transistor.

- Also due to its unipolar nature it is more stable during changes of temperature.

The main disadvantages of an FET over its bipolar counterpart are as follows.

- Its very high input impedance renders it susceptible to internal damage from static electricity.

- Its voltage gain for a given bandwidth is lower. Although this may be a disadvantage at low frequencies (below 10 MHz), at high frequencies the low noise amplification that an FET achieves is highly desirable. This facet of FET operation, though, is usually only exploited in radio and TV applications, where very small high frequency signals need to be amplified.

- The FET cannot switch from its fully on to its fully off condition as fast as a bi-polar transistor. It is for this reason that digital logic circuits employing MOSFET technology are slower than bipolar equivalents, although faster switching speeds are being achieved as FET production technology continues to advance.

Remember

The main reason why the FET has such a differing characteristic from the bipolar transistor is because current flow through the device is controlled by an electric field, which is not the case with the bipolar transistor. It is for this reason that the FET is considered to be a voltage operated device rather than current operated

Figure 4.76 illustrates the basic construction of the FET, which consists of a channel of n-type semiconductor material with two connections, source (S) and drain (D). A third connection is made at the gate (G), which is made of p-type material to control the n-channel current. The symbol is shown in Figure 4.77.

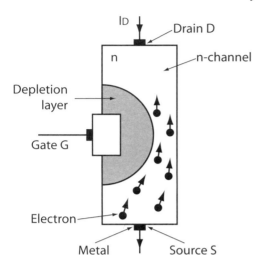

Figure 4.76 Basic construction of field effect transistor (FET)

Figure 4.77 Field effect transistor (FET) symbol

In theory, the drain connection is made positive with respect to the source, and electrons are attracted towards the D terminal. If the gate is made negative there will be reverse bias between G and S, which will limit the number of electrons passing from S to D.

The gate and source are connected to a variable voltage supply, such as a potentiometer, and increasing or lowering the voltage makes G more or less negative, which in turn reduces or increases the drain current.

Diodes

Diodes allow electricity to flow in one direction only. The arrow of the circuit symbol shows the direction in which the current can flow. Diodes are the electrical version of a valve and early diodes were actually called valves. Figure 4.78 shows a diode and its symbol.

Diodes must be connected the correct way round. They may be labelled **a** or + for anode and **k** or – for cathode. The cathode is marked by a line painted on the body of the diode. Diodes are labelled with their code in small print. You may need a magnifying glass to read this on small signal diodes!

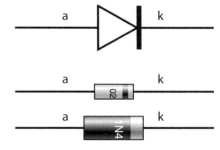

Figure 4.78 A diode and its symbol

Rectifier diodes are used in power supplies to convert a.c. to d.c. This process is called rectification. Diodes can also be used elsewhere in circuits. In places where a large current must pass through the diode, but only in one direction, the diode will block current from the opposite direction.

All rectifier diodes are made from silicon and therefore have a forward voltage drop of 0.7 V. Table 4.08 shows maximum current and maximum reverse voltage for some diodes. The 1N4001 is suitable for most low voltage circuits with a current of less than 1 A.

Diode	Maximum current	Maximum reverse voltage
1N4001	1 A	50 V
1N4002	1 A	100 V
1N4007	1 A	1000 V
1N5401	3 A	100 V
1N5408	3 A	1000 V

Table 4.08 Maximum current and reverse voltage for some diodes

Zener diodes

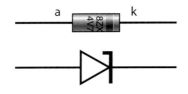

Figure 4.79 Zener diodes

These diodes can be distinguished by the code printed on them, for example BZX or BZY. As with an ordinary diode, a line is painted around the body to indicate the position of the cathode.

Zener diodes are rated by their breakdown voltage and maximum power.

- The minimum voltage available is 2.7 V.
- Power ratings of 400m W and 1.3 W are common.

Integrated circuits

Integrated circuits are complete electronic circuits housed within a plastic case (known as the **black box**). The chip contains all the components required, which may include diodes, resistors, capacitors and transistors etc.

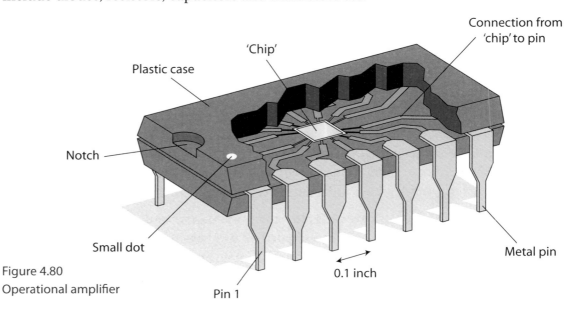

Figure 4.80
Operational amplifier

There are several categories which include analogue, digital and memories. The basic layout is shown in Figure 4.79, which is an operational amplifier (**dual in-line IC**).

The plastic case has a notch at one end and, if you look at the back of the case with the notch at the top, Pin 1 is always the first one on the left-hand side, sometimes noted with a small dot. The other pin numbers follow down the left-hand side, 2, 3 and 4 then back up the right-hand side from the bottom right to the top, 5, 6, 7 and 8. You can get some chips with up to 32 pins or more.

Definition

Dual in-line ICs – type of IC with the pins lined up down each side

Photo cell and light-dependent resistor

The photo cell shown in Figure 4.81 changes light (also infrared and ultraviolet radiation) into electrical signals and is useful in burglar and fire alarms as well as in counting and automatic control systems. Photoconductive cells or light-dependent resistors make use of the semiconductors whose resistance decreases as the intensity of light falling on them increases. The effect is due to the energy of the light setting free electrons from donor atoms in the semiconductor, making it more conductive. The main use of this type of device is for outside lights along the streets, roads and motorways. There are also smaller versions for domestic use within homes and businesses.

Figure 4.81 Photo cell and its circuit symbol

Photodiode

A photodiode is a p–n junction designed to be responsive to optical input. As a result they are provided with either a window or optical fibre connection that allows light to fall on the sensitive part of the device.

Photodiodes can be used in either zero bias or reverse bias. When zero biased, light falling on the diode causes a voltage to develop across the device, leading to a current in the forward bias direction. This is called the photovoltaic effect and is the principle of operation of the solar cell, this being a large number of photodiodes. The use of the solar cell can then be seen in providing power to equipment such as calculators, solar panels and satellites orbiting Earth.

When reverse biased, diodes usually have extremely high resistance. This resistance is reduced when light of an appropriate frequency shines on the junction. When light falls on the junction, the energy from the light breaks down bonds in the 'lattice' structure of the semiconductor material, thus producing electrons and allowing current to flow. Circuits based on this effect are more sensitive to light than ones based on the photovoltaic effect. Consequently, the photodiode is used as a fast counter or in light meters to measure light intensity.

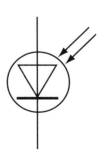

Figure 4.82 Circuit symbol for photodiode

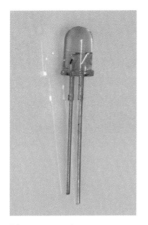

Phototransistors

Phototransistor

Phototransistors are solid-state light detectors that can be used to provide either an analogue or digital signal. They are a photodiode amplifier integrated onto a single silicon chip. They have been designed to overcome photodiode faults caused by unity gain, and are used when fast counting of units is required.

Phototransistors are available in a wide range of styles including epoxy coated, transfer moulded, cast hermetic and chip form. They can be used with almost any visible or near infrared light source such as IREDs, neons, fluorescent, incandescent bulbs, lasers, flames sources and sunlight. For sizes and types it would be necessary to refer to manufacturers catalogues.

Opto-coupler

The opto-coupler, also known as an opto-isolator, consists of an LED combined with a photodiode or phototransistor in the same package, as shown in Figure 4.83 and the photograph.

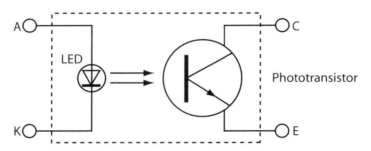

Figure 4.83 Opto-coupler circuit

Opto-coupler package

The opto-coupler package allows the transfer of signals, analogue or digital, from one circuit to another when the second circuit cannot be connected electrically to the first, for example due to different voltages.

Light (or infrared) from the LED falls on the photodiode/transistor which is shielded from outside light. A typical use for one of these is in a VCR, to detect the end or start of the tape.

Fibre optic link

The simplified block diagram in Figure 4.84 shows a system of communication which can be several thousand kilometres in length. On the far left, information such as speech or visual pictures is input as electrical signals. They are then pulse code modulated in the coder and changed into equivalent digital light signals by the optical transmitter via a miniature laser or LED at the end of the fibre optic cable.

Figure 4.84 Simplified block diagram of fibre optic link

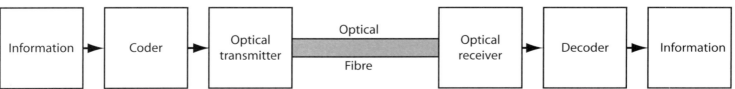

The light is transmitted down the cable to the optical receiver, which uses a photodiode or phototransistor to convert the incoming signals back to electrical signals before they are decoded back into legible information.

The advantages of this type of link over a conventional communication system are:

- high information-carrying capacity
- free from the noise of electrical interference
- greater distance can be covered, as there is no volt drop
- the cable is lighter, smaller and easier to handle than copper
- cross-talk between adjacent channels is negligible
- it offers greater security to the user.

The fibre optic cable

The fibre optic cable (see Figure 4.85) has a glass core of higher refractive index than the glass cladding around it. This maintains the light beam within the core by total internal reflection at the point where the core and the cladding touch.

This is similar to the insulation of the single core cable preventing the current leaking from the conductor. The beam of light bounces off the outer surface of the core in a zigzag formation along its length.

There are two main types of cable: multimode and singlemode (monomode).

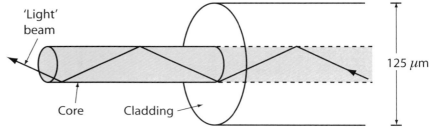

Figure 4.85 Fibre optic cable

Multimode

The wider core of the multimode fibre (Figure 4.86) allows the infrared to travel by different reflected paths or modes. Paths that cross the core more often are longer and take more time to travel along the fibre. This can sometimes cause errors and loss of information.

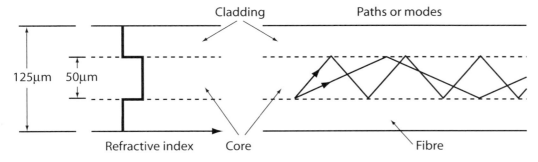

Figure 4.86 Multimode fibre optic cable

Singlemode

The core of the singlemode fibre (Figure 4.87) is about one-tenth of the multimode and only straight through transmission is possible.

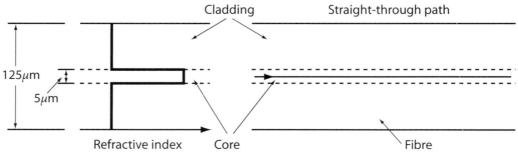

Figure 4.87 Singlemode fibre optic cable

Solid state temperature device (thermistors)

Thermistors monitor the changes in temperature of components, which could become damaged due to excessive heat. As the temperature rises there is a reaction within the semiconductor for its resistance to either rise (positive temperature coefficient) or fall (negative temperature coefficient) depending upon its make up. There are two types of thermistor:

- ptc – made from barium titanate
- ntc – made from oxides of nickel, manganese, copper and cobalt.

Lighting

On completion of this topic area the candidate will be able to describe incandescent lighting and discharge lighting and will have a knowledge of the application of the inverse square law, cosine law and lumen method of calculation.

Electric lamps

Illumination by electricity has been available for over 100 years. In that time it has changed in many ways, though many of the same ideas are still in use. The first type of electric lamp was the 'arc lamp', which used electrodes to draw an electrode through the air; this is now known as discharge lighting. This was quite an unsophisticated use of electricity and many accidents and fires resulted from it. Regulations had to be developed to control discharge lighting installations, and it is interesting that the first edition of the Regulations, introduced in 1882, was entitled 'Rules and Regulations for the Prevention of Fire Risks Arising from Electric Lighting'.

Remember

Electric lamps create light by heating a fine filament of wire connected across an electrical supply. GLS lamps use a tungsten filament. The filament wire in an incandescent lamp reaches a temperature of about 2500–2900°C

Incandescent lamps and, in particular, GLS lamps (or light bulb as they are commonly known) were introduced at Level 2. In this section we will focus on further examples of incandescent lighting.

Tungsten halogen lamps

These types of lamps were introduced in the 1950s. For their operation the tungsten filament is enclosed in a gas-filled quartz tube together with a carefully controlled amount of halogen such as iodine. Figure 4.87 illustrates the linear tungsten halogen lamp.

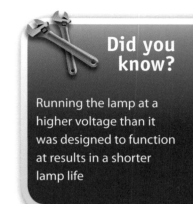

Did you know?

Running the lamp at a higher voltage than it was designed to function at results in a shorter lamp life

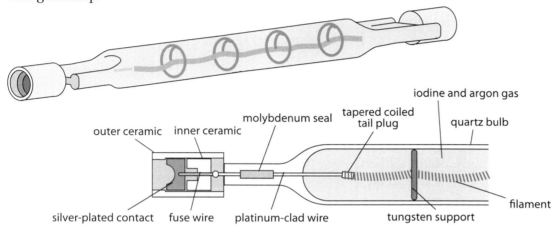

outer ceramic   inner ceramic   molybdenum seal   tapered coiled tail plug   iodine and argon gas   quartz bulb

silver-plated contact   fuse wire   platinum-clad wire   tungsten support   filament

Figure 4.87 Linear tungsten lamp

Operation of tungsten halogen lamps

The inclusion of argon and iodine in the quartz tube allows the filament to burn at a much higher temperature than the incandescent lamp. The inclusion of halogen gas produces a regeneration effect, which prolongs the life of the lamp.

As small particles of tungsten fall away from the filament, they combine with the iodine passing over the face of the quartz tube, thereby forming a new compound. Convection currents in the tube cause this new compound to rise, passing over the filament. The intense heat of the filament causes the compound to separate into its component parts and the tungsten is deposited back on the filament.

The lamp should not be touched with bare fingers as this would deposit grease on the quartz glass tube. This would lead to small cracks and fissures in the tube when the lamp heats rapidly, causing the lamp to fail. If accidentally touched on installation, the lamp should be cleaned with methylated spirit before use. The linear type of lamp must be installed within 4 degrees of the horizontal to prevent the halogen vapour migrating to one end of the tube, causing early failure.

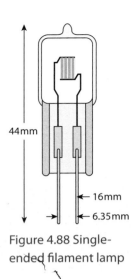

44mm

16mm

6.35mm

Figure 4.88 Single-ended filament lamp

These types of lamps have many advantages, which include:

- increased lamp life (up to 2000 hours)
- increase in efficacy (up to 23 lumens per watt)
- reduction in lamp size.

There are two basic designs of halogen lamps: the double-ended linear lamp and the single-ended lamp, with both contacts embedded in the seal at one end (see Figure 4.88). This type of lamp has been produced to work on extra-low voltages and is used extensively in the automobile industry for vehicle headlamps; it may also be used for display spotlights where extra-low voltage is desirable. The single-ended lamp may be supplied from an in-built 230-volt/12-volt transformer.

Lamp caps

To connect the lamps to the supply there are several different caps in use. These can be divided into three groups: bayonet-fitting, screw-fitting and plug-fitting. In general the lamp number is the diameter of the cap or the distance between pins, in millimetres. The Goliath Edison Screw (GES) design is reserved for lamps of a higher wattage, such as 300 and 500 watt lamps, used for floodlighting purposes.

Discharge lighting

The principles of discharge lighting were introduced at Level 2. In that section low-pressure mercury vapour lamps were described. These use a heated fluorescent tube containing a gas (such as krypton or argon) and an arc of mercury vapour. The heat leads to a flow of electrons, which ionises the gas. This reaction is maintained by the mercury. This emits ultra-violet light, which is absorbed by the phosphor coating and transformed into visible light. The high voltage needed to create this reaction is usually achieved by a **transformer** or a **choke**.

In this section we will look at further examples of discharge lighting:

- the glow-type starter circuit
- semi-resonant starting
- high frequency
- stroboscopic effect
- other methods of starting the fluorescent tube
- other discharge lamps.

The glow-type starter circuit

In the starter, a set of normally open contacts is mounted on bi-metal strips and enclosed in an atmosphere of helium gas. When switched on, a glow discharge takes place around the open contacts in the starter, which heats up the bi-metal strips causing them to bend and touch each other. This puts the electrodes at either end of the fluorescent tube in circuit and they warm up, giving off a cloud of electrons; simultaneously an intense magnetic field is building up in the choke, which is also in circuit.

The glow in the starter ceases once the contacts are touching so that the bi-metal strips cool down and they spring apart again. This momentarily breaks the circuit, causing the magnetic field in the choke to rapidly collapse. The high back-e.m.f. produced provides the high voltage required for ionisation of the gas and enables the main discharge across the lamp to take place. The voltage across the tube under running conditions is not sufficient to operate the starter and so the contacts remain open.

The resistance of the ionised gas gets lower and lower as it warms up and conducts more current. This could lead to disintegration of the tube. However, the choke has a secondary function as a current-limiting device: the impedance of the choke limits the current through the lamp, keeping it in balance. This is one reason why it is often referred to as ballast.

This type of starting may not succeed first time and can result in the characteristic flashing on/off when initially switching on.

Remember

Once the tube strikes, current no longer passes through the starter; therefore the starter takes no further part in the circuit

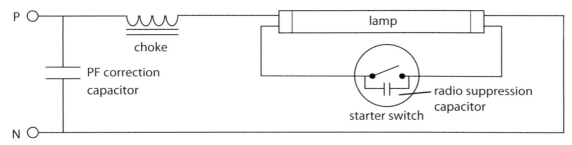

Figure 4.89 Glow-type starter circuit

Semi-resonant starting

In this circuit the place of the **choke** is taken by a specially wound transformer. Current flows through the primary coils to one cathode of the lamp and then through the secondary coil, which is wound in opposition to the primary coil. A fairly large **capacitor** is connected between the secondary coil and the second cathode of the lamp (the other end of which is connected to the neutral).

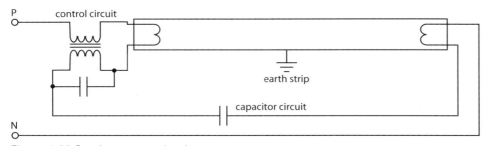

Figure 4.90 Semi-resonant circuit

The current that flows through this circuit heats the cathodes and, as the circuit is predominantly capacitive, the pre-start current leads the main voltage. Owing to the primary and secondary windings being in opposition, the voltages developed across them are 180 degrees out of phase, so that the voltage across the tube is increased, causing the arc to strike. The primary windings then behave as a choke, thus stabilising the current in the arc. The circuit has the advantage of high power factor and easy starting at low temperatures.

High frequency

Standard fluorescent circuits operate on a mains supply frequency of 50 Hz; however, high-frequency circuits operate on about 30,000 Hz. There are a number of advantages to high frequency:

- higher lamp efficacy
- first-time starting
- noise-free
- the ballast shuts down automatically on lamp failure.

The higher efficacy for this type of circuit can lead to savings of at least 10 per cent, and in some large installations savings may be as high as 30 per cent. Many high-frequency electronic ballasts will operate on a wide range of standard fluorescent lamps. The high-frequency circuit will switch on the lamp within 0.4 seconds and there should be no flicker (unlike glow-type starter circuits). Stroboscopic effect is not a problem as this light does not flicker on/off due to its high frequency.

However, a disadvantage of this type of circuit is that supply cables installed within the luminaire must not run adjacent to the leads connected to the ballast output terminals as interference may occur. Also, the initial cost of these luminaires is greater than traditional glow-type switch starts.

Stroboscopic effect

A simple example of the stroboscopic effect is when watching the wheels rotate on a horse-drawn cart on television – it appears that the wheels are stationary or even going backwards. This phenomenon is brought about by the fact that the spokes on the wheels are being rotated at about the same number of revolutions per second as the frames per second of the film being shot. This effect can also be produced by fluorescent lighting.

The discharge across the electrodes is extinguished 100 times per second, producing a flicker effect. This flicker is not normally observable but it can cause the stroboscopic effect. For example, if rotating machinery is illuminated from a single source it will appear to have slowed down, changed direction of rotation or even stopped. This is a potentially dangerous situation to any operator of rotating machines in an engineering workshop.

Did you know?

Certain frequencies of stroboscopic flash can induce degrees of drowsiness, headache, eye fatigue and, in extreme cases, epileptic fit

However, the stroboscopic effect can be harnessed, for example to check the speed of a CD player or the speed of a motor vehicle for calibration purposes.

By using one of the following methods, the stroboscopic effect can be overcome or reduced. The first three maintain light falling on the rotating machine. The fourth makes the effective flicker at a different frequency from the operating frequency of the machines.

(i)　Tungsten filament lamps can be fitted locally to lathes and pillar-drilling machines etc. This will lessen the effect but will not eliminate it completely.

(ii)　Adjacent fluorescent fittings can be connected to different phases of the supply. Because in a three-phase supply the phases are 120 degrees out of phase with each other, the light falling on the machine will arrive from two different sources. Each of these will be flickering at a different time and will interfere with each other, reducing the stroboscopic effect.

(iii)　Twin lamps can be wired on lead-lag circuits, thus counteracting each other. The lead-lag circuit, as the name implies, is a circuit that contains one lamp in which the **power factor** leads the other – hence the other lags. Using the leading current effect of a capacitor and the lagging current effect of an inductor produces the lead-lag effect. The lagging effect is produced naturally when an inductor is used in the circuit as shown in Figure 4.91. The leading effect uses a series capacitor, which has a greater effect than the inductor in the circuit. When these two circuits are combined as shown there is no need for further power factor correction as one circuit will correct the other.

(iv)　The use of high-frequency fluorescent lighting reduces the effect by about 60 per cent.

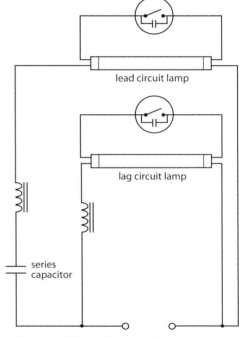

Figure 4.91 Lead lag circuit

Other methods of starting the fluorescent tube

Quick start

The electrodes of this type of circuit are rapidly pre-heated by the end windings of an autotransformer, so that a quick start is possible. The method of ionisation of the gas is the same as in the semi-resonant circuit. Difficulties may occur in starting if the voltage is low.

Thermal starter circuit

This type of circuit has waned in popularity over the years. However, there are still thousands of these fittings in service, so it is worth describing them. In this starter, the normally closed contacts are mounted on a bi-metal strip. A small heater coil heats one of these when the supply is switched on. This causes the strip to bend and the contacts to open, creating the momentary high voltage and starting the circuit discharge. The starter is easily recognised as it has four pins instead of the usual two, the extra pins being for the heater connection.

Other discharge lamps

These include:

- high-pressure mercury vapour
- high-pressure sodium
- low-pressure sodium.

Typically these lamps consist of a glass (arc) tube, in which an electrode has been fitted at either end. Rare metals such as lanthanum, lutetium or mercury may be added. The tube is pressurised with gases such as argon and neon before sealing.

Lamps such as these will not normally start using a raw mains voltage but require an external igniter to provide a high-voltage injection in order to set up an electromagnetic field. This excites ions present in the gas, causing them to collide with molecules of gas, resulting in the emission of visible light.

Regulations concerning lighting circuits

Although a number of regulations have been covered already in this book, there are several others that are applicable to the installation of lighting circuits.

Figure 4.92
Lighting circuit
regulations

Regulations concerning lighting circuits

(i) Where flexible conductors enter a luminaire as, for example, when a bulkhead fitting or batten lampholder is used, the conductors should be able to withstand any heat likely to be encountered or sleeved with heat resistant sleeving; see Regulation 522.2.2.

(ii) A ceiling rose, unless specially designed for the purpose, should have only one flexible cord; see Regulation 559.6.1.3.

(iii) A ceiling rose shall not be installed in any circuit operating at a voltage normally exceeding 250 volts: see Regulation 559.6.1.2.

(iv) Where a flexible cord supports or partly supports a luminaire, the maximum mass supported shall not impair the safety of the installation (Table 4F3A).

(v) For circuits on a TN or a TT system, the outer contact of every Edison screw or single-centre bayonet cap-type lamp holder shall be connected to the neutral conductor; see Regulation 559.6.1.8.

(vi) Semi-conductors used in dimmer controls may be used for functional switching (not isolators) provided that they comply with Sections 512 and 537 of BS 7671.

(vii) When installing lighting circuits, the current assumed is equivalent to the connected load, with a minimum of 100 watts per lampholder; see Table 1A of the *IEE On-Site Guide*. However, it should be noted that diversity could be applied to lighting circuits in accordance with Table 1B of the *IEE On-Site Guide*.

(viii) Final circuits for discharge lighting (this includes fluorescent luminaires) shall be capable of carrying the total steady current. Where this information is not available the demand in volt-amperes can be worked out by multiplying the rated watts by 1.8; see Table 1B of the *IEE On-Site Guide*.

On the job: Fluorescent fault

Your employer asks you to visit the owners of an engineering manufacturing factory to discuss a problem with them. Staff have reported that some rotating machines in the main factory area seem to be slowing down. The manager has investigated this and in actual fact the machines are not. The main factory area is lit by single tube fluorescent lighting fittings hung from chains.

1. How would you go about investigating the problem?

2. What do you think the problem might be?

Terms used to describe illumination calculations

Before we can look and apply calculations to illumination problems, it is important to have an understanding of the terms used. Below is a list of such terms.

Term	Symbol	Measurement	Description
Luminous intensity	I	Candela (cd)	The brightness and power of a light.
Luminous flux	F	Lumen (lm)	The amount of light emitted from a source.
Illuminance	E	Lux (lx) and measured in lumen/m²	The amount of light falling on a surface.
Luminous efficacy	K	Lumens per Watt (lm/w)	The ratio of luminous flux to power input.
Maintenance factor	Mf	Average of 0.8 when no specific data exists	This allows for gathering of dust on fittings, which reduces I. It can be calculated by dividing lumens on the work surface by the lumens on the same surface before dust gathered.
Co-efficient of utilisation	C of U	Average of 0.6 used	The illumination efficiency of the room, determined by the colours in the room, height of the ceiling and the type of luminaires. A list of measurements is published by the Illumination Engineering Society.
Space-height ratio		A ratio of 1.5:1 must not be exceeded	The relation between the rows of lights and their height. If even lighting is required, the space in the last row of lamps should not be more than half the space between the rows.

Table 4.09 Terms used to describe illumination problems

The inverse square law

The inverse square law recognises that as a light source is moved further away from a surface on which it is shining, the level of illumination of the surface is reduced by a factor of four. Look at the diagram below to see how this law works in practice.

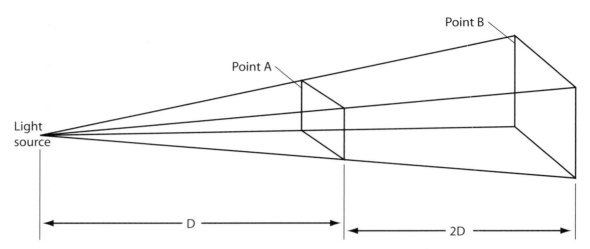

Figure 4.93 The inverse square law

The illumination at point A is E lumens/m²

The illumination at point B is $\dfrac{E \text{ lumens/m}^2}{4}$

The illumination of a surface is proportional to the intensity of the light source and inversely proportional to the square of the distance between the surface and the light source.

This can be expressed as: $E = \dfrac{I}{d^2}$

Where:

- I is the intensity in candela (cd)
- d is the distance in metres
- E is the illuminance in lux.

Example 1

A light source of 900 candela (cd) is situated 3m above a work surface.
Calculate:

(a) the illuminance directly below the lamp
(b) the illuminance if the lamp were to be moved to a height of 4 m above the work surface

(a) $E = \dfrac{I}{d^2}$ $E = \dfrac{900}{3^2} = \textbf{100 lux}$

(b) $E = \dfrac{I}{d^2}$ $E = \dfrac{900}{4^2} = \textbf{56.25 lux}$

Example 2

An incandescent lamp is suspended at 1.5m above a bench and gives an illuminance of 150 Lux on the surface of the bench, directly below the lamp. Calculate:

(a) the luminous intensity of the lamp
(b) the new illuminance on the bench, if the lamp were moved 0.5m lower.

(a) $E = \dfrac{I}{d^2}$ Therefore $I = E \times d^2$ $I = 150 \times 1.5^2 = \textbf{337.5 cd}$

(b) $E = \dfrac{I}{d^2}$ Therefore $E = \dfrac{337.5}{1^2} = 337.5\ lx$

Cosine law

The cosine law states the relationship in a normal triangle between the lengths of its sides and the cosine of one of its angles. The cosine is the ratio of the adjacent side to the hypotenuse of a right-angled triangle.

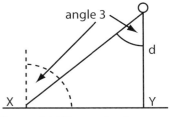

Figure 4.94 The cosine law

From the diagram it is clear that the distance from the light source to point X is further than the distance on the light source to point Y. Therefore the illuminance at point X is less than at point Y.

The level of illumination at point X is affected by the angle of the light falling on the surface.

We know that that level of illumination at point Y can be calculated by:

$$E = \dfrac{I}{d^2}$$

At point X the level of illumination is calculated using the cosine of the angle 3. Therefore:

$$Ex = \dfrac{I \times Cos\theta^3}{d^2}$$

Example

A 250 W Sodium vapour lamp emits a light of 22,500 candela and is situated 5m above the road. Calculate the illuminance:

(a) directly below the lamp

(b) at a position 6m along the road

(a) Illumination at point A directly below the lamp:

$$E = \frac{I}{d^2} \qquad E = \frac{22500}{25} = \textbf{900 lx}$$

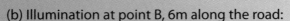

(b) Illumination at point B, 6m along the road:

$$E = \frac{I \times \cos\theta^3}{d^2} \qquad E = \frac{22,500 \times \cos\theta^3}{5^2}$$

We do not know the angle θ therefore we have to work it out:

$$\tan\theta = \frac{Opp}{Adj} \qquad \text{therefore} \qquad \tan\theta = \frac{6}{5} = 1.2$$

Now convert the tangent of the angle into an actual angle using inverse tan on the calculator.

Tan of 1.2 inversed = 50.2^0

Therefore the cosine of the angle, from the calculator = 0.64

Now complete the calculation:

$$E = \frac{I \times \cos\theta^3}{d^2} \qquad E = \frac{22500 \times 0.64^3}{5^2} \qquad = \textbf{236 lx}$$

The lumen method of calculation

Calculating the number of luminaires required in an area is an exacting science, and would require much more time and space than we have here to explain.

The lumen method of calculation forms the basis of this very complicated task. In its simplest form all that is needed is to establish the level of illumination required. Once this is known divide the level of illumination by the output of the fittings, making some allowance for the dirt and dust that will appear over a period of time (maintenance factor) and allowing for the decor of the room (coefficient of utilisation).

There are other factors that would also be taken into account for more accurate lighting levels, such as the mounting heights of the light fittings and the height of work surfaces and desks, as well as the performance factors of the luminaires to be fitted, polar curves for the light distribution and the performance of the individual lamps.

Below is an example that we will work through to try and explain the basic calculation.

Example

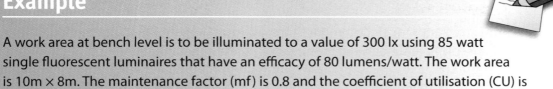

A work area at bench level is to be illuminated to a value of 300 lx using 85 watt single fluorescent luminaires that have an efficacy of 80 lumens/watt. The work area is 10m × 8m. The maintenance factor (mf) is 0.8 and the coefficient of utilisation (CU) is 0.6. Calculate the number of fittings required.

$$\text{Total lumens required (F)} = \frac{E\,(lx) \times Area}{mf \times CU}$$

$$F = \frac{300 \times (10 \times 8)}{0.8 \times 0.6}$$

Total lumens Required (F) = 50,000 lm

The efficacy of the lamps is 80 lm/W

$$\text{Therefore the power required} = \frac{Total\ lumens}{Efficacy} = \frac{50000}{80} = 625\ watts$$

$$\text{Therefore the number of lamps required} = \frac{Total\ power}{Rating\ of\ the\ lamps}$$

$$\text{No. of Lamps} = \frac{625}{85} = 7.35\ luminaires$$

As it is not possible to have 0.35 of a lamp go to the next whole number above 7.35. **Therefore the number of luminaires required is 8.**

FAQ

Q What is e.m.f.?

A E.m.f stands for electro-motive force or, in other words, the force to move the electrons. It is measured in volts and is given the symbol V.

Q What is an ampere?

An ampere (or Amp) is a measure of the flow rate in a circuit, the number of electrons per second passing a fixed point. If you measured the number of electrons per second you would end up with a huge number, so we call them amps instead. The symbol for amps is I.

Q What is resistance?

A Resistance is often described as 'the opposition to the flow of current' in a circuit. The more resistance a circuit has, the less current will flow in it (for a given voltage). Resistance is measured in Ohms and given the symbol R.

Q What is impedance?

A Impedance is very similar to resistance and can also be described as 'the opposition to the flow of current' in a (usually a.c.) circuit. In a given circuit, impedance is higher than resistance because it includes 'extra resistance' called reactance. Impedance is given the symbol Z and is measured in ohms.

Q What would happen if I connect an electrolytic capacitor the wrong way around?

A Don't do it! It usually goes bang and makes a mess!

Q I still don't understand what phasor diagrams are all about?

A Phasor diagrams are all about the 'time shift' between the voltage and current flowing in a circuit. Essentially, in a resistive circuit the voltage appears at the same time as the current flows, so there is no time shift and the lines are drawn on top of one another. In an inductive circuit the current flows after the voltage appears (remember coil current lags) so the lines are drawn with a phase angle (or time shift) between them. In a capacitive circuit the current flows before the voltage appears, so the current line is drawn (leading) or before the voltage line.

Q Does a thyristor or triac control device actually save energy?

A Yes. Unlike a variable resistor, which simply 'diverts' energy away from the device it controls, an electronic switch 'chops up' the waveform and prevents the energy from leaving the supply.

Q How does a transistor amplify?

A A transistor is simply an electronic variable switch which uses a small current (the base current) to control a larger one (the collector current). The result is an output current that is the same shape but larger than the input current.

Q What are logic gate circuits used for?

A Logic gates are a very important part of computer circuitry and allow an electric circuit to make a decision based on certain conditions. In its simplest form, even two light switches in series are a form of 'decision-making' circuit. If both switch 1 AND switch 2 are ON, then the lamp will light.

Q Why are discharge lamps much more efficient than filament or incandescent lamps?

A Mainly because they don't get anywhere near as hot, so they don't waste energy heating up the atmosphere!

Activity

1. Obtain a GLS Lamp of 60W and work out what the resistance of the filament should be using the power and ohms law formulae. Now use a multi-meter to measure the actual resistance of the lamp filament. Compare the two values and explain any difference between them.

2. Using a multi-meter, measure the resistance through a diode in each direction. Decide which end of the diode should be marked as positive.

3. Connect up a thyristor or triac controlled lamp dimming or speed control circuit and measure the average voltage across the load as the control knob is adjusted. If you have access to an oscilloscope you may like to look at the wave-shape by connecting it across the load.

Knowledge check

1. State the units of measurement and symbols for voltage, current and resistance.

2. State the formula for self-inductance.

3. What is the correct formula for inductive reactance?

4. List and describe the different types of dielectric found in capacitors and name the three factors that the value of capacitance depends on.

5. Capacitors of 10 μF, 12 μF and 40 μF are connected in series and then in parallel. Calculate the total capacitance in each case.

6. What is the correct formula for calculating impedance in a circuit containing resistance and inductance in series?

7. Describe with the aid of diagrams the process for testing diodes.

8. Describe the characteristics of a zener diode.

9. Describe with the aid of diagrams, voltage amplification in a transistor circuit.

10. Draw the lamp dimmer circuit and describe the operation of the individual components.

11. Name the four coloured bands found in the colour coding system for resistors and capacitors.

12. State a rhyme for remembering the colours in the resistor colour code system, then name the colours.

13. State the efficacy and average life of a linear tungsten halogen lamp.

14. State three methods of overcoming stroboscopic effects of discharge lighting.

chapter 5

Supply systems, protection and earthing

Unit 1 outcome 5

In order to successfully complete the construction of an electrical system, it is essential to be fully informed on where electricity comes from. The chapter will look at how electricity is generated and supplied, both to private homes and major industries.

This chapter will also explore systems used to protect electricians and users from the dangers of electrical supply. The principles of these are covered at Level 2. This chapter builds on concepts first explored in earlier chapters and at Level 2.

On completion of this chapter the candidate will be able to describe:

- **electricity supply systems**
- **industrial distribution systems**
- **transformers**
- **switchgear**
- **earthing systems**
- **protection systems.**

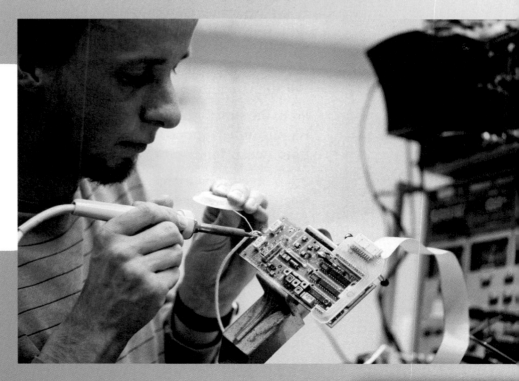

The distribution of electricity

On completion of this topic area the candidate will be able to describe electrical supply systems.

Generation, transmission and distribution

This topic area is covered in detail at Level 2. Please refer back to this title for information. For this chapter, we will briefly recap the principles of the process of electricity production.

Electricity is generated in power stations through the turning of the shaft of a three-phase alternator. This is often powered by steam, heated through either coal, gas, oil or nuclear power. From the alternator electricity goes to a transformer. The output of most alternators is 25,000 V (25 kV) and needs to be transformed to:

- 400 kV or 275 kV for the super grid
- 132 kV for the original national grid
- 66 kV and 33 kV for secondary transmission
- 11 kV for local sub-station distribution and industry
- 400 V for commercial consumer supplies
- 230 V for domestic consumer supplies.

Electricity is transmitted at very high voltage in order to compensate for power losses in power lines. When voltage is increased, current reduces for a given value of power. The effect of volt drop can be calculated using the formula:

$$Vd = IR$$

So, for example, if a supply cable with resistance 2 Ω were carrying 1000 amps, then the volt drop along its length would be 2000 volts. If the supply cable was carrying 100 amp, then the volt drop would be reduced to 200 volts.

From the super grid, electricity is transmitted to the original national grid. Electricity is transmitted around these grids via steel-cored aluminium conductors, suspended from steel pylons.

From here sub-stations transform the grid supply down to 11 kV and distribute the electricity to a series of local sub-stations. These transform the supply to 400 V/230 V and distribute the power to the customer. The supply arrives at the customers' **main intake position**. These contain an overcurrent device and energy metering system, and are controlled by a main switch.

Did you know?

The National Grid is a network of nearly 5000 miles of overhead and underground power lines, all of which are interconnected. This means that, should a fault develop in a power station, power can be requested from another station in the grid

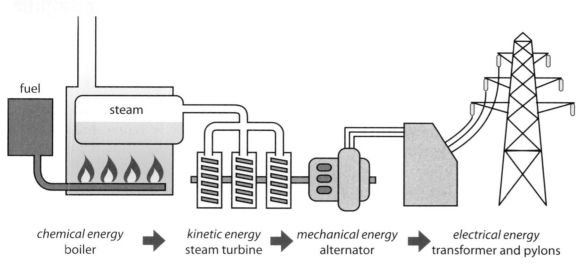

chemical energy ➡ kinetic energy ➡ mechanical energy ➡ electrical energy
boiler   steam turbine   alternator   transformer and pylons

Figure 5.01 The basic components of electricity generation systems

Three-phase supplies

Again, these were introduced at Level 2. However, we will briefly recap the principles behind three phase supplies.

Three-phase systems use three or four conductors, depending on the type of load being used. To get the phases, we spin three loops inside a magnetic field, with each loop on the same rotating shaft kept 120 degrees apart. This is shown in Figure 5.02. Each loop creates an identical sinusoidal waveform – in other words three identical voltages.

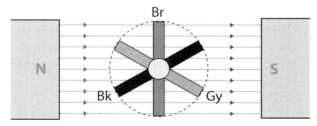

Figure 5.02 Loops spinning inside magnetic field

These are connected either in **star** or **delta**.

- **Delta** – the line voltage between any two phases is equal to the phase voltage; the line current is $\sqrt{3}$ times the phase current. We tend to use this when we have a balanced load.

- **Star** – the line voltage between any two phases is the same as the current in the line connected to it; this is $\sqrt{3}$ times voltage induced in any one phase coil in the current in each phase coil. With star a fourth wire can be added, giving the option of having both a single and three phase supply. We tend to use this when we have an unbalanced load.

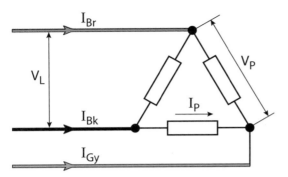

Figure 5.03 Delta connection

Delta connection

Figure 5.03 shows a three-phase load. You can see that each leg of the load is connected across two of the lines, e.g. Br–Gy, Gy–Bk and Bk–Br. We refer to the connection between phases as being the line voltage and have shown this on the drawing as V_L.

Equally, if each line voltage is pushing current along, we refer to these currents as being line currents, which are represented on the drawing as I_{Br}, I_{Gy} or I_{Bk}. These line currents are calculated as being the phasor sum of two phase currents, which are shown on the drawing as I_P and represent the current in each leg of the load. Similarly, the voltage across each leg of the load is referred to as the phase voltage (V_P).

In a delta connected balanced three-phase load, we are then able to state the following formulae:

$$V_L = V_P \qquad I_L = \sqrt{3} \times I_P$$

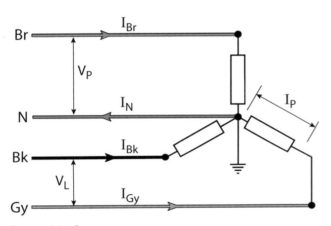

Figure 5.04 Star connection

Star connection

Figure 5.04 shows a star connection. All three star connected loops are connected to a central point, from where we take the neutral connection. This is a three-phase four wire system. In Figure 5.04 we refer to the connection made between phases as the line voltage and have shown this on the drawing as V_L. However, unlike delta, the **phase voltage** exists between any phase conductor and the neutral conductor, and we have shown this as V_P. Our line currents have been represented by I_{Br}, I_{Gy} and I_{Bk} with the phase currents being represented by I_P.

In a star connected load, the line currents and phase currents are the same, but the line voltage (400 V) is greater than the phase voltage (230 V).

In a star-connected load, we are therefore able to state the following formulae:

$$I_L = I_P \qquad V_L = \sqrt{3} \times V_P$$

Three-phase power

To find the power in a single-phase a.c. circuit we use the following formula:

$$\text{Power} = V \times I \times \cos \emptyset$$

Logically, you might assume that for three-phase power we could multiply this formula by three. Although this is not far from the truth we must remember that this could only apply where we have a balanced three-phase load. We can therefore say that for any three-phase balanced load, the formula to establish power is:

$$\text{Power} = \sqrt{3} \times (V_L \times I_L \times \cos \varnothing)$$

However, in the case of an unbalanced load, we need to calculate the power for each separate section and then add them together to get total power.

Example 1

Three coils of resistance 40 Ω and inductive reactance 30 Ω are connected in delta to a 400 V 50 Hz three-phase supply. Calculate the following.

(a) The current in each coil.

(b) Line current.

(c) Total power.

(a) Current in each coil

We must first find the impedance of each coil:

$$Z = \sqrt{R^2 + X_L^2} = \sqrt{40^2 + 30^2} = \sqrt{2500} = 50\ \Omega$$

The current in each coil (I_p) can then be found by applying Ohm's Law:

This gives $I_p = \dfrac{V}{Z} = \dfrac{400}{50} = \textbf{8 A}$

(b) Line current

For a delta connected system

$$I_L = \sqrt{3} \times I_p$$

Therefore $I_L = 1.732 \times 8 = \textbf{13.86 A}$

(c) Total power

We must first find the power factor using the formula:

$$\cos \varnothing = \frac{R}{Z}$$

This gives us $\cos \varnothing = \dfrac{40}{50} = 0.8$

And for a delta system $V_L = V_p$

Therefore we can now use the power formula of:

$P = \sqrt{3} \times V_L \times I_L \times \cos \varnothing$
$P = 1.732 \times 400 \times 13.86 \times 0.8$
$P = \textbf{7682 W or 7.682 kW}$

Example 2

A small industrial estate is fed by a 400 V, three-phase, 4-wire TN-S system. On the estate there are three factories connected to the system as follows.

Factory A taking 50 kW at unity power factor
Factory B taking 80 kVA at 0.6 lagging power factor
Factory C taking 40 kVA at 0.7 leading power factor

Calculate the overall kW, kVA, kVar and power factor for the system.

To clarify, we are trying to find values of P (true power), S (apparent Power) and Q (reactive power). First, we need to work out the situations for each factory.

Factory A

We know that power factor

$$\cos \varnothing = \frac{\text{True power (P)}}{\text{Apparent power (S)}}$$

We also know that the power factor is 1.0 and that P = 50 kW

Therefore, by transposition:

$$S = \frac{P}{\cos \varnothing} = \frac{50}{1} = 50 \text{ kVA}$$

And with unity power factor for Factory A, Q = 0

Factory B

Using the same logic, we need to find true power and reactive power.

Therefore P = cos Ø x S = 0.6 × 80 kW = 48 kW

Reactive component (Q) = S × sin Ø = 80 × 0.8 = 64 kVAr

Factory C

P = S × cos Ø = 40 kW × 0.7 = 28 kW

Q = S × sin Ø = 40 × 0.714 = 28.6 kVAr

We can now find the **total kW** by addition: 50 + 48 + 28 = **126 kW**

We can find the total **kVAr** as the difference between the reactive power components, the larger one, Factory B, lagging and the smaller one, Factory C, leading:

64 kVAr − 28.6 kVAr = **35.4 kVAr** lagging

We can use Pythagoras' Theorem to find the total kVA

$$S = \sqrt{P^2 + Q^2} = \sqrt{126^2 + 35.4^2} = \textbf{131 kVA}$$

Consequently, the overall power factor will be:

$$\cos \varnothing = \frac{P}{S} = \frac{126}{131} = \textbf{0.96 lagging}$$

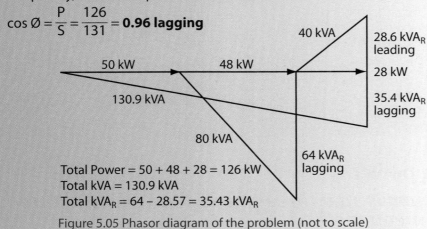

Total Power = 50 + 48 + 28 = 126 kW
Total kVA = 130.9 kVA
Total kVA$_R$ = 64 − 28.57 = 35.43 kVA$_R$

Figure 5.05 Phasor diagram of the problem (not to scale)

Neutral currents

In a three-phase four-wire system, we are saying that each line (Br, Gy, Bk) will have an unequal load and therefore the current in each line can be different. It therefore becomes the job of the neutral conductor to carry the out-of-balance current. If we used **Kirchhoff's law** in this situation, we would find that the current in the neutral is normally found by the phasor addition of the currents in the three lines.

Example

For a three-phase four-wire system, the line currents are found to be B_R = 30 A and in phase with V_{Br}, I_{Gy} = 20 A and leading V_{Gy} by 20° and I_{Bk} = 25 A and lagging V_{Bl} by 10°. Calculate the current in the neutral by phasor addition.

The phasor diagram for this example has been provided. However, you should note that, in order to establish the current in the neutral (I_N), you would need to draw two parallelograms. The first should represent the resultant currents I_{Br} and I_{Gy}. The second, I_N, should be drawn between this resultant and the current in I_{Bk}.

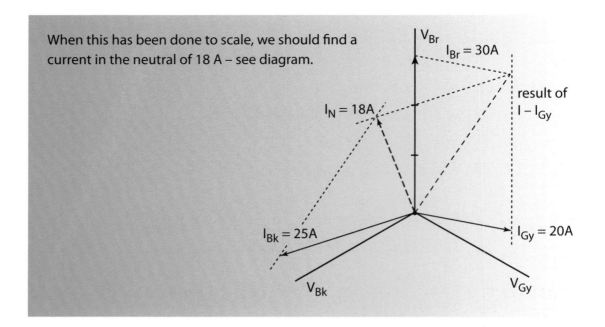

When this has been done to scale, we should find a current in the neutral of 18 A – see diagram.

V_{Br}
$I_{Br} = 30A$
result of $I - I_{Gy}$
$I_N = 18A$
$I_{Bk} = 25A$
$I_{Gy} = 20A$
V_{Bk}
V_{Gy}

Measuring power

Measuring power in a three-phase, four-wire balanced load – one-meter method

A wattmeter can be used to measure power in this circuit. In a three-phase, four-wire balanced load the total power will be equal to three times the value measured on the meter.

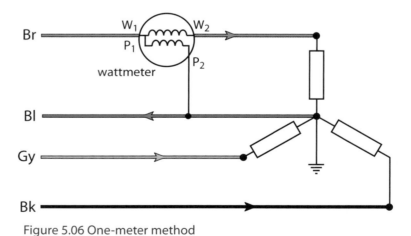

Figure 5.06 One-meter method

Measuring power in a three-phase, three-wire balanced load – two-wattmeter method

In the two-wattmeter method, the total power is found by adding the two values together. At unity power factor the instruments will read the same and be half of the total load. For other power factors the instrument readings will be different; the difference in the reading could then be used to calculate the power factor.

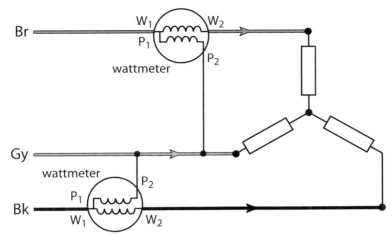

Figure 5.07 Three-wire two-meter method

Figure 5.08 Four-wire three-meter method

Measuring power in an unbalanced three-phase four-wire circuit – three-wattmeter method

Here the total power will be the sum total of the readings on the three-wattmeters.

This allows the power to be measured in a situation where the load is unbalanced, such as the three-phase supply to a large building where it is impossible to balance the load completely.

Measuring power factor

To measure power factor, there are a number of purpose made instruments available. All of these meters use voltage and current measurement to determine power factor. Most designs are based on the clamp meter idea where the meter clamps around the current-carrying conductor. However, additional leads are then required to connect the meter across the supply. Many meters are also combined with the ability to measure all aspects of power, i.e. kW, kVA and kVAr.

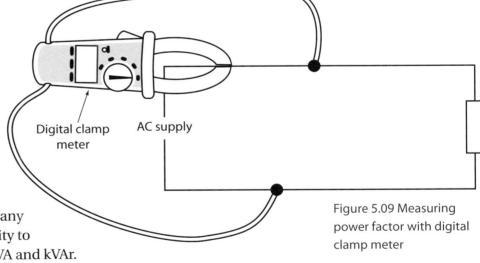

Figure 5.09 Measuring power factor with digital clamp meter

Figure 5.09 illustrates the connections of a power factor measuring instrument.

There is also an alternative method. The calculation to establish power factor is to divide true power by the apparent power (VA). As we know that a wattmeter will give us the true power, by adding an ammeter and a voltmeter to our circuit we can establish the apparent power (VA) and therefore establish the power factor. Figure 5.10 shows this arrangement.

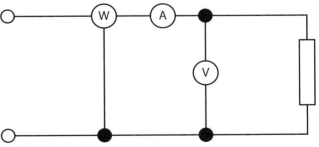

Figure 5.10 Circuit with wattmeter, ammeter and voltmeter connected

Industrial distribution systems and equipment

On completion of this topic area the candidate will be able to describe industrial distribution systems and be able to list the types of material and equipment used.

We first looked at a number of these installations at Level 2.

Steel and PVC conduit installations

Annealed mild steel tubing, known as **conduit**, is widely used as a commercial and industrial wiring system. PVC-insulated (non-sheathed) cables are run inside the steel tubing. This offers excellent mechanical protection to the wiring. The British Standard covering steel conduit and fittings is BSEN 61386. The two most commonly used types of steel conduit are black enamel conduit (used indoors) and galvanized conduit (used in damp conditions). In this section the following areas will be looked at:

- screwed conduit
- bending machines
- types of bend
- bending conduit
- conduit fixings
- cutting and screwing conduit
- running coupling
- termination of conduit
- use of non-inspection elbows and tees
- wiring conduit
- conduit and cable capacities
- plastic conduit
- flexible conduit
- miscellaneous points.

Screwed conduit

Screwed steel conduit can be either seam-welded or solid-drawn. Solid-drawn is stronger but much more expensive. At one time solid-drawn conduit was the only conduit that could be used in hazardous areas. More recently, because of manufacturing improvements, seam-welded conduit can now also be used in hazardous areas. The thread used on steel conduit is not used on any other pipe and special conduit dies are therefore required.

Prior to 1970 conduit sizes were imperial, and of course many of these conduits are still around. Since 1970 metric sizes have been used. Where imperial and metric conduits must be joined together, adapters may be required. Adapters may have external imperial threads and internal metric threads or vice versa.

Bending machines

Bending machines used to be considered an expensive item but are easily available today. They give consistent results every time and require the minimum of practice.

To position the stand as shown in the photograph, swing the rear leg (E) to its maximum. Place the safety pin (D) through the hole beneath where the pin hangs, locking the rear leg in place. The machine should now be standing with the swivel arm (A) hanging downwards. (B) illustrates the conduit guide and (C) is the adjusting arm for the conduit guide.

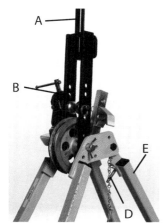

Bending machine set up

Bending machine with a piece of conduit inserted, which prevents the swivel arm from hanging downwards

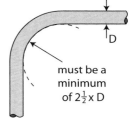

must be a minimum of $2\frac{1}{2} \times$ D

Figure 5.11 Minimum-bending radius allowed

Types of bend

Sharp bends must be avoided. The minimum radius of steel conduit is laid down in BS 7671 as twice the outside diameter of the conduit. See Figure 5.11.

Right-angled bend	This is used to go around a corner or change direction by 90 degrees. When bending, measurements may be taken from the back, centre or front of the bend. Allowance should be made for the depth of the fixing saddle bases.	
Set	The set is used when surface levels change or when terminating into a box entry. Sets should be parallel and square, not too long and not too short so that the end cannot be threaded. Where there are numerous sets together all sets must be of the same length. The double set is used when passing girders or obstacles, as shown.	Set Double set
Kick	The kick is used when a conduit run changes direction by less than 90 degrees.	
Bubble set or saddle set	The bubble set or saddle set is used when passing obstructions, especially pipes or roof trusses etc. The centre of the obstruction should be central to the set.	

Table 5.1 Types of conduit bend

Bending conduit

Making a 90 degree bend from a fixed point

Figure (a) shows the required bend.

Figure 5.12 a–c Conduit bending sequence

(a)

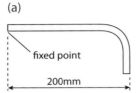

Mark the conduit as shown in diagram (b), 200mm from the fixed point. If the distance is given to the inside or centre of the tube, simply add on either the diameter or half the diameter respectively to give the back bend measurement and follow the same procedure as for outside measurement.

(b)

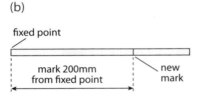

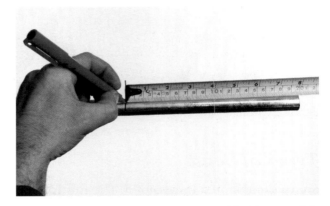

Marking conduit

Place the tube in the 'former' with the fixed point to the rear. Position the tube so that a square held against the tube at the fixed point touches and forms a tangent to the leading edge of the former.

(c)

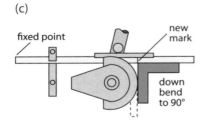

Conduit in former with set square

Where the remaining length of tube from the measured point is too long to down-bend and where it is not convenient or possible to up-bend using the method described, then the problem can be overcome by using the following method.

Figure 5.12 d–f Conduit bending sequence (continued)

(d)

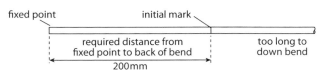

fixed point ⌐ initial mark

required distance from
fixed point to back of bend

200mm

too long to
down bend

Deduct three times the outside diameter of the tube from the initial mark.

(e)

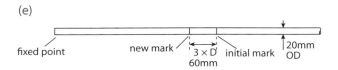

fixed point

new mark

3 × D
60mm

initial mark

20mm
OD

Place the tube in the former with a fixed point to the front, with the mark at 90 degrees to the edge of the former. This will give a 90 degree bend at the required distance from the fixed point to the back of the bend as shown.

Conduit bent to shape in former

(f)

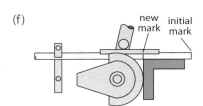

new
mark

initial
mark

Finished 90° bend

Making double sets

Figure 5.12 shows the required double set.

Figure 5.13

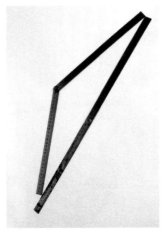

Angled rule

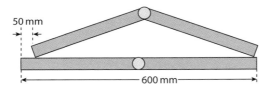

Figure 5.14

To ensure re-entry of the bent tube into the bending machine to complete the return set, a determined angle of initial bend is required. Determine the distance of the set at 50mm and deduct this distance from a 600mm rule.

The tube can be bent using the angled rule to indicate the angle of the first set.

Angle of the first set

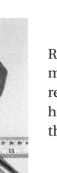

Remove the tube from the bending machine and mark the tube for the return set, making sure to measure the height of the obstacle or accessory from the inside of the tube.

Marking the tube for the return set

Reposition the tube in the machine, ensuring that the mark on the tube forms a tangent to the edge of the former. The final set can be made parallel with the first.

Conduit in former showing marking

Second bend in former

Making bubble sets

To obtain the correct angle for the first set, multiply the external diameter of the obstacle – (e.g. 50mm tube) by 3 (i.e. 50 × 3 = 150mm) as shown.

Stagger the legs of a 600mm folding rule between the 150mm and 600mm marks on a second rule.

Having marked the centre of the set on the tube, the tube is positioned with the mark vertically above a mark on the former. This is determined by bisecting the angle of the rule when placed as shown.

Place the conduit over the obstacle; measure 50mm from the inside of the first set to a straight edge and mark the tube at A and B as shown.

Position the tube in the bending machine so that the mark A forms a tangent to the edge of the former. Bend down until the top edge of the tube is level and in line with mark B.

Reverse the tube in the former and position as for mark A. Down-bend until the top edges of the tube are in line.

Figure 5.15 a–f Making a bubble set

(a)

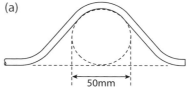

50mm

(b)

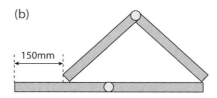

150mm

(c)   tangent line   mark

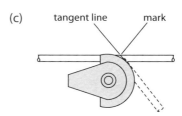

(d)

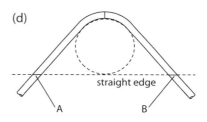

straight edge

A   B

(e)

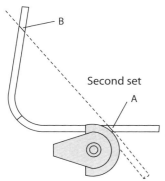

B

Second set

A

(f)

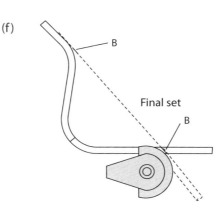

B

Final set

B

Conduit fixings

Conduits must be securely fixed in accordance with the following distances between supports.

Nominal size of conduit	Maximum distance between supports					
	Rigid metal		Rigid insulating		Pliable	
	Horizontal m	Vertical m	Horizontal m	Vertical m	Horizontal m	Vertical m
Not exceeding 16 metres	0.75	1.00	0.75	1.00	0.30	0.50
Exceeding 16 but not exceeding 25 metres	1.75	2.00	1.50	1.75	0.40	0.60
Exceeding 25 but not exceeding 40 metres	2.00	2.25	1.75	2.00	0.60	0.80
Exceeding 40 metres	2.25	2.50	2.00	2.00	0.80	1.00

Table 5.2 Conduit fixing parameters (Table 4C, *On-Site Guide*)

Methods of supporting conduit

Conduit fixings

Strap saddle

- A **'strap saddle'** or **'half saddle'** is used for fixing conduit to cable tray or steel framework.

- The spacer bar saddle is used when fixing to an even surface; it gives a clearance of 2mm.

- The distance saddle is used if the surface is uneven and where brick on concrete can give rise to heavy condensation.

- The hospital saddle is used where it is necessary to clean around the conduit fixing.

- The multiple saddle strip is used to fasten multiple runs of conduit together.

- The girder clamp will fix conduit to girders and I-beams without having to drill a hole in the girder.

- A pipe hook or crampet is used when conduits are secured to a wall or cast in concrete.

Cutting, screwing and terminating conduit

Conduit should be cut with a hacksaw. The cut should be square and the full length of the blade should be used, taking steady strokes. Hold the conduit in a pipe vice, not a bench vice. The vice should be secured but not so tight that it cuts into the pipe.

Pipe vice and method of cutting

Before threading, the conduit should be chamfered with a file to help the die start. Screwing is carried out using stocks and dies. Another part of the stock and die is the guide, which ensures the screw cut is square. Stocks and dies should be kept clean and any lubricant or steel shavings should be removed after cutting. The cut is made by placing the stock and die on the conduit and then turning clockwise while applying forward pressure; sometimes a great deal of pressure may be required. Once the cut is started the stock and die are removed so that a cutting agent can be applied. Having applied the cutting agent the stock is placed on the conduit again and the threading begins. The stock and die is turned back every rotation to clean out the cuttings.

Cutting the thread

When the thread is finished the stock and die is removed and the inside of the conduit is cleaned and reamed. This removes all burrs and sharp edges, which would cut the cables (if not removed) when they were installed. Reaming can be carried out with a reamer or round file. The standard length of thread for a normal joint is half a coupling length.

The cut thread   Reaming the thread

All couplings, bushes and conduit boxes must be fully tightened before installation. Where possible, couplings, bushes and boxes should be tightened while the conduit is held firmly in a pipe vice.

Tightening the conduit

Running coupling

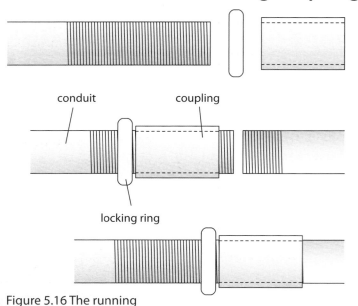

conduit   coupling

locking ring

Figure 5.16 The running coupling

Sometimes two conduits must be joined together and neither can be turned. This may be due to one conduit coming through a wall or ceiling, or long runs combined with bends making turning impossible. In these cases a running coupling must be used. Running couplings are made by having one thread a normal half-coupling length and the other thread the length of a coupling plus locking ring.

The coupling and locking ring are fixed on the long thread side and the two conduits are then butted together. The coupling is then removed from the long thread to the shorter thread and finally rests across the two sides. After tightening, the coupling is locked. A locking ring must be used because locknuts get caught on the ceiling in tight situations.

Because the coupling is traversing two threads simultaneously, the thread must be very clean and well cut. Reversing the dies and running them over the thread can help this. This is particularly important where the running coupling is in an awkward position (as it often is).

Termination of conduit

There are several methods available for terminating conduit, three of which are illustrated.

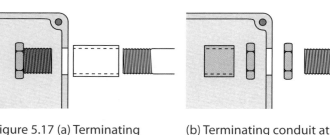

Figure 5.17 (a) Terminating conduit at a box using a conduit coupling and brass male bush

(b) Terminating conduit at a box using lockouts and a brass female bush

(c) Flanged coupling washer and brass male bush method for use with PVC box

Fitting the bush

Tightening the bush

Use of non-inspection elbows and tees

The main consideration here is that damage to the cables does not occur during installation. Non-inspection elbows are only used adjacent to an outlet box or inspection-type fitting. One solid elbow may be used if positioned less than 500mm from an accessible outlet in a conduit run of less than 10m that has other bends which are not more than the equivalent of one right angle. Figure 5.18 illustrates this use.

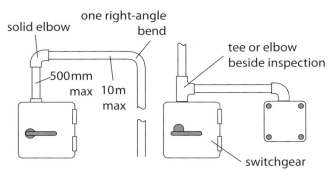

Figure 5.18 Non-inspection elbows and tees

Wiring conduit

Cables must not be drawn into a conduit system until the system is complete. If a large number of cables are to be drawn into a conduit system at the same time, the cable reels should be arranged on a stand or support so they can revolve freely. If the reels cannot be supported, then sufficient cable must be run off them before drawing, taking care to avoid them tangling.

In new buildings, cables should not be drawn in until the conduit is dry and free from moisture. If in any doubt, a draw tape with a swab at the end should be drawn through the conduit to remove any moisture that may have accumulated.

It is usual to commence drawing in cables from a mid-point in the conduit system to minimise the length of cable that has to be drawn in. A steel tape should be used from one draw-in point to another. The draw tape should not be used for drawing in cables, as it may become damaged. A steel tape should only be used to pull through a draw wire. The ends of the cables must be stripped for a distance of approximately 75mm and threaded through a loop in the draw wire.

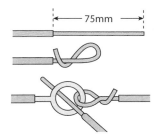

Figure 5.19 Drawing in cables

When drawing in a number of cables, they must be fed in very carefully at the delivery end while someone pulls them at the receiving end. Care should be taken to feed into the conduit in such a manner as to prevent any cables crossing. Always leave some slack cable in all draw-in boxes, and make sure that cables are fed into the conduit so as not to finish up with twisted cables at the draw-in point.

This operation requires care and there must be synchronisation between the person who is feeding and the person who is pulling. When they are in sight of each other, this can be achieved by some pre-arranged signal; if within earshot, by word of command given by the person feeding the cables. If the two people are not within earshot or sight of each other the process is more difficult. A good plan is for the individual feeding the cables to give pre-arranged signals by tapping the conduit.

Remember

If the cables are allowed to spiral off the reels they will become twisted and this would cause damage to the insulation

Remember

If cables are not drawn in carefully as described, they will almost certainly become crossed, and this might result in them becoming jammed inside the conduit. This would prevent one or more cables being drawn out of the conduit should this become necessary later

Drawing in cables

1. Pass draw tape through and between outlets.

2. Fasten a draw wire securely to the draw tape.

3. Feed draw wire into the conduit while withdrawing the draw tape. Ensure that the draw wire is long enough and strong enough for the job.

4. Fasten cables to the draw wire. At least 75mm of insulation should be stripped away and secured as illustrated. Fasten each cable separately.

5. Where possible, cables should be drawn into the conduit directly off the cable drum.

6. First make sure that there is sufficient cable for the job.

7. Feed cables into the conduit using the fingers of one hand to feed the cables and the other hand to keep cables straight.

8. Ensure that no crossed or kinked cables enter the conduit.

9. Keep hands close to the conduit entry and feed only short lengths of cable at a time.

Cables attached to draw wire

Cable drum

Conduit and cable capacities

The number of cables that can be drawn into or laid in any enclosure of a wiring system must be such that no damage can occur to the cables or the enclosure during installation. The number of cables that can be used is the overall sum of the cables' cross-sectional area (csa) compared to the overall csa of the conduit. This is expressed as a percentage and should not exceed 45 per cent; this is already taken into account by the standard sizes of cable and enclosures in the *Amicus Tables*. This section looks at the cable capacities of conduit – in other words, how many cables can be safely installed into what size conduit.

The tables that will be referred to in this section can be found in Appendix 5 of the *IEE On-Site Guide*. These tables only give guidance to the maximum number of cables that should be drawn in. The sizes should ensure an easy pull with low risk of damage to cables and enclosures.

The electrical effects of grouping are not taken into account. Therefore, as the number of circuits increases, the current-carrying capacity of the cables will decrease. Cable sizes would have to be increased, with a consequent increase in cost of cable and conduit.

The following wiring systems are covered by the appropriate tables in the *IEE On-Site Guide*, Appendix 5 and Amicus Guide pages 69–70:

- straight runs of conduit not exceeding 3m in length (Tables 5A and 5B)

- straight runs of conduit exceeding 3m in length, or in runs of any length incorporating bends or sets (Tables 5C and 5D).

Example

A lighting circuit for a school extension requires the installation of a conduit system with a conduit run of 8m with two right-angle bends. The number of cables required is twelve × 1.5mm stranded, PVC-insulated. What size of conduit should be used for this installation?

Step 1   Select the correct table for cable runs over 3m with bends. (Table 5C)

Step 2   Obtain the factor for 1.5mm stranded cable = 22

Step 3   Multiply the number of cables by the factor = 22 × 12 = 264

Step 4   Select the correct table for conduit systems with runs in excess of 3m with bends. (Table 5D)

Step 5   Obtain the factor for a length of run which is greater than 264. The table gives a factor of 292 for an 8 metre run in conduit with two bends.

Step 6   The answer is 25mm conduit.

Plastic conduit (PVC)

Plastic conduit is made from polyvinyl chloride (PVC), which is produced in both flexible and rigid forms. It is impervious to acids, alkalis, oil, aggressive soils, fungi and bacteria, and is unaffected by sea, air and atmospheric conditions. It withstands all pests and does not attract rodents. PVC conduit is preferable for use in areas such as farm milking parlours. PVC conduit may be buried in lime, concrete or plaster without harmful effects.

Plastic conduit advantages	Plastic conduit disadvantages
• Light in weight • Easy to handle • Easy to saw, cut and clean • Simple to bend • Does not require painting • Minimum condensation due to low thermal conductivity in walls • Quick to install	• Care must be taken when applying glue to the joints to avoid forming a barrier across the inside of the conduit • If insufficient adhesive is used the joints may not be waterproof • PVC expands around five times as much as steel and this expansion must be allowed for

Table 5.3 Advantages and disadvantages of plastic conduit

Working with PVC conduit

The techniques required for working with PVC conduit differ considerably from those in steel conduit installations. PVC conduit is easily cut using a junior hacksaw. Any roughness of cut and burrs should be removed by simply wiping with a cloth. The most common jointing procedure uses a PVC solvent adhesive. The adhesive should

Safety tip

Care must be taken when using these adhesives as they are volatile liquids; the lid must be replaced on the tin immediately after use. Take care when using in a confined space. Always read the manufacturer's instructions be applied to the female part of the joint and the conduit twisted into it to ensure a total coverage. Generally the joint is solid enough for use after two minutes, although complete adhesion takes several hours. In order to ensure a sound joint, the tube and fittings must be clean and free from dust and oil.

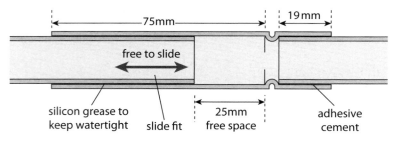

75mm

19mm

free to slide

silicon grease to keep watertight

slide fit

25mm free space

adhesive cement

Figure 5.20 Expansion provision in conduits

Where expansion is likely and adjustment is necessary a mastic adhesive should be used. This is a flexible adhesive which makes a weatherproof joint, ideal for surface installations and in conditions of wide temperature variation. It is also advisable to use mastic adhesive where there are straight runs on the surface exceeding 6m in length.

PVC conduit expands considerably more than metal conduit with an increase in temperature. The expansion can be ignored where the conduit is buried in concrete or plaster. In surface work precautions must be taken to prevent such expansion from causing the conduit to bow. Usually where bends and sets are close together these take up any expansion. Where longer runs of conduit occur in conditions of varying temperatures, some provision for expansion must be made, using expansion couplers as shown in Figure 5.20.

PVC conduits not exceeding 25mm diameter can be bent cold by using a spring. The bend is then made by either the hands or across the knee. In order to achieve the angle required the original bend should be made at twice the angle required and the tube allowed to return to the correct angle. Under no circumstances should an attempt be made to force the bend back with the spring inserted, as this can damage the spring. It is easier to withdraw the spring by twisting it in an anticlockwise direction. This reduces the diameter of the spring, making it easier to withdraw.

In cold weather it may be necessary to warm the conduit slightly at the point where the bend is required. One of the simplest ways is to rub the conduit with the hand or a cloth. The PVC will retain the heat long enough for the bend to be made. In order that the bend is maintained at the correct angle the conduit should be fastened to the surface with a saddle as soon as possible.

Remember

A good guide to the use of expansion couplers is one coupler per 6m in straight runs

Flexible conduit

This type of conduit comes in a role and, as its name implies, is flexible. It is used in increasing quantities. A typical example of an installation in which a flexible conduit is used would be within metal stud partitioning, above false ceilings and below the suspended floors of office blocks etc. Another use for flexible conduit is to make connections between cable management systems and any equipment that may vibrate, such as motors.

In some cases flexible conduit comes with corrugated PVC sides but others have an extruded PVC coating over the flexible tubing. Termination into boxes and enclosures is by means of glands and locking nuts. Where flexible conduit is used, a separate circuit protective conductor (CPC) must always be installed.

Miscellaneous points for steel and plastic conduits

- Ample capacity must be provided at junctions employed for cable connections.

- Where a steel conduit is used as a circuit-protective conductor it must be tested in accordance with BS 7671.

- Where a steel conduit forms the protective conductor, a separate conductor must be used to connect from any socket outlet to its back box.

- Where flush switch boxes and switch grids are used a CPC is required from an earthing terminal in the box to the earthing terminal in the switch grid.

- Cable capacities should be calculated in accordance with the relevant tables found in the *On-Site Guide.*

- Conduits run overhead should be run in accordance with Table 4B in the *On-Site Guide.*

- Where conduits pass through walls the hole shall be made good during installation with a fire-resistant material.

Trunking

Trunking is a fabricated casing for cables, normally of rectangular cross-section, one side of which may be removed or hinged back to permit access. It is used where a number of cables follow the same route, or in circumstances where it would otherwise be expensive to install a large number of separate conduits or runs of mineral-insulated cable. Trunking is commonly installed, for example, in factories where the introduction of new equipment and the relocation of existing equipment may involve frequent modification of the installation.

Surface pattern trunking is available in 3m lengths and a wide variety of sizes, ranging from 38mm^2 up to 225mm × 100mm. This section will look at the following areas:

- square steel trunking with cover and coupling

- floor trunking

- multi-compartment trunking

- flush cable trunking

- overhead trunking

- skirting trunking

- busbar trunking (overhead/horizontal)

- feed units

- busbar trunking (rising mains)

- site-built trunking accessories

- regulations concerning trunking

- trunking capacities.

Did you know?

Zinc-coated and enamel trunking is generally used internally. Outside, or in situations where atmospheric conditions are likely to be arduous, a galvanised or electro-zinc plated trunking would be necessary

Square steel trunking with cover and coupling

To facilitate connections and terminations etc. and in order to run the trunking with the contours of the building in which it is installed, a wide number of trunking fittings and accessories are available, a selection of which are illustrated below.

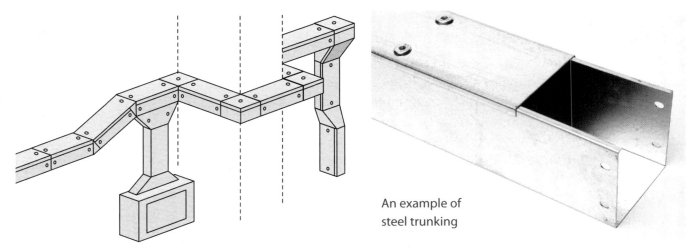

An example of steel trunking

Figure 5.21
Steel trunking

Floor trunking

Floor trunking or ducting is also available in a wide range of sizes and fittings. It is used extensively in schools, hospitals and industrial situations and can either be laid below the floor surface, where access is by means of a number of inspection covers, or installed just below floor level and covered by a steel chequer plate.

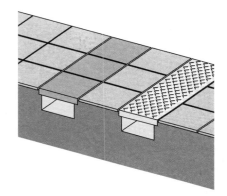

Figure 5.22 Floor trunking

Multi-compartment trunking

This is used where segregation is desired. It is normally a broad, flat trunking with internal steel fillets. The fillets are normally spot welded to the trunking.

Flush cable trunking

Flush cable trunking is used where a neat and unobtrusive cable trunking is desired. It is available in a range of sizes. The covers can be supplied in a number of finishes and have a large overlap on each side.

Skirting trunking

Skirting trunking is used mainly in offices, schools and colleges to provide a large number of socket outlets for 230 V and telecommunications equipment. It is made with an internal fillet for segregation. It can be installed before or after plastering and it eliminates the need for conventional skirting.

Busbar trunking (overhead horizontal)

Busbar trunking is a popular means of three-phase power distribution in machine shops, laboratories and many industrial situations. It consists of a broad, flat trunking in which three or four busbars are rigidly fixed on to moulded block insulators. The conductors used in busbar trunking are generally made of copper or sometimes aluminium. They can be sleeved with insulating material or left bare. The size of conductor will vary with the current-carrying capacity, and its shape can be either round, oval or rectangular. A complete range of fittings is available for right-angle bends, tee pieces and crossovers. These fittings are self-contained assemblies complete with busbars and couplings (see Figure 5.23).

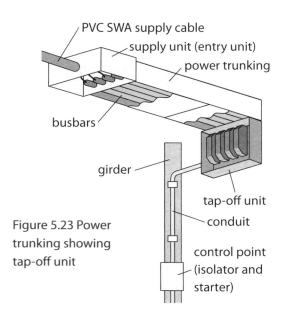

Figure 5.23 Power trunking showing tap-off unit

Feed units

Feeding the busbars can be done in a number of ways. For smaller runs of busbar trunking a screwed conduit or cable trunking could be used with PVC-insulated conductors of appropriate rating. In the case of a large installation, or where an armoured cable would be more suitable, a feed unit with a cable-sealing chamber would be used. The busbar trunking can be centre-fed, the centre feed being more popular, as this method makes full use of the busbar capacity.

Fused outlet boxes can either be fixed or of the plug-in type; they can be side-mounted or under-slung. Sets of drillings are provided at regular intervals along the busbar trunking to facilitate the tap-off boxes, and blanking-off plates are supplied to cover off the apertures not initially used. HBC (High breaking capacity) fuses must always be used and never replaced with rewirable fuse links.

Busbar trunking (rising mains)

The increasing number of multi-storey developments for hospitals, offices and flats etc. poses certain problems for electrical power distribution. In some cases a rising main busbar trunking system can provide a compact and economical solution. In this system, purpose-made busbar trunking is run vertically through the walls of the building. It is fed and controlled usually from the bottom at the service entry, and has a fixed fuse box mounted at each floor.

The essential difference between horizontal-type busbar trunking and vertical rising main-type busbar trunking is the substantial insulated support rack at the base of each riser. It is designed to carry the full weight of the copper conductors, which are then free to expand upwards. Figure 5.24 illustrates rising mains trunking.

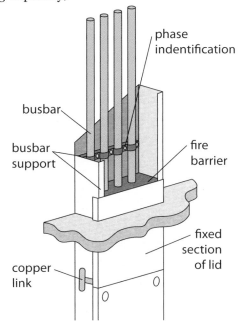

Figure 5.24 Rising mains busbar trunking

Site-built trunking accessories

Prefabricated bends and sets will typically be used to install trunking systems because they are quicker and cheaper. However, there will be times when the prefabricated bends and sets are either not suitable or are not available. In this situation you will need to be able to fabricate your own.

The accessories that are considered here are:

- right-angle internal bend
- right-angle vertical bend
- trunking sets
- tee junction.

Right-angle internal bend

Safety tip

Remove any rough or jagged edges with a medium-cut file in order to prevent damage to the cables

Select a short section of trunking between 900mm and 1000mm in length. Using a soft pencil and a reliable set square, draw a line (called a datum line) around the outside (periphery) of the trunking. This should be done at the mid-point positions as shown.

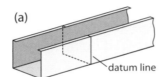
(a)
datum line

Check the width of the workpiece and transfer this measurement to either the top left or right-hand side of the central datum line, as shown.

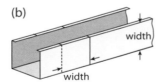
(b)
width
width

Using an adjustable set square as a guide, draw a pencil line from the marked trunking to the bottom of the centre datum line as shown in (c). Repeat this guideline on the opposite vertical side.

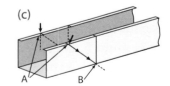
(c)
A   B

At this stage there will be a right-angled triangle drawn on each outer side of the trunking. Remove the two triangles using a hacksaw with a blade fitted with 25–30 teeth per 25mm of blade. Also remove sufficient flange so that it will not restrict bending up of the trunking. Once removed, file smooth all rough or jagged edges as these may damage the cables.

Figure 5.25 a–d Right angle internal bend

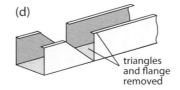
(d)
triangles and flange removed

Cut a wooden block with a good square edge on one side able to be fitted comfortably across the internal width of the trunking. Place the wooden block to the vertically cut side as shown in the diagram.

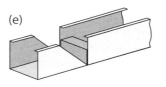

Figure 5.25 e–f Right angle internal bend (continued)

Hold the block firmly in place and with the other hand push up the side of the trunking adjoining the angled cut. Allow the vertical sides to be sandwiched between the angled trunking sides. The wooden block will help provide a sharper edge at the bending point. Once completed, dress the bend with a hammer and remove the wood. Check for squareness and strengthen with pop rivets, nut and bolt or spot weld.

(f)

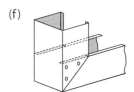

Right-angle vertical bend

Mark out the position of the bend on all sides of the trunking. (The two lengths marked Y must be equal.)

(a)

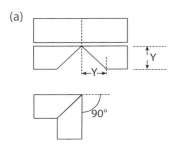

Drill small holes in the corners at the point of the bend to stop the metal from folding. Then place wooden blocks inside the trunking for support. Cut the sides of the trunking with an appropriate hacksaw.

(b)

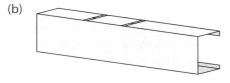

The edge of the trunking can be cut with a file (as shown in (c) opposite) and the waste broken off.

(c)

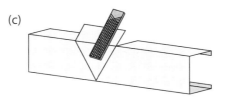

Cut away the back of the trunking using a suitable hacksaw. Then file all the rough and jagged edges and bend the trunking to shape as shown.

(d)

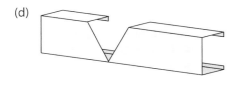

(e)

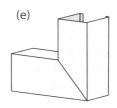

Figure 5.26 a–e Right angle vertical bend

Figure 5.26 f–g Right angle
vertical bend (continued)

Make a fishplate out of some scrap trunking and drill in some fixing holes.

(f)

Finally, mark out the trunking from the holes in the fishplate and drill. Secure the assembly with nuts and bolts or pop rivets. Alternatively, the joint may be spot-welded.

(g)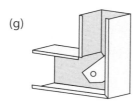

Trunking sets

A trunking set and a trunking return set are both constructed in a similar fashion when making a right-angled bend. Figure 5.27(a) illustrates a trunking set (A) and a trunking return set (B).

(a)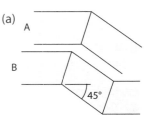

Select a section of trunking that can be worked comfortably yet is long enough to accommodate either a set or return set. Draw a datum line using a soft pencil and set square as a guide.

(b)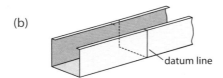

Measure the width of the trunking and transfer half of this measurement to either the top left or top right-hand side of the datum line. Draw a line from this mark to the base of the datum line A.

(c)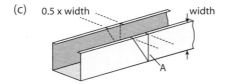

Cut out both triangular shapes from each side of the trunking.

(d)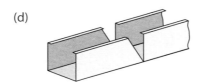

Figure 5.27 a–d
Trunking set

Making a right-angle vertical bend (alternative method)

1. Mark the trunking for cutting.

2. Place wooden block inside trunking to secure vice.

3. Use hacksaw to cut trunking.

4. Bend trunking into right-angled bend.

5. Secure with rivets.

6. Final result.

Cut a section of a wooden block which will fit comfortably across the internal diameter of the trunking. Place the piece of wood as shown in (e) so that it is at the side adjacent to the vertically cut datum line.

(e)

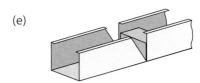

(f)

Secure the wooden block with one hand while gently bending up the remaining half of the workpiece until the set is formed. Check that the required angle is correct and secure using pop rivets, nuts and bolts or spot welding. (f) illustrates the completed set.

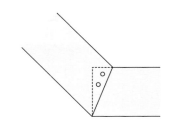

If a return set is involved, then extra work is required. Draw a line of reference on a suitable flat surface and offer the shortest section of the set to the line as shown. Measure the depth of the required set (A) and mark on the trunking at (B). (C) represents the shortest leg.

(g)

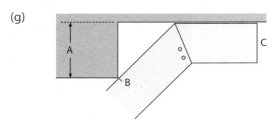

Prepare the trunking as shown in (h). Cut out the complete left-hand side of the centre line comprising two side triangles bridged by a rectangular base section. To provide electrical continuity the trunking flange must remain unbroken.

(h)

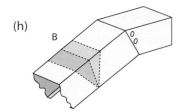

A 2mm diameter V-shape should be cut along both sides of the bottom edge of the trunking from the vertical centre line to the angled dotted line, as shown. This will act as a supplementary lip and help stabilise the return set when it is assembled.

(i)

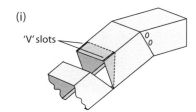

'V' slots

Figure 5.27 e–i Trunking set (continued)

Gently bend the workpiece and unite both sections of the return set, allowing the vertically cut centre edges to be sandwiched between the angled sides of the trunking as shown in the diagram. Check that the set has been worked to meet the required measurement. Secure using pop rivets, nuts and bolts or spot welding. Dress the supplementary base flange to accommodate the changed angle and secure as necessary.

Figure 5.27 j Trunking set (continued)

(j)

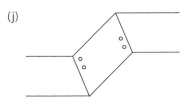

Tee junction

Mark out the position of the tee using a second piece of trunking to gauge the width. Cut out a space for the tee as shown in (a). Use blocks of wood to support the sections being cut. File all rough edges to protect the cables.

Cut away the section to leave two lugs (b). These can be bent in a vice using a hammer to give a clean edge. File all edges smooth and check the fit as necessary.

Mark out the holes, drill and secure with nuts and bolts or pop rivet, or alternatively spot weld (c).

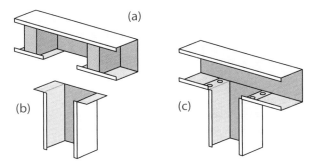

Figure 5.28 a–c Tee junction

Regulations concerning trunking

- **Regulation 521.5.2:** Cables of a.c. circuits installed in ferromagnetic enclosures shall be arranged so that the conductors of all lines and the neutral conductor and the circuit-protective conductor of each circuit are contained in the same enclosure (duct or trunking). This is to prevent the occurrence of eddy (induced) currents.

- **Regulation 521.5.1:** Every conductor or cable shall have adequate strength and be so installed as to withstand the electromagnetic forces caused by any current, including fault current, it may have to carry in service.

- **Regulation 521.6:** Where applicable, trunking, ducting and their fittings shall comply with BSEN 50085. Cable conduits shall comply with BSEN 61386, cable tray and ladder systems with BSEN 61537.

The regulations also lay down the following guidelines.

- Ducts and metallic trunking must be securely fixed and protected from mechanical damage.

- The number of cables to be installed in ducts shall be such that a space factor of 35 per cent must not be exceeded.

- The number of cables installed in trunking shall be such that a space factor of 45 per cent must not be exceeded.

- During installation where conduit, ducts or trunking passes through walls, floors, ceilings or partitions, the surrounding hole must be made good with a non-combustible material, and internal fire barriers are to be installed within trunking at these positions.

- Copper links are used across joints in metallic conduit systems in order to maintain the continuity of the exposed parts.

- Cable entries must be protected with grommets to prevent damage or abrasions to cables.

- Straight runs of trunking are best joined by using a joining section and the nuts and bolts supplied.

- When terminating metal trunking at a distribution board it is essential to ensure that the junction between trunking and distribution board will not cause abrasion to the cables.

- Grommet strips should be fitted over the edges of any holes drilled in trunking to prevent damage to cables.

- To prevent the effects of eddy currents (electromagnetic effects), cables of an a.c. circuit should pass through a single hole.

Trunking capacities

The number of cables that can be drawn into or laid in any enclosure of a wiring system must be such that no damage can occur to the cables or the enclosure during installation. The number of cables that can be used is the overall sum of the cables, cross-sectional area (csa) compared to the overall csa of the trunking. This is expressed as a percentage and should not exceed 45 per cent; which is already taken into account by the standard sizes of cable and enclosures in the *Amicus Tables*.

The following wiring systems are covered by the appropriate tables in the *IEE On-Site Guide* Appendix 5 and *Amicus Guide* pages 34–37.

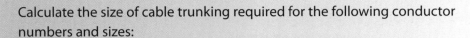

Example 1

Calculate the size of cable trunking required for the following conductor numbers and sizes:

$25 \times 1.5mm^2$ $20 \times 2.5mm^2$ $6 \times 4.0mm^2$ $2 \times 10mm^2$ $2 \times 16mm^2$

Assume that all the conductors are stranded PVC.

Step 1

Obtain the 'Term' values from the *Amicus Guide* page 70 or Appendix 5 *IEE On-Site Guide*.

Conductor size	Term value	Number of conductors	Total term value
1.5mm	8.6	25	215
2.5mm	12.6	20	252
4mm	16.6	6	99.6
10mm	35.3	2	70.6
16mm	47.8	2	95.6

Table 5.4 Term values by conductor size

Step 2

Multiply the term values by the number of conductors.

Step 3

Add the term values together.

Total terms = 732.8

Step 4

From the tables, select the nearest trunking term value (equal to or larger than that calculated), which in this case would be:

767 terms, which would give a trunking size of 50mm × 38mm

Or:

738 terms, which would give a trunking size of 75mm × 25mm

Whichever is selected will depend on the designer's choice.

Example 2

Steel trunking is to be installed as the wiring system for 80 single-phase circuits, each having a design current of 15 A. Single-core PVC stranded cables of 4mm² will be used. 20 A BS 88 fuses will be used to protect the circuit, and PVC-insulated copper cables will be installed. Determine the size of trunking required, assuming voltage drop requirements have been met.

Select term values from the *Amicus Guide* page 70.or Appendix 5 of the *IEE On-Site Guide*.

The term value for 4mm² is 16.6.

The number of conductors is 160 (80 circuits each of 2 V conductors).

Multiply $16.6 \times 160 = 2656$ terms.

The nearest trunking sizes are:

150mm × 50mm, which has a term value of 3091

Or:

150mm × 38mm, which has a term value of 2999

Whichever is selected will depend on the designer's choice.

PVC trunking

Essentially the same range of accessories and installation techniques are used for the installation of PVC trunking. As well as being lighter and therefore easier to handle, PVC trunking is easier on the eye than metal when installed for data cabling or computer supplies in locations such as shops and offices. Two examples are shown below.

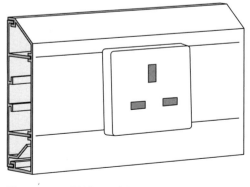

Figure 5.29 PVC trunking

PVC trunking

Cable tray

On large industrial and commercial installations, where several cables take the same route, cable trays may be used. We will look at cable tray designs, as well as their advantages and disadvantages, in greater depth in Chapter 7. Meanwhile in this section we will look at:

- types of accessories
- installing cable tray
- fabricating on site – the tray bending machine
- bending cable tray by hand
- fabricating a flat 90° bend
- fabricating a flat 90° bend from three pieces of tray
- forming a cable tray reduction
- fixing cable tray.

Types of accessories

A wide variety of factory-made accessories is available to suit both standard and return-flange trays of various sizes.

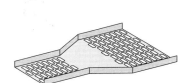  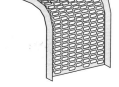 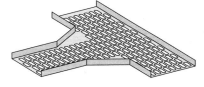

Figure 5.30 (a) Straight reducer   (b) 90° flat bend   (c) 90° outside riser   (d) Equal tee

Installing cable tray

When installing cable tray it is essential that it is well supported and secured. It is usually possible to complete the installation by making use of the wide range of accessories and fittings generally available, although it may sometimes be necessary to fabricate joints, bends or fittings to meet particular requirements.

Cable trays can be joined in a number of ways. Different manufacturers will supply a variety of patent couplings and fasteners. Links or fish plates are commonly used, and some cable trays are designed with socket joints. In some circumstances a welded joint may be required. If this method is used care must be taken to restore the finish around the weld to prevent corrosion.

Most methods of jointing cable tray involve the use of nuts and bolts. A round-headed or mushroom-headed bolt (roofing bolt) or screw should be used, and this should be installed with the head inside the tray. This reduces the risk of damage to the cables being drawn along the tray.

Remember

When installing a cable tray run, the following points should be considered:

- ease of installation
- economy of time and materials
- facility for extending the system to take additional cables

Fabricating on site (the bending machine)

It is sometimes necessary to fabricate joints, bends and fittings to meet particular requirements or where factory-made accessories are not available. Careful measurement and marking out is required. Cable trays can be cut quite easily with a hacksaw but care should be taken to remove any sharp edges or burrs.

Bends can be formed by hand after a number of cuts have been made in the flange to accommodate the bend, although a far better job can be made by using a crimping tool or a tray-bending machine. Bending machines are available from various manufacturers. Where a lot of cable tray work is to be installed, machine bending is quicker and more practical. The machines are made to accommodate the various widths and gauges of cable tray. They may also be used to bend and form flat strips of metal, and have a vice to hold the length of cable tray being worked upon. Cable-tray chain vices are also available for this purpose.

'Making good' cannot normally match the protection qualities of the original factory-applied finish, but the absence of any protection at all can seriously reduce the effectiveness of the original finish through corrosion spreading from this point. Primed tray should have a proper finishing paint applied over the primer as soon as possible. The purpose of the finish is to protect the cable tray from corrosion.

Figure 5.31 (a) Cable-tray bending machine

(b) Sharp-radius bend being formed

(c) Large upward-radius bend being formed

Bending cable tray by hand

Light-duty cable tray may be bent by hand with the aid of a crimping tool. This can be made from a piece of 6mm mild steel bar.

Figure 5.32 a
Bending cable tray

(a)

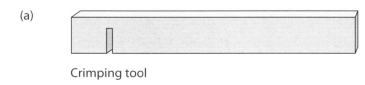

Crimping tool

To make an inside bend, first determine the radius (r) of the bend from the diameter of the largest cable to be installed using Table 4E in the *On-Site Guide*. Mark the points at which the bend will begin and end, and the centre of the bend on the piece of tray to be bent.

Figure 5.32 b–f Bending cable tray (continued)

(b)

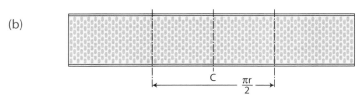

Using the crimping tool, crimp evenly on each side of the tray. Work slowly, checking the form of the bend.

(c)

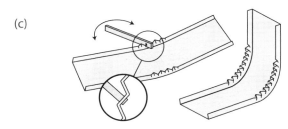

To make an 'outside' bend, saw through the flange and bend the tray in a vice. Mark out the bend, as before, taking care that the radius of the bend accepts the largest size cable as stated in Table 4E from the *On-Site Guide*.

Mark off along both flanges on the cable tray a series of equal distance points. Make appropriate hacksaw cuts that are equal to the depth of the tray flange on either side of the centre line as shown.

(d)

(e)

Grip the tray with wooden blocks in a vice and bend gradually, moving the tray along and checking the bend for evenness. (f) illustrates the completed bend.

(f)

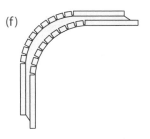

Figure 5.33 a–d Flat
90º bend

Fabricating a flat 90° bend

Measure and mark the mid-point of the bend.

Mark off X when $X = \sqrt{2 \times W}$, where W = width of tray.

(a)

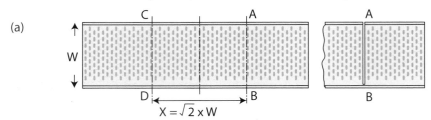

Cut through the flange with a hacksaw at point A and along line A–B but do not cut through the opposite flange. Make a similar cut at point C and along line C–D. Bend the two outer sections of the tray together to form a 90° angle.

(b)

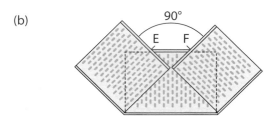

Mark the flange at the points of overlap E, F and E1, F1. Cut through the flanges at these points and bend these flanges flat. Cut away the tongues at both slots A and C as shown.

(c)

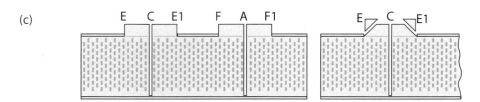

Remove all sharp edges and burrs with a file. Make up the assembly as shown and secure with round-headed bolts. Ensure that the bolt heads are uppermost.

(d)

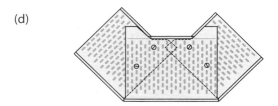

Fabricating a flat 90° bend from three pieces of cable tray

Place two pieces of tray together to form a 90° angle at the point of contact marked X.

(a)

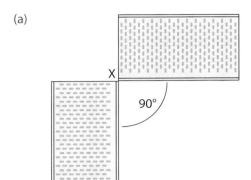

Figure 5.34 a–e Flat 90° bend made from three pieces of cable tray

(b)

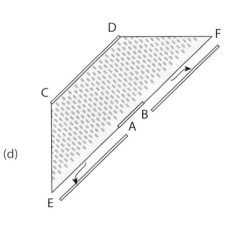

Place a third piece of tray over these and mark points of contact A, B, C, D, E, F.

Cut away flanges XA and XB as shown.

(c)

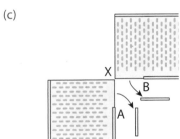

Cut along CE and DF. Cut away inner flange AE, BF.

Remove all sharp edges and burrs with a file. Assemble as shown, using round-headed bolts.

(d)

(e)

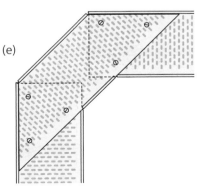

Forming a cable-tray reduction

Overlap the two segments of cable tray by at least 100mm. Mark the tray width and reduction angle as shown.

Figure 5.35 a–e
Cable-tray reduction

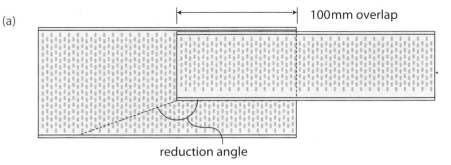

(a)

100mm overlap

reduction angle

Cut with a hacksaw and cold chisel but do not cut through the flange.

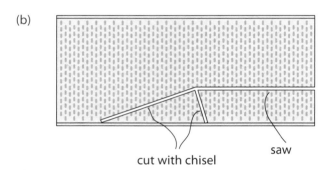

(b)

cut with chisel   saw

Bend the cable tray into shape. Remove all sharp edges and burrs with a file.

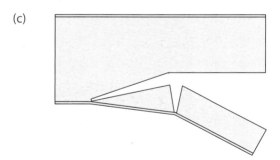

(c)

Assemble the joint using round-headed bolts. Ensure that the bolt heads are uppermost to prevent any damage to the cables.

(d)

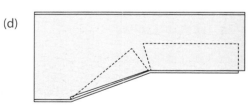

(e)

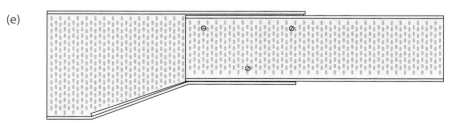

Fixing cable tray

Cable tray is fixed to building surfaces using three basic methods, as illustrated.

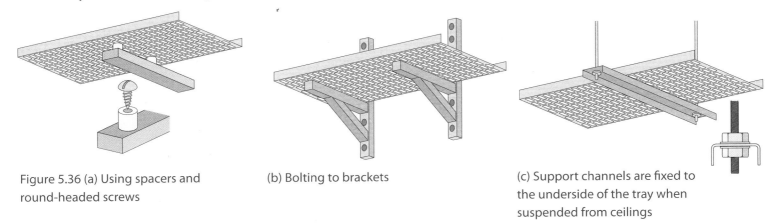

Figure 5.36 (a) Using spacers and round-headed screws

(b) Bolting to brackets

(c) Support channels are fixed to the underside of the tray when suspended from ceilings

Armoured cables

PVC/SWA/PVC cable

PVC-insulated SWA (steel wire armoured) and PVC-sheathed cables, commonly known as PVC/SWA/PVC cables, are used extensively for mains and sub-mains, and for wiring circuits in industrial installations. The cable consists of multi-core PVC-insulated conductors made of copper or aluminium with a PVC-sheathed steel wire armour and PVC sheath overall. This type of cable has many advantages: it is more pliable and easier to handle than paper-insulated lead-covered cables, and termination is also relatively easy. This section will cover the installation and termination of various armoured cables.

A typical example of PVC/SWA/PVC cable is shown below.

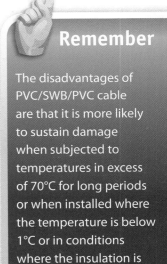

Remember

The disadvantages of PVC/SWB/PVC cable are that it is more likely to sustain damage when subjected to temperatures in excess of 70°C for long periods or when installed where the temperature is below 1°C or in conditions where the insulation is likely to be split

1. Shape stranded copper conductor
2. PVC insulation
3. Extruded bedding
4. Galvanised steel wire armour
5. PVC oversheath

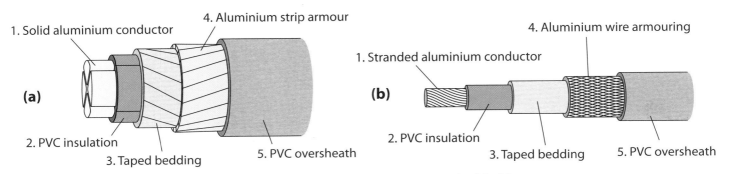

1. Solid aluminium conductor
2. PVC insulation
3. Taped bedding
4. Aluminium strip armour
5. PVC oversheath

(a)

1. Stranded aluminium conductor
2. PVC insulation
3. Taped bedding
4. Aluminium wire armouring
5. PVC oversheath

(b)

Figure 5.37 Four-core cable aluminium strip armoured (a) and single-core sectoral cable (b)

XLPE/LSF/SWA/LSF

This low smoke and fume SWA cable appears to be the same as a PVC/SWA/PVC cable in its construction. However, it is designed to give off small amounts of smoke and fumes in the event of a fire. This (copper) BS 6724 cable has plain annealed stranded copper conductors with cross-layered polyethylene insulation (XLPE). Other versions include the low smoke and fume extruded bedding (LSF), steel wire armoured and LSF sheath.

This gives the cables the following combustion characteristics: HCL emission 0.5 per cent in accordance with BS 6425 part 1 and low smoke classification based on the metre cube test. Completed cables comply with the requirements of fire test BS 4066 part 3.

These cables are used as distribution cables in environments where the public has access, such as train stations and airports etc.

Terminating PVC/SWA/PVC cable

The cable consists of PVC-insulated conductors with an overall covering of PVC. Between this covering and the outer PVC sheath the galvanised steel wire armouring is embedded. The armouring is used as a circuit protective conductor and special glands are employed to ensure good continuity between this and the metal work of the equipment to which you are connecting.

These glands vary a little from one manufacturer to another and their design also depends on the environment in which they are to be used. An earthing tag (bonding ring) provides earth continuity between the armour of the cable and the box or panel. If PVC/SWA/PVC cable is used to connect directly to an electric motor mounted on slide rails, a loop should be left in the cable adjacent to the motor to permit necessary movement.

Terminations are made by stripping back the PVC sheathing and steel wire armouring then fitting a compression gland, which is terminated into the switchgear or control gear housing.

For correct termination, start by measuring the length of armouring required to fit over the cable clamp; make a note of this measurement. Then establish how long the conductors need to be in order to connect to your equipment (make a note of this) and then mark that distance on the cable by measuring from the end of the cable (position A) to position B, as shown in (a). This then represents the length of conductor required.

Follow this by marking the length required for the cable clamp from B to C as shown in(a). At this stage some people will strip off the PVC outer sheathing. However, this is best left on as it will hold the steel wire armouring in place for you. You must remember to place the shroud over the cable. This is often forgotten – a sign of poor workmanship.

Remember

From 1 April 2004 the colours changed as follows: red changed to brown; black changed to blue. From 1 April 2004 the colours for a three-core cable changed as follows: red changed to brown; yellow changed to black; blue changed to grey

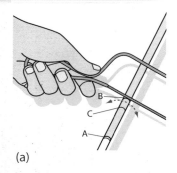

(a)

Figure 5.38 a
Terminating PVC/SWA/
PVC cable

Next, taking a junior hacksaw, cut through the PVC outer sheathing and partly through the armouring at point B. The PVC outer sheath can now be cut away as shown (b).

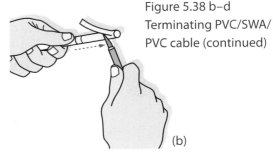

Figure 5.38 b–d Terminating PVC/SWA/ PVC cable (continued)

(b)

Taking each strand of the armouring in turn, snap them off at the point where they are partly cut through.

Then, using either a hacksaw or a knife, cut neatly around the PVC outer sheath at point C and remove the remaining pieces of outer sheathing; this will leave the cable as shown (c).

(c)

The gland can now be fitted onto the cable. First slide the backnut and compression rings, if any, on to the cable. Then, taking the gland body, slide this on to the cable, making sure that it fits under all the strands of armouring as shown (d).

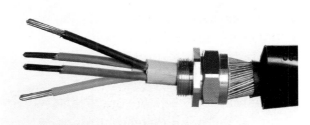

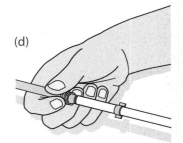

(d)

Finally, slide up the backnut and screw it on to the gland body, thus clamping the armouring tightly. The inner PVC sheath can be stripped off like any other PVC cable and the gland is then ready for connecting to your equipment.

It is important to clean any paintwork from the area of contact before tightening up the locknut and securing the gland. Bonding rings, or earthing tags as they are sometimes called, can be used to provide better contact with surrounding metal work. A CPC (circuit-protective conductor) should be fitted between the bolt securing the earthing tag and the earthing terminal of the equipment.

Installation

Cables can be laid directly in the ground, in ducts or fixed directly on to walls using cable cleats. If several cables are to follow the same route they are best supported on cable trays or racks. When several cables are installed in enclosed trenches the current ratings will be reduced due to their disposition. The correction factors for cables run under these conditions are found in BS 7671 Table 4B3.

Safety tip

If, while carrying out an installation of underground cables, you realise that excavations are still taking place you should inform your supervisor before continuing with the installation

Cleats

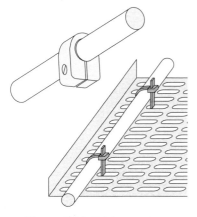

Figure 5.39 Cable tray and cleat

Installation is relatively easy for the smaller size cables but it will become necessary to employ an installation team to handle the bigger sizes or multi-core cables. For the most part, one-hole cable cleats constructed of solid PVC will be used. In the case of cables installed on a cable tray, cable ties will be used.

For the bigger cables, cleats made of die-cast aluminium are used. These are often designed to be slotted into steel channels so that, once a piece of channel has been fixed, multiple runs of cable can be accommodated. The minimum radius that these cables should be bent to is eight times the outside diameter, and the spacing of the cleats should be as recommended in Table 4A of the *On-Site Guide*. Where these cables are to be installed directly into the ground they should be marked with cable covers or suitable PVC marking tape to indicate their presence. Regulation 522.8.10 states that they should be buried at a sufficient depth to avoid being damaged by any reasonably foreseeable disturbance of the ground.

Conductor colour identification for standard 600/1000 V armoured cables to BS 6346, BS 5467 or BS 6724 from 1 April 2004

- Single core was red or black; it is now brown or blue.

- Two core was red, black; it is now brown, blue.

- Three core was red, yellow, blue; it is now brown, black, grey.

- Four core was red, yellow, blue, black; it is now brown, black, grey, blue.

- Five core was red, yellow, blue, black, green and yellow; it is now brown, black, grey, blue, green and yellow.

PVC/GSWB/PVC cable

PVC/Galvanised Steel Wire Braided(GSWB)/PVC cable consists of individual conductors within an aluminium screen surrounded by an inner sheath then a steel braid (similar to a basket weave) underneath an outer sheath of PVC. The braid makes this cable more flexible than SWA.

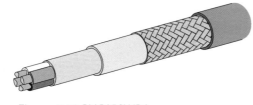

Figure 5.40 PVC/GSWB/ PVC cable

This durable cable is used in many instrumentation applications or where shielding is required for signal applications.

Installation techniques are virtually identical to SWA, but cable glands need either to be of the universal type or to have a special **olive** inside them to hold the thin braid.

Cat (Category) 5 cable

This cable is used extensively for data transfer in computer networks and telephone systems. It has four pairs of wires that transmit and receive data along them at very high frequencies, typically 350 MHz. Special termination ends are required for these cables.

There are three basic types of cabling used in data systems: coaxial, fibre-optic and Unshielded Twisted Pair (UTP). Coaxial is widely installed in older networks but is not recommended for new network installations. Fibre-optic is used for high-speed networks and to connect networking devices separated by large distances. However, UTP is currently the most common and recommended cabling type.

UTP is inexpensive, flexible and can transmit data at high speeds. Most new installations are currently installed with Cat5 UTP cabling and components.

Collector columns

Collector columns supply electrical current to a rotating unit from a fixed source. They can be easily adapted to an endless variety of applications, including slewing cranes, sewage treatment plant, test rigs and rotating displays.

The columns usually contain a mounting frame, main body and slipring cover manufactured from aluminium, and are fitted with phosphor-bronze sliprings and dual leg brush gear, incorporating copper graphite brushes. The slipring assembly is mounted on a mild steel shaft rotating in self-lubricating bearings. The slipring cover then incorporates a location pocket to accept the customer's drive or anchor pin.

Collector column

Trailing cables

Other types of installation might require the use of trailing cables. The main characteristics of a trailing cable are its ability to flex and bend and also to withstand varying levels of mechanical stress. Invariably these cables contain stranded conductors to give them this flexibility.

For example, in certain construction sites or in mining, heavy plants may have their operating cables trailed across a site where vehicles drive over them and have to withstand adverse weather conditions. Such cables are made with heavy inner and outer sheathing and will usually contain wire-braid armouring or flexible steel-wound armouring. They will normally be terminated using special couplers and have to withstand regular bending as part of their normal use. However, cables feeding light equipment may only risk being dragged or scraped (e.g. when using a drilling machine), and their construction may therefore only consist of heavy-duty rubber sheathing.

These types of cable are also used in factories and manufacturing processes where production machinery is constantly moving, e.g. robotic equipment.

Structure of a trailing cable

Fibre-optic cables

Hair-thin fibres make up fibre-optic cables

If you wanted to see down a dark corridor, you might shine a torch down it. But what if the corridor had a bend in it? You could probably put a mirror in just the right place at just the right angle and shine the light round the corner. But what if the corner had lots of bends? Well, what if I made the entire corridor walls out of mirrors, then I wouldn't need to put them in just the right place or angle. The light would be able to bounce around all the mirrors along the walls.

This is the theory behind fibre-optics, as the glass core is essentially a mirror wound into a thin tube. Some 10 billion digital bits can be transmitted per second along an optical fibre link in a commercial network, enough to carry tens of thousands of telephone calls. The hair-thin fibres consist of two concentric layers of high-purity silica glass, the core and the cladding, which are enclosed by a protective sheath.

MICC cable

Mineral-insulated copper cables (MICC) consist of high-conductivity copper conductors insulated by a highly compressed white powder (magnesium oxide). A seamless copper sheath encapsulates the conductors and powder. The basic properties of MICC cables are covered at Level 2.

MICC cable

In this section we will look at the following:

- terminating MICC cables
- practical termination of MICC cables
- running of cables
- testing and fault finding
- FP 200.

Terminating MICC cables

Mineral-insulated cables must be sealed at each end, otherwise the magnesium oxide insulator will absorb moisture resulting in a low insulation-resistance reading. A complete termination comprises two sub-assemblies, each performing a different function.

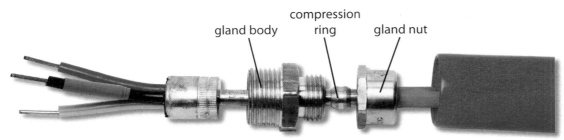

MICC gland and seal

The purpose of the seal is to exclude moisture from the cable. The seal consists of a brass pot with a disc to close the mouth, and sleeves to insulate the conductor tails. Gland body compound is used to fill the pot. This is used to connect or anchor the cable; the gland consists of three brass components as shown.

Practical termination of MICC cables

The first thing that must be done when terminating MI cables is to strip off the copper sheath. If the cable has a PVC oversheath, then this must be removed before the copper sheath. Some electricians recommend that the cable should be indented with a ringing tool prior to sheath stripping. However, it may, be better to carry out ringing nearer the end of the stripping process. Stripping off the copper sheath can be carried out in several ways, although all of them require practice before the task can be carried out in reasonable time.

Stripping MICC cables

Using side cutters

When using side cutters or pliers it is essential that the points of the blades are in good condition, otherwise difficulties can arise. First make a small tear in the sheath then peel off the sheath with the side cutters, working around the cable in a clockwise direction, keeping the side cutters at an angle of approximately 45 degrees to the cable.

Using a stripping bar (fork-ended stripper)

First the cable sheath is started or broken away with side-cutting pliers as above. Then the torn-off sheath portion is flattened and the slot of the stripping bar pushed into the torn-off sheath. The stripping bar is then rotated while keeping the bar at 45° to the cable.

Using rotary stripping tools

Rotary stripping tools are available in many shapes and forms. They are much easier to use than either side cutters or the stripping bar but can be difficult to set up. This is not a problem if the same size cables are being used all the time.

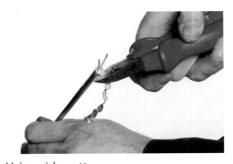

Using side cutters

Using a stripping bar

Rotary stripping tool

Ringing cable sheaf

Rotary strippers usually require a good square end to the cable. The stripper is then pushed over the cable, a little pressure applied and then rotated. If the sheath is to be stripped for a longer distance, then the sheath being removed should be cut away at intervals to avoid fouling the tool. Another advantage of the stripping tool is that a ringing tool is not required. When the stripping tool has reached a required position, pliers are applied to the cable sheath and the stripper is further turned.

Ringing cable sheath

Before the pot can be fixed the sheath must have a good, clean circular end. To achieve this a ringing tool is used which makes an indent in the sheath. This will be required when the sheath is stripped with either side cutters or the stripping bar, but is not necessary (as previously mentioned) when rotary strippers are used.

Actions prior to fitting the pot

Before fitting the pot a check should be made that the gland nut and olive are in position. If the cables have the PVC oversheath then a shroud must be fitted.

Fitting the pot

Fitting the pot

The following describes the fitting of a screw-on pot. Pots are best fitted using a pot wrench. This ensures that the pot goes on square. The pot has an internal thread, which screws on to the copper sheath. The pot should be screwed on to the copper sheath until it just lines up inside the pot.

Finally the pot is tapped to empty it of any filings or loose powder. Do not be tempted to blow into the pot as the moisture in your breath could reduce the insulation resistance of the magnesium oxide, causing a short circuit at a later stage.

Actions prior to sealing

Before sealing, the conductors should be wiped with a clean, dry rag to remove any loose powder. If the cables have become twisted they should be straightened by being firmly pulled with a pair of pliers.

MICC pot and seal

The cable insulation and disc should be prepared. The insulation is usually neoprene rubber sleeving and this is cut to the required length. The sleeving is then fitted to the conductors and disc. The disc also ensures the conductors are kept apart while the compound is inserted into the pot.

Figure 5.41 Tapping the pot empty of powder

Sealing the pot

Filling the pot with compound affects the sealing. It is important that the compound is filled from one side only, directed towards and in between the conductors. This prevents air locks, which could lead to condensation problems. Hands should be clean when using the compound and the compound should be kept covered to avoid entry of dirt.

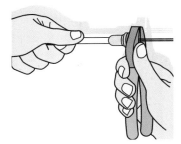

Figure 5.42 Sealing the pot

Crimping

Before the crimping tool is applied any excess compound should be removed. The crimping tool is applied and gradually tightened while keeping it straight. The operation should be stopped at intervals to allow the compound to seep out.

Crimping the pot

Finished seal and identification

When crimping is completed the seal should be tested to ensure a high value of insulation resistance both between cores and from cores to the copper sheath. Having done this, the cable can be run out and the other end sealed. The whole cable is then insulation tested and the conductors are given a continuity test to indicate which conductor is which before putting tags or coloured sleeves on them for identification.

Gland connection

When the cable has been tested and conductors identified, the gland must be tightened up and connected into the equipment or accessory in which it is being terminated. After the conduit thread has been tightened into the accessory, the back nut must also be tightened. This will compress the olive ring inside the gland, thereby providing earth continuity. Once this has been tightened it is very difficult to remove.

Types of gland

Standard glands have their sizes stamped on them, e.g. 2L1 meaning 2core, light-duty, 1mm^2 csa (cross-sectional area) or 4H6 meaning 4-core, heavy-duty 6mm^2 csa. The size of the conduit thread will depend on the overall size of the cable being used.

Running of MICC cables

One of the advantages of MI cables is that they can be run on the surface as well as being run under plaster. When they are run on the surface, for neat appearance it is imperative that they are run straight. Lines should be run out before fixings are made and care should be taken not to twist the cable unnecessarily before installation. Fixings are made either with one-hole clips or two-hole saddles. Screws are usually brass round-heads.

Where a large number of cables are run together an adjustable saddle is available that will take many cables. Often MI cables are run on cable tray. This is used to provide easier fixing, neat appearance and cleanliness. It also means that a great number of cables can be run together. Where several cables leave or enter an enclosure, e.g. at a distribution board, the cable should be distributed equally. If there are too many cables to fit into one row, two staggered rows of holes should be drilled.

Putting bent cable in roller   Closing roller   Running roller along cable

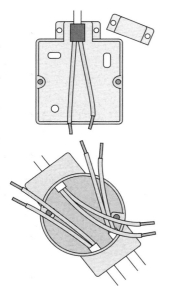

Figure 5.43 Types of plaster-depth termination boxes

MI cables are usually terminated using standard conduit boxes. Where connections have to be made, connectors should be used. Where cables are run buried in plaster, plaster-depth boxes are available. With this type of box no gland is required. It is also important to note that the earth clamp over the pot is secured to give continuity. Earth-tailed pots are also available. These are pots that have an earth conductor manufactured as part of the pot, thus giving good earth continuity. If this cable is to be installed directly into a motor then a loop should be made in the cable to prevent mechanical stress at the termination due to excessive vibration.

Testing and fault-finding

The seal should be visually inspected for obvious defects. If there is a minor fault, e.g. incomplete crimping, it may be practicable just to repeat the operation. However, it may be necessary to remove the seal and re-terminate. After both ends of the cable have been terminated with permanent seals, the cable should be subjected to an insulation-resistance test using a voltage appropriate to the intended operating voltage in accordance with BS 7671. The purpose of this initial test is to check for major faults, e.g. short circuit within the pot, in which case the fault should be located and rectified. The insulation resistance should be noted and compared with the value measured at least 24 hours later. The second reading should be at least 100 MΩ and should have risen from the initial value.

FP 200 cable

Construction of FP 200 Gold

FP 200 Gold has either solid or stranded copper conductors covered with a fire- and damage-resistant insulation (Insudite). Electrostatic screening is provided by a laminated aluminium-tape screen, which is applied longitudinally and folded around the cores to give an overlap. The aluminium-tape screen is applied metal side down and in contact with the uninsulated cpc. The sheath is a robust thermoplastic low-smoke, zero-halogen sheath, which is an excellent moisture barrier.

Types of FP 200 Gold

FP 200 Gold is available in 2, 3, 4, 7, 12 and 19 cores, with conductor sizes that vary from 1mm² to 4mm². The cable is used in fixed installations in dry or damp premises in walls, on boards, in channels or embedded in plaster for situations in which prolonged operation is required in the event of fire. It is primarily intended for use in fire-alarm and emergency-lighting circuits and has a low voltage rating of 300/500 volts.

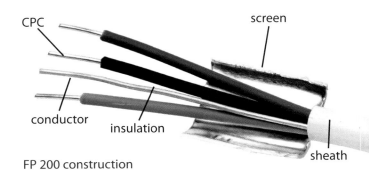

FP 200 construction

The sheath colour is available in red or white. Core colours are:

- two-core: brown and blue

- three-core: brown, blue and green-yellow

- four-core: brown, black, grey and green-yellow.

Advantages	Disadvantages
• Easy to handle • Designed for use with conventional terminations • No special tools required • Age-resistant insulation • Bends easily without the use of bending tools	• Can be damaged if the bending radii are not observed • Cable lacks mechanical strength • Easily damaged if not correctly terminated

Table 5.5 Advantages and disadvantages of FP 200 Gold

Remember

From 1 April 2004 the colours changed as follows: red changed to brown; black changed to blue. From 1 April 2004 the colours for a three-core changed as follows: red changed to brown; yellow changed to black; blue changed to grey

Installing FP 200 Gold

When the cable is required to maintain circuit integrity during a fire, it is important that any clips or ties used to support the cable can also withstand that fire. They should be copper, steel or copper-coated clips or ties suitable for fixing a fire-rated cable; standard plastic or nylon clips or ties should not fix the cable.

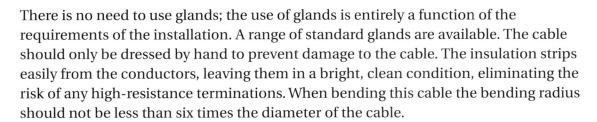

FP 200 Gold termination

There is no need to use glands; the use of glands is entirely a function of the requirements of the installation. A range of standard glands are available. The cable should only be dressed by hand to prevent damage to the cable. The insulation strips easily from the conductors, leaving them in a bright, clean condition, eliminating the risk of any high-resistance terminations. When bending this cable the bending radius should not be less than six times the diameter of the cable.

Termination

FP 200 does not possess the same mechanical strength as other cables. It is important to terminate the cable according to the manufacturer's instructions. The following sequence explains a recognised method of termination.

Scoring the sheath

Score around the sheath with a knife or suitable cable-stripping tool, taking care not to cut right through to the aluminium tape. Flex the cable gently at the point of scoring until the sheath yields.

Pull off the sheath, twisting gently to follow the lay of the cores.

Pulling the sheath

FP 200 Gold does not require a ferrule; the cable may enter wiring accessories or fittings through a simple grommet. When installed in wet conditions or outdoors, a standard waterproof gland incorporating a PCP sealing ring must be used. Earthing of the aluminium screen is achieved automatically by correctly terminating the bare-tinned copper circuit-protective conductor. Two-, three- and four-core cables have a full size cpc whereas 7-, 12- and 19-core cables have a drain wire which has a short-circuit rating of 75 A for 1 second; suitable protective equipment should therefore be provided.

In order to prevent electrical faults occurring from phase to earth when terminating FP cables, care should be taken to avoid damage to the insulation by not bending the cores sharply over the end of the sheath.

Remember

When terminating cables in enclosures where space is limited (e.g. plaster-depth boxes), sleeving should be applied over the insulated core and earth conductors to prevent 'pinching' of the insulation and consequent damage

Transformers

On completion of this topic area the candidate will be able to describe the operation of transformers, perform simple calculations relating to input, output and losses and explain rating.

The transformer is one of the most widely used pieces of electrical equipment. It can be found in situations such as electricity distribution, construction work and electronic equipment. Its purpose, as the name implies, is to transform something – the something in this case being the voltage, which can enter the transformer at one level (input) and leave at another (output).

When the output voltage is higher than the input voltage this is called a step-up transformer. When the output voltage is lower than the input, this is called a step-down transformer.

We covered the operation of transformers in great detail at Level 2. Please refer back to that section for more information on transformers.

Calculations involving transformers

The number of turns in each winding will affect the induced e.m.f., with the number of turns in the primary being referred to as N_p, and those in the secondary referred to as N_s. We call this the turns ratio. When voltage V_p is applied to the primary winding, it will cause a changing magnetic flux to circulate in the core. This changing flux will cause an e.m.f. V_s to be induced in the secondary winding.

Assuming that we have no losses or leakage (i.e. 100 per cent efficiency) then power input will equal power output and the ratio between the primary and secondary sides of the transformer can be expressed as follows:

$$\frac{V_P}{V_S} = \frac{N_P}{N_S} = \frac{I_S}{I_P}$$

(where I_p represents the current in the primary winding and I_s the current in the secondary winding).

The losses that occur in transformers can normally be classed under the following categories.

Copper losses

Although windings should be made from low resistance conductors, the resistance of the windings will cause the currents passing through them to create a heating effect and subsequent power loss. This power loss can be calculated using the formula:

$P_c = I^2 \times R$ watts

(where I is the current flowing in amps and R is the resistance of the winding in ohms).

Remember

Transformers make use of an action known as **mutual inductance**. Two coils, primary and secondary, are placed side by side, but not touching each other. The primary coil is connected to an a.c. supply and the secondary coil is connected to a load, such as a resistor. Current flow creates a magnetic field in the primary coil, creating a changing magnetic flux. This produces an e.m.f. in the secondary coil which then starts flowing through the load

Remember

No transformer can be 100 per cent efficient. There will always be power losses

Example

Calculate the copper loss of a secondary winding in a small step-down transformer that has a resistance of 0.35 Ω the connected external load is drawing 10 Amp.

$P_c = I^2 \times R$ therefore $P_c = 10^2 \times 0.35 = 100 \times 0.35 =$ **35 Watts**

What are the copper losses in the primary winding of a step up transformer, where the resistance of the winding is 0.02 Ω and the transformer is drawing 6 amps from the supply?

$P_c = I^2 \times R$ therefore $P_c = 6^2 \times 0.02 = 36 \times 0.02 =$ **0.72 Watts**

Definition

Hysteresis – a generic term meaning a lag in the effect of a change of force

Iron losses

These losses take place in the magnetic core of the transformer. They are normally caused by eddy currents (small currents which circulate inside the laminated core of the transformer) and **hysteresis**. To demonstrate this, let's say that you push on some material and it bends. When you stop pushing, does the material return completely to its original shape without being pushed the other way? If it doesn't the material is demonstrating hysteresis. Let's look at this in context.

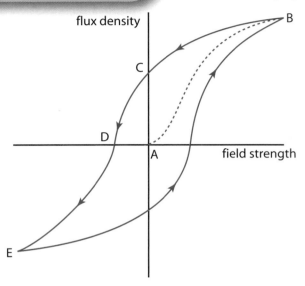

Figure 5.44 Double wound, core-type transformer

Figure 5.44 shows the effect within ferromagnetic materials of hysteresis. Starting with an unmagnetised material at point A, both field strength and flux density are zero. The field strength increases in the positive direction and the flux begins to grow along the dotted path until we reach saturation at point B. This is called the initial magnetisation curve.

If the field strength is now relaxed, instead of retracing the initial magnetisation curve the flux falls more slowly. In fact, even when the applied field has returned to zero, there will still be a degree of flux density (known as the **remanence**) at point C. To force the flux to go back to zero (point D), we have to reverse the applied field. The field strength that is necessary to drive the field back to zero is known as the **coercivity**. We can then continue reversing the field to get to point E, and so on. This is known as the hysteresis loop.

As we have already said, we can help reduce eddy currents by using a laminated core construction. We can also help to reduce hysteresis by adding silicon to the iron from which the transformer core is made.

The version of the double wound transformer that we have looked at so far makes the principle of operation easier to understand. However, this arrangement is not very efficient, as some of the magnetic flux being produced by the primary winding will not react with the secondary winding and is often referred to as 'leakage'. We can help to reduce this leakage by splitting each winding across the sides of the core (see Figure 5.45).

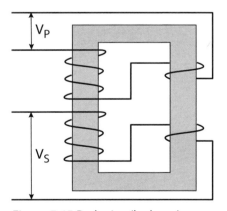

Figure 5.45 Reducing 'leakage'

Transformer ratings

Transformers are rated in kVA (kilovolt-amps) rather than in watts. This rating has to be used as the transformer is made of copper coils. When a conductor is formed into a coil and connected to an a.c. supply, the coil takes on the property of an inductor. In this state it is not the pure resistance of the conductor that opposes the flow of current but the combined values of inductance and resistance – the impedance (Z).

We also know that in an a.c. circuit:

Power = IVCosθ.

We therefore know the output voltage of the transformer but not the current or its power factor. Both are affected by the load connected to the transformer. Therefore, the VA rating is used to give an indication of the transformer's performance.

The kVA rating can be used to determine the current that a transformer can deliver for a known voltage.

Example 1

What is the current that can be drawn from a 100 VA transformer if it has an output voltage of 20 V?

$$A = \frac{VA}{V} \qquad A = \frac{100}{20} \qquad A = 5\,A$$

If a transformer has a maximum output voltage of 200 V and supplies a current of 12 A what is the kVA rating of the transformer?

$$kVA = \frac{VA}{1000} \qquad kVA = \frac{200 \times 12}{1000} = 2.4\,kVA$$

Calculate the line current delivered by an 11 kVA three-phase transformer.

$$kVA = \frac{\sqrt{3}V_L I_L}{1000} \qquad \text{therefore } 11 = \frac{\sqrt{3} \times 400 \times I_L}{1000}$$

$$\text{Transpose to find } I_L = \frac{11 \times 1000}{\sqrt{3} \times 400} = 15.88\,A$$

Example 2

A transformer having a turns ratio of 2:7 is connected to a 230 V supply. Calculate the output voltage.

When giving transformer ratios they appear in the order primary then secondary. Therefore, in this example we are saying that for every two windings on the primary winding there are seven on the secondary. Therefore using our formula:

$$\frac{V_P}{V_S} = \frac{N_P}{N_S}$$

We should now transpose this to get:

$$V_S = \frac{V_P \times N_S}{N_P}$$

However, we do not know the exact number of turns involved. But do we need to if we know the ratio? Let us find out.

The ratio is 2:7, meaning for every two turns on the primary there will be seven turns on the secondary. Therefore if we had six turns on the primary this would give us 21 turns on the secondary but the ratio of the two has not changed. It remains 2:7.

This means we can just insert the ratio rather than the individual number of turns into our formula:

$$V_S = \frac{V_P \times N_S}{N_P} = \frac{230 \times 7}{2} = \textbf{805 V}$$

Now, to prove our point about ratios, let us say that we know the number of turns in the windings to be 6 in the primary and 21 in the secondary (which is still giving us a 2:7 ratio). If we now apply this to our formula we get:

$$V_S = \frac{V_P \times N_S}{N_P} = \frac{230 \times 21}{6} = \textbf{805 V}$$

The same answer.

Example 3

A single-phase transformer, with 2000 primary turns and 500 secondary turns, is fed from a 230 V a.c. supply. Find:

(a) the secondary voltage
(b) the volts per turn.

Secondary voltage:

$$\frac{V_P}{V_S} = \frac{N_P}{N_S}$$

Using transposition, re-arrange the formula to give:

$$V_S = \frac{V_P \times N_S}{N_P}$$

$$V_S = \frac{230 \times 500}{2000} = \textbf{57.5V}$$

Volts per turn:

This is the relationship between the volts in a winding and the number of turns in that winding. To find volts per turn we simply divide the voltage by the number of turns.

Therefore, in the primary:

$$\frac{V_P}{N_P} = \frac{230}{2000} = \textbf{0.115 volts per turn}$$

In the secondary:

$$\frac{V_S}{N_S} = \frac{57.5}{500} = \textbf{0.115 volts per turn}$$

Example 4

A single-phase transformer is being used to supply a trace heating system. The transformer is fed from a 230 V 50 Hz a.c. supply and needs to provide an output voltage of 25 V. If the secondary current is 150 A and the secondary winding has 50 turns, find:

(a) the output kVA of the transformer
(b) the number of primary turns
(c) the primary current
(d) the volts per turn.

The output kVA:

$$kVA = \frac{\text{volts} \times \text{amperes}}{1000} = \frac{V_S \times I_S}{1000} = \frac{25 \times 150}{1000} = \textbf{3.75 kVA}$$

The number of primary turns:

If: $\dfrac{V_P}{V_S} = \dfrac{N_S}{N_P}$ then by transposition:

$$NP = \frac{V_P \times N_S}{V_S} = \frac{230 \times 50}{25} = \textbf{460 turns}$$

The primary current:

If: $\dfrac{V_P}{V_S} = \dfrac{I_S}{I_P}$ then by transposition: $I_P = \dfrac{V_S \times I_S}{V_P} = \dfrac{25 \times 150}{230} = \mathbf{16\ A}$

The volts per turn:

In the primary: $\dfrac{V_P}{N_P} = \dfrac{230}{460} = \mathbf{0.5\ volts\ per\ turn}$

In the secondary: $\dfrac{V_S}{N_S} = \dfrac{25}{50} = \mathbf{0.5\ volts\ per\ turn}$

Example 5

A step-down transformer, having a ratio of 2:1, has an 800-turn primary winding and is fed from a 400 V a.c. supply. The output from the secondary is 200 V and this feeds a load of 20 Ω resistance. Calculate:

(a) the power in the primary winding   (b) the power in the secondary winding.

We know that the formula for power is: $P = V \times I$

We also know that we can use Ohm's law to find current: $I = \dfrac{V}{R}$

Therefore, if we insert the values that we have, we can establish the current in the secondary winding:

$$I_S = \frac{V_S}{R_S} = \frac{200}{20} = \mathbf{10A}$$

Now that we know the current in the secondary winding, we can use the power formula to find the power generated in the secondary winding:

$$P = V \times I = 200 \times 10 = 2000\ W = \mathbf{2\ kW}$$

We now need to find the current in the primary winding. To do this we can use the following formula:

$$\frac{V_P}{V_S} = \frac{I_S}{I_P}$$

However, we need to transpose the formula to find. This would give us:

$$I_P = \frac{I_S \times V_S}{V_P}$$

Which, if we now insert the known values, gives us:

$$I_P = \frac{I_S \times V_S}{V_P} = \frac{10 \times 200}{400} = \mathbf{5\ A}$$

Now that we know the current in the primary winding, we can again use the power formula to find the power generated in the secondary winding:

$$P = V \times I = 400 \times 5 = 2000\ W = \mathbf{2\ kW}$$

Switchgear

On completion of this topic area the candidate will be able to describe the function of circuit breakers, switches and isolators.

Any installation must be controlled. This is usually done by a combination of devices such as circuit breakers, switches and isolators. The aim of these devices is to prevent serious damage to circuits in fault conditions.

Circuit breakers

A circuit breaker is designed to break a circuit under full fault conditions, such as an overload in a circuit. This means ending the current flow. Unlike a fuse it can then be reset after operation rather than replaced. The contacts of a circuit breaker are closed against spring pressure and held closed by a latch arrangement. A small movement of the latch will release the contacts, which will open quickly against the spring pressure to break the circuit. Circuit breakers come in a variety of sizes, from small devices within household appliances to large switchgear that can serve to protect whole cities.

The latch can be activated by a variety of methods.

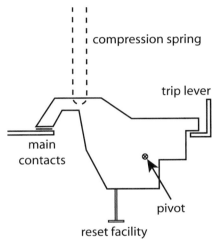

Figure 5.46 Compression spring

- **Magnetic** – some circuit breakers use a magnet whose pull increases with the current. When the current passes a certain point, the solenoid's magnetic pull opens the latch.

- **Thermal** – some circuit breakers have a bimetallic strip, which heats up with increased current. This causes it to bend. Once this bending reaches a certain point it is no longer able to keep the latch closed.

- **Thermomagnetic** – these incorporate both of the above methods, with the magnet responding to large surges (short circuits) and the strip responding to long-term conditions (overcurrent).

We will cover the tripping of these devices in more detail in Chapter 7.

As contacts move apart to break the circuit, it is possible for the current to jump the gap. This is called an **arc** (spark). The arc will remain in place until the contacts move far enough apart that the current can no longer jump the gap. This movement apart takes time. During this period when the arc is still active, it is important to remember two things.

- The fault condition within the device is still operating and causing damage.

- The arc generates heat and therefore the contacts will start to suffer the effects of this heat. We can calculate this using the formula $P = I^2R$.

Figure 5.47 Circuit breaker

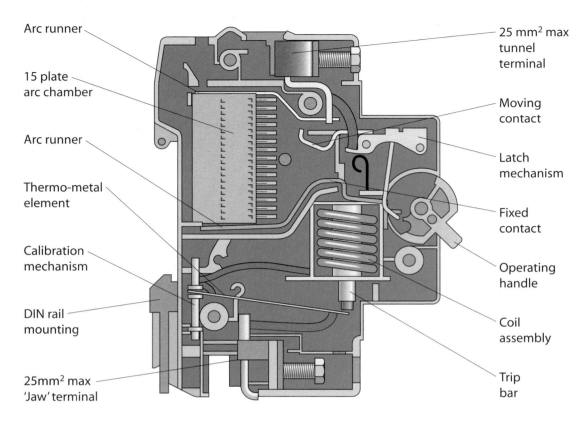

Arc runner

15 plate arc chamber

Arc runner

Thermo-metal element

Calibration mechanism

DIN rail mounting

25mm² max 'Jaw' terminal

25 mm² max tunnel terminal

Moving contact

Latch mechanism

Fixed contact

Operating handle

Coil assembly

Trip bar

Circuit breakers incorporate features to reduce this arc. In low voltage breakers, an arc runner made of metal plates or ceramic ridges cools the arc, while a blowout coil redirects the arc away from the circuit.

Of course, under fault conditions the current values can be hundreds of amps and sometimes thousands.

Low voltage (not exceeding 1000 volts)

In installations the final circuits are protected by Miniature Circuit Breakers (MCBs). They operate through thermal or thermo-magnetic systems. MCBs are selected for:

- current rating of the circuit to be protected, i.e. 6, 16 and 32 amp
- the prospective fault current rating, for example M6 (6000) M16 (16,000) is the value of current they will be able to break when it sustains damage.

Due to improved design and performance, the modern MCB now forms an essential part of the majority of installations at the final distribution level. From about 1970 onward the benefits of current-limiting technology have been incorporated into MCBs, thus providing the designer/user with predictable high performance overcurrent devices.

Circuit breakers

Distribution circuits, supplying main and sub-main cables, are quite commonly protected by Moulded Case Circuit Breakers (MCCBs) as well as MCBs. These operate up to 1000 V and use a thermal or thermo magnetic connection. These are also selected for:

- current rating of the circuit to be protected, i.e. 16 through to 800 amp
- the prospective fault current rating, i.e. up to 80 kA.

Both single and three-phase units are available. Some versions have adjustable tripping sensitivities. This means that the amount of current they are designed to break the circuit at can be altered, depending on the intended use of the circuit breaker.

The function, selection and operation of MCBs will be looked at in greater detail in Chapter 7.

Distribution circuits

High voltage (HV)

On high voltage distribution systems the circuit breakers are physically much bigger. The potential for arcing, and the damage associated with it, is considerably higher as the values of fault current involved are many tens of thousands of amps.

To combat this, larger circuit breakers use different methods to dispel the arc. The arcs may be immersed in oil or a vacuum may be created around the switchgear. This cools and slows the progress of the arc, preventing it from inflicting huge damage on the circuit.

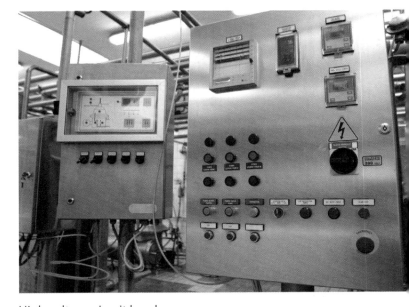

High voltage circuit breakers

Isolators

An isolator switch is often found in industrial applications. Like a circuit breaker, it is designed to break a circuit in the event of a fault. However, an isolator is a switch that isolates circuits that are still powered. As such they are used by electricians when working on a circuit. Isolators are off-load devices rather than on-load devices (circuit breakers). We covered the reasons for isolation and good practice when using isolator devices at Level 2.

Earthing systems

On completion of this topic area the candidate will be able to describe the need for earthing low and high voltage systems and identify typical systems.

We covered the basics of earthing at Level 2, and the need for earthing low and high voltage systems. We will cover the typical systems in more detail in chapter 7. For now we will merely state what these systems are:

- TT system
- TN-S system
- TN-C-S system

- Protective Multiple Earthing (PME)
- TN-C system
- IT system.

Protection systems

On completion of this topic area the candidate will be able to describe detection systems for short-circuits, earth leakage and overload.

We covered the basics of electric shock at Level 2. Here we will briefly recap the main features. The abnormal conditions that could arise, leading to the need for protection systems, are:

- dangerous currents due to short circuits
- overload in a circuit.
- earth leakage.

At Level 2 we looked at how short circuiting can cause dangerous currents and how to prevent overload through circuit breakers and fuses. Please look back at this section to refresh your memory.

Later in this book (chapter 7) we will make an in-depth exploration of differing types of fuses and circuit breakers and their application.

FAQ

Q   Does a three-phase system always need a neutral conductor?

A   No. In a three-phase system the neutral conductor is only used to carry any 'out of balance' currents. If the currents in the phases are 'balanced' (the same), then there will be no neutral current and therefore no need for a neutral conductor.

Q   Why is plastic conduit used in farm installations rather than steel conduit?

A   Although steel conduit is mechanically stronger than plastic it is not resistant to corrosion. Farms can be very wet places!

Q   Why can't I keep adding cables to some trunking or conduit until it is full?

A   Two reasons. One is that if you fill conduit or trunking up to the maximum there is a much higher risk that the cables will be damaged during installation. The second reason is that the more cables you group together, the more heat will be generated and there will be less air-space to dissipate the heat. Your cables may overheat and cause a fire.

Q   Is it best to fabricate your own joints in metal wiring systems?

A   It is normally much more economic to purchase proprietary joints. You should only make your own joints if it is strictly necessary for that particular job.

Q   If an mcb trips out, can I just switch it back on again?

A   In theory yes, after the fault has been cleared, you could just switch it back on. But, if the MCB has operated on a large fault, such as a short circuit, the internal workings of the MCB may have been damaged. If you suspect this is the case, manufacturers recommend the mcb is replaced to be on the safe side.

Activity

1. Obtain some manufacturers' catalogues containing conduit and trunking accessories. Imagine you were going to install a simple wiring system to add a single circuit in a building . Sketch out the route that the wiring system would have to take and draw up a list of the parts needed to complete the job.

2. Draw up a list of tools and their part numbers that you would need to purchase in order to terminate MICC cable.

Knowledge check

1. Draw a simple diagram of a bending machine and label the main components.

2. List six different fixing devices for steel conduit and describe their use.

3. What is the minimum space factor allowed in trunking?

4. List the different types of cable tray that are commercially available.

5. Describe in words how to construct a 90-degree inside bend using a crimping tool.

6. Describe what magnesium oxide is and where it can be found in the electrical industry.

7. What is the purpose of the MICC seal?

8. What is the name of the important test to be carried out after terminating MICC cable?

9. List the relative advantages and disadvantages of MICC and FP 200 cable.

10. A single-phase transformer with 2000 primary turns and 500 secondary turns is fed from a 230V a.c. supply. What would the secondary voltage be?

11. Why does an MCB have both thermal and magnetic devices?

12. What does the 'M' number on an MCB mean (e.g. M6)?

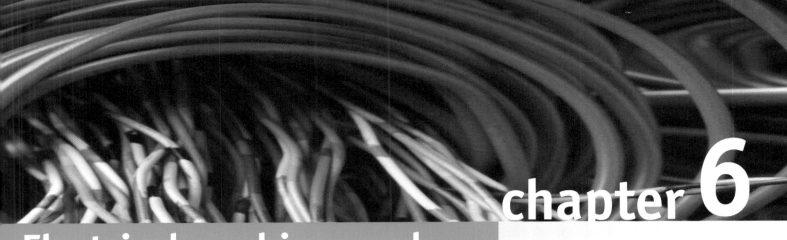

chapter 6

Electrical machines and motors

Unit 1 outcome 6

The principles of magnetism are central to many of the tasks you will carry out as an electrician. Magnetism, like gravity, is a fundamental force. It arises due to the movements in electrical charge and is seen whenever electrically charged particles are in motion.

The concepts of magnetism are crucial to understanding how machines and motors work. In essence there are two different categories of motor: those that run on direct current (d.c.) and those that run on alternating current (a.c.). As well as magnetic fields, it is also important to remember how current flow and induced motion operate, as together these make a motor rotate.

This chapter builds on topics already covered at Level 2.

On completion of this chapter, the candidate will be able to:

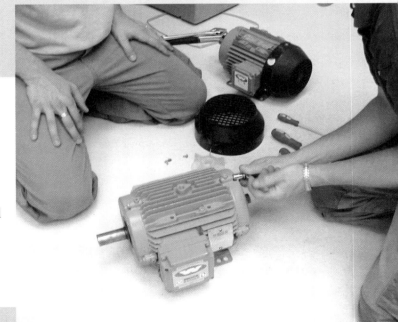

- describe the principles of operation of electrical rotating machines
- identify three-phase induction motors, distinguishing between cage and wound rotors
- identify single-phase a.c. motors, distinguishing between series wound, split phase, capacitor and capacitor start/run
- identify basic principles of starting and speed control.

Electrical rotating machines

On completion of this topic area the candidate will be able to describe the principle of the operation of electrical rotating machines, identify three-phase and single-phase induction motors and the basic principles of starting and speed control.

Basic a.c. and d.c. motors and generators

The principles of a.c. and d.c. motors were introduced at Level 2. Here, we will briefly recap the principles of these devices, before identifying the elements of cage and wound rotors.

The d.c. motor and generators

Figure 6.01 shows a d.c. motor with a conductor formed into a loop and pivoted at its centre. When a d.c. current is fed into this system, it causes a magnetic field to be produced in the conductor that interacts with the magnetic field between the two poles (**pole pair**) of the magnet. This causes a bending or stretching of the lines of force between the two poles. These lines attempt to return to the shortest length, exerting a force on the conductor, causing the loop to rotate.

The loops are wound around the **armature**. In order to keep the amature turning the current is reversed every half-rotation by using a **commutator** to switch the polarity as shown in Figure 6.02. Practical d.c. motors use electromagnets, as these allow the poles to reverse as well as allowing the user to control the flow of current.

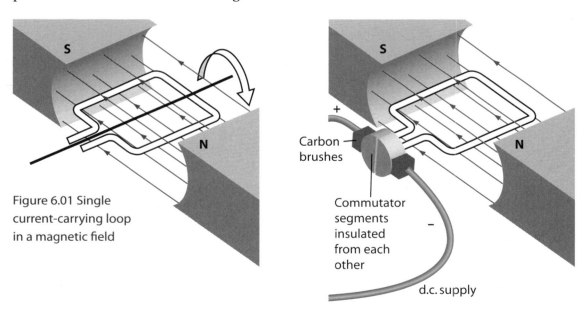

Figure 6.01 Single current-carrying loop in a magnetic field

Figure 6.02 Single loop with commutator

There are three basic forms of d.c. motor: series, shunt and compound. They are very similar to look at, the difference being the way in which the field coil and armature coil circuits are wired.

- The series motor has the field coil wired in series with armature. It is also called a universal motor because it can be used in both d.c. and a.c. situations, has a high starting torque (rotational force) and a variable speed characteristic. The motor can therefore start heavy loads but the speed will increase as the load is decreased.

- The shunt motor has the armature and field circuits wired in parallel, and this gives constant field strength and motor speed.

- The compound motor combines the characteristics of both the series and shunt motors and, thus, has high starting torque and fairly good speed torque characteristics. However, because it is complex to control, this arrangement is usually only used on large bi-directional motors.

If mechanical power is applied to rotate the armature, then an e.m.f. is created within the wire and a current will flow if connected to an external circuit. Hence, we have produced a generator. The current still alternates but the commutator ensures that it is always in the same direction; and, with many turns on the armature set at different angles, the effective output is a fairly steady current.

The a.c. motor and generator

There are many types of a.c. motor, each one having a specific set of operating characteristics such as **torque**, speed, single-phase or three-phase, and this determines their selection for use. We can essentially group them into two categories: single-phase and three-phase.

An a.c. motor does not need a commutator to reverse the polarity of the current. Instead, the a.c. motor changes the polarity of the current running through the stator (stationary part of the motor).

The operating principle of the a.c. generator is much the same as that of the d.c. version. However, instead of the loop ends terminating at the commutator, they are terminated at slip rings, which enable the output current to remain alternating.

A simple a.c. motor

The series-wound (universal) motor

We will begin the discussion of a.c. motors by looking at the series-wound (universal) motor because it is different in its construction and operation from the other a.c. motors considered here. It is constructed with field windings, brushes, commutator and an armature.

As can be seen in Figure 6.03 (page 234), because of its series connection, current passing through the field windings also passes through the armature. The turning motion (torque) is produced as a result of the interaction between the magnetic field in the field windings and the magnetic field produced in the armature.

For this motor to be able to run on an a.c. supply, modifications are made both to the field windings and armature **formers**. These are heavily laminated to reduce eddy currents and I^2R losses, which reduces the heat generated by the normal working of the motor thereby making the motor more efficient.

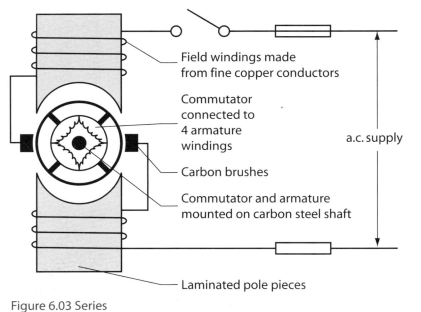

Field windings made from fine copper conductors

Commutator connected to 4 armature windings

a.c. supply

Carbon brushes

Commutator and armature mounted on carbon steel shaft

Laminated pole pieces

Figure 6.03 Series universal motor

This type of motor is generally small (less than a kilowatt) and is used to drive small hand tools such as drills, vacuum cleaners and washing machines.

A disadvantage of this motor is that it relies on contact with the armature via a system of carbon brushes and a commutator. It is this point that is the machine's weakness, as much heat is generated through the arcing that appears across the gap between the brushes and the commutator. The brushes are spring-loaded to keep this gap to a minimum, but even so the heat and friction eventually cause the brushes to wear down and the gap to increase. These then need to be replaced, otherwise the heat generated as the gap gets larger will eventually cause the motor to fail.

The advantages of this machine are:

- more power for a given size than any other normal a.c. motor
- high starting torque
- relative cheapness to produce.

Three-phase induction motors

Three-phase induction motor showing component parts

Induction motors operate because a moving magnetic field induces a current to flow in the rotor. This current in the rotor then creates a second magnetic field, which combines with the field from the stator windings to exert a force on the rotor conductors, thus turning the rotor.

Production of the rotating field

Figure 6.04 shows the stator of a three-phase motor to which a three-phase supply is connected. The windings in the diagram are in star formation and two windings of each phase are wound in the same direction.

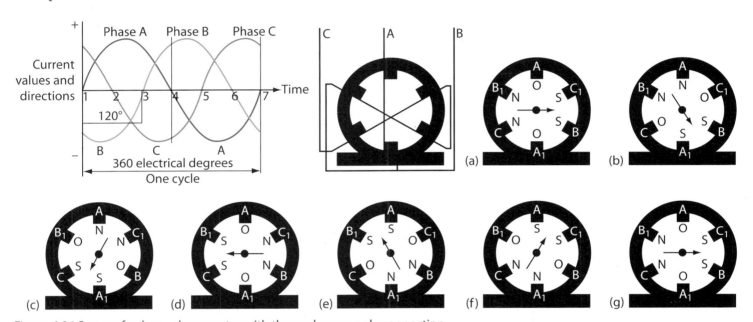

Figure 6.04 Stator of a three-phase motor with three-phase supply connection

Each pair of windings will produce a magnetic field, the strength of which will depend upon the current in that particular phase at any instant of time. When the current is zero, the magnetic field will be zero. Maximum current will produce the maximum magnetic field.

As the currents in the three phases are 120 degrees out of phase (see graph in Figure 6.04) the magnetic fields produced will also be 120 degrees out of phase. The magnetic field set up by the three-phase currents will therefore give the appearance of rotating clockwise around the stator.

The resultant magnetic field produced by the three phases is at any instant of time in the direction shown by the arrow in the diagram, where diagrams (a) to (g) in Figure 6.04 show how the direction of the magnetic field changes at intervals of 60 degrees through one complete cycle. The speed of rotation of the magnetic field depends upon the supply frequency and the number of 'pole pairs', and is referred to as the **synchronous** speed. We will discuss synchronous speed later in this chapter.

The direction in which the magnetic field rotates is dependent upon the sequence in which the phases are connected to the windings. Reversing the connection of any two incoming phases can therefore reverse rotation of the magnetic field.

Stator construction

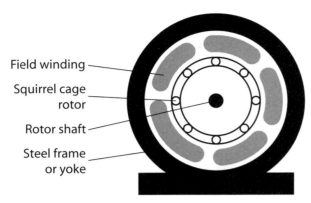

Field winding

Squirrel cage rotor

Rotor shaft

Steel frame or yoke

Figure 6.05 Stator construction

Stator field winding

As shown in Figure 6.05, the stator (stationary component) comprises the field windings, which are many turns of very fine copper wire wound on to formers. These are then fixed to the inside of the stator steel frame (sometimes called the yoke).

The formers have two roles.

1. To contain the conductors of the winding.

2. To concentrate the magnetic lines of flux to improve the flux linkage.

The formers are made of laminated silicon steel sections to reduce eddy currents, thereby reducing the I^2R losses and reducing heat. The number of poles fitted will determine the speed of the motor.

Rotor construction and principle of operation

Essentially there are two main types of rotor:

1. Squirrel-cage rotor.

2. Wound rotor.

Squirrel-cage rotor

In the squirrel cage (see Figure 6.06), the bars of the rotor are shorted out at each end by 'end rings' to form the shape of a cage. This shape creates numerous circuits within it for the induced e.m.f. and resultant current to flow and thus produce the required magnetic field.

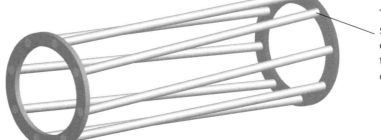

Tinned bars shorted out at each end by a tinned copper end ring

Figure 6.06 Squirrel-cage rotor

Figure 6.07 shows the cage fitted to the shaft of the motor. The rotor bars are encased within many hundreds of very thin laminated (insulated) segments of silicon steel and are skewed to increase the rotor resistance.

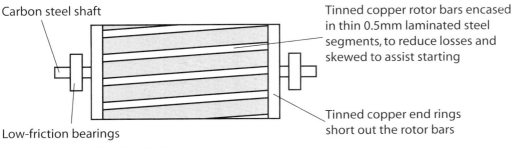

Carbon steel shaft

Tinned copper rotor bars encased in thin 0.5mm laminated steel segments, to reduce losses and skewed to assist starting

Tinned copper end rings short out the rotor bars

Low-friction bearings

Figure 6.07 Cage fitted to shaft and motor

On the shaft will be two low-friction bearings which enable the rotor to spin freely. The bearings and the rotor will be held in place within the yoke of the stator by two end caps which are normally secured in place by long nuts and bolts that pass completely through the stator.

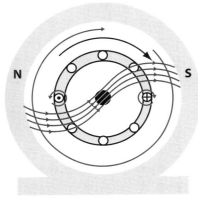

Figure 6.08 Rotating field

When a three-phase supply is connected to the field windings, the lines of magnetic flux produced in the stator, rotating at 50 revolutions per second, cut through the bars of the rotor, inducing an e.m.f. into the bars.

Faraday's law states that 'when a conductor cuts, or is cut, by a magnetic field, then an e.m.f. is induced in that conductor the magnitude of which is proportional to the rate at which the conductor cuts or is cut by the magnetic flux.'

Did you know?

Squirrel cage induction motors are sometimes referred to as the 'workhorse' of the industry, as they are inexpensive and reliable and suited to most applications

The e.m.f. produces circulatory currents within the rotor bars, which in turn result in the production of a magnetic field around the bars. This leads to the distortion of the magnetic field as shown in Figure 6.08. The interaction of these two magnetic fields results in a force being applied to the rotor bars, and the rotor begins to turn. This turning force is known as a torque, the direction of which is always as Fleming's left-hand rule indicates.

Wound rotor

In the wound rotor type of motor, the rotor conductors form a three-phase winding which is starred internally. The other three ends of the windings are brought out to slip rings mounted on the shaft. Thus it is possible through brush connections to introduce resistance into the rotor circuit, although this is normally done on starting only to increase the starting torque. This type of motor is commonly referred to as a slip-ring motor.

Figure 6.09 shows a completed wound-rotor motor assembly. Although it looks like a squirrel-cage motor, the difference is that the rotor bars are exchanged for heavy conductors that run through the laminated steel rotor, the ends then being brought out through the shaft to the slip rings on the end.

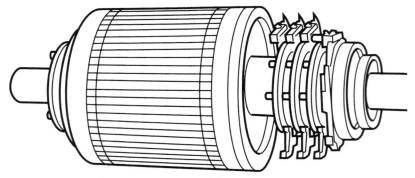

Figure 6.09 Wound-rotor motor assembly

The wound-rotor motor is particularly effective in applications where using a squirrel-cage motor may result in a starting current that is too high for the capacity of the power system. The wound-rotor motor is also appropriate for high-inertia loads having a long acceleration time. This is because you can control the speed, torque and resulting heating of the wound-rotor motor. This control can be automatic or manual. It is also effective with high-slip loads as well as adjustable-speed installations that do not require precise speed control or regulation. Typical applications include conveyor belts, hoists and elevators.

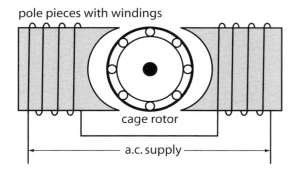

Figure 6.10 Single-phase induction motor

Single-phase induction motors

If we were to construct an induction motor as shown in Figure 6.10, we would find that, on connecting a supply to it, it would not run. However, if we were then to spin the shaft with our fingers, we would find that the motor *would* continue to run. Why is this?

When an a.c. supply is connected to the motor, the resulting current flow, and therefore the magnetic fields produced in the field windings, changes polarity, backwards and forwards, 100 times per second.

However, if we spin the rotor we create the effect of the rotor bars cutting through the lines of force, hence the process starts and the motor runs up to speed. If we stop the motor and then connect the supply again, the motor will still not run. If this time we spin the rotor in the opposite direction, the motor will run up to speed in the new direction. So how can we get the rotor to turn on its own?

If we think back to the three-phase motor we discussed previously, we did not have this problem because the connection of a three-phase supply to the stator automatically produced a rotating magnetic field. This is what is missing from the motor in Figure 6.10: we have no rotating field.

The split-phase motor (induction start/induction run)

We can overcome this problem if we add another set of poles, positioned 90 degrees around the stator from our original wiring, as shown in Figure 6.11.

Now when the supply is connected, both sets of windings are energised, both windings having resistive and reactive components to them. This is the resistance that every conductor has and also inductive reactance because the conductors form a coil. These are known as the 'start' and 'run' windings.

The start winding is wound with fewer turns of smaller wire than the main winding, so it has a higher resistance. This extra resistance creates a small phase shift that results in the current in the start winding lagging the current in the run winding by approximately 30 degrees, as shown in Figure 6.12.

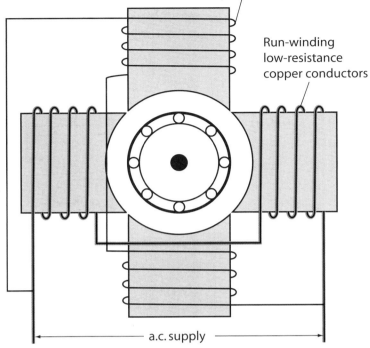

Start-winding high-resistance smaller csa conductors compared with the run winding

Run-winding low-resistance copper conductors

a.c. supply

Figure 6.11 Split-phase motor induction

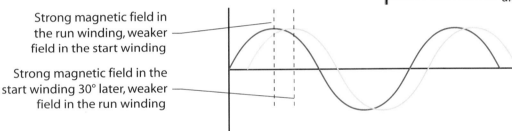

Strong magnetic field in the run winding, weaker field in the start winding

Strong magnetic field in the start winding 30° later, weaker field in the run winding

Figure 6.12 Run- and start-winding phases

Thus the magnetic flux in each of the windings is growing and collapsing at different periods in time, so that, for example, as the run winding has a strong north/south on the face of its pole pieces, the start winding will only have a weak magnetic field.

In the next instance, the run winding's magnetic field has started to fade but the start winding's magnetic field is now strong, presenting to the rotor an apparent shift in the lines of magnetic flux.

In the next half of the supply cycle, the polarity is reversed and the process repeated so there now appears to be a rotating magnetic field. The lines of force cutting through the rotor bars induce an e.m.f. into them, and the resulting current flow now produces a magnetic field around the rotor bars. The interaction between the magnetic fields of the rotor and stator is established and the motor begins to turn.

It is because the start and run windings carry currents that are out of phase with each other that this type of motor is called 'split-phase'.

Once the motor is rotating at about 75 per cent of its full load speed, the start winding is disconnected by the use of a device called a centrifugal switch, which is attached to the shaft; see Figure 6.13.

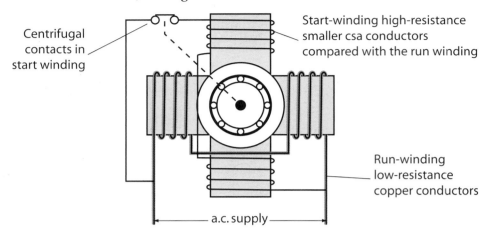

Centrifugal contacts in start winding

Start-winding high-resistance smaller csa conductors compared with the run winding

Run-winding low-resistance copper conductors

a.c. supply

Figure 6.13 Split-phase induction motor with centrifugal switch

This switch works by centrifugal action in that sets of contacts are held closed by a spring, and this completes the circuit to the start winding. When the motor starts to turn, a little weight gets progressively thrown away from the shaft, forcing the contacts to open and thus disconnecting the start winding. It is a bit like the fairground ride known as 'the rotor', in which you are eventually held against the sides of the ride by the increasing speed of the spinning wheel. Once the machine has disconnected the start winding, the machine continues to operate from the run winding.

The split-phase motor's simple design makes it typically less expensive than other single-phase motors. However, it also limits performance. Starting torque is low at about 150–175 per cent of the rated load. Also, the motor develops high starting currents of about six to nine times the full load current. A lengthy starting time can cause the start winding to overheat and fail, and therefore this type of motor should not be used when a high starting torque is needed. Consequently it is used on light-load applications such as small hand tools, small grinders and fans, where there are frequent stop/starts and the full load is applied after the motor has reached its operating speed.

Reversal of direction

If you think back to the start of this section, we talked about starting the motor by spinning the shaft. We also said that we were able to spin it in either direction and the motor would run in that direction. It therefore seems logical that, in order to change the direction of the motor, all we have to concern ourselves with is the start winding. We therefore need only to reverse the connections to the start winding to change its polarity, although you may choose to reverse the polarity of the run winding instead. The important thing to remember is that, if you change the polarity through both the run and start windings, the motor will continue to revolve in the same direction.

The capacitor-start motor (capacitor start/induction run)

Normally perceived as being a wide-ranging industrial motor, the capacitor-start motor is very similar to the split-phase motor. Indeed, it probably helps to think of this motor as being a split-phase motor but with an enhanced start winding that includes a capacitor in the circuit to facilitate the start process. If we look at the photograph we can see the capacitor mounted on top of the motor case.

In this motor the start winding has a capacitor connected in series with it, and since this gives a phase difference of nearly 90 degrees between the two currents in the windings, the starting performance is improved. We can see this represented in the sine waves shown in Figure 6.14.

Capacitor-start motor

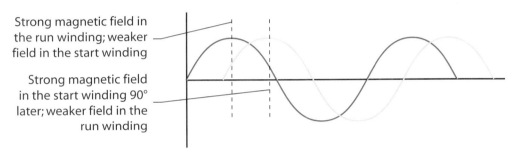

Strong magnetic field in the run winding; weaker field in the start winding

Strong magnetic field in the start winding 90° later; weaker field in the run winding

Figure 6.14 Magnetic field in capacitor-start motor

In this motor the current through the run winding lags the supply voltage due to the high inductive reactance of this winding, while the current through the start winding leads the supply voltage due to the capacitive reactance of the capacitor. The phase displacement in the currents of the two windings is now approximately 90 degrees. Figure 6.15 shows the winding connections for a capacitor-start motor.

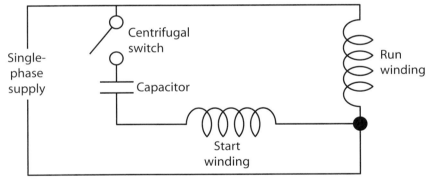

Figure 6.15 Winding connections for capacitor-start split-phase motor

The magnetic flux set up by the two windings is much greater at starting than in the standard split-phase motor, and this produces a greater starting torque. The typical starting torque for this type of motor is about 300 per cent of full-load torque, and a typical starting current is about five to nine times the full-load current.

The capacitor-start motor is more expensive than a comparable split-phase design because of the additional cost of the capacitor. But the application range is much wider because of higher starting torque and lower starting current. Therefore, because of its improved starting ability, this type of motor is recommended for loads that are hard to start, so we see this type of motor used to drive equipment such as lathes, compressors and small conveyor systems.

As with the standard split-phase motor, the start windings and the capacitor are disconnected from the circuit by an automatic centrifugal switch when the motor reaches about 75 per cent of its rated full-load speed.

Reversal of direction

Reversing the connections to the start winding will only change its polarity, although we may choose to reverse the polarity of the run winding instead.

Permanent split capacitor (PSC) motors

Permanent split capacitor (PSC) motors look exactly the same as capacitor-start motors. However, a PSC motor does not have either a starting switch or a capacitor that is strictly used for starting. Instead, it has a run-type capacitor permanently connected in series with the start winding, and the second winding is permanently connected to the power source. This makes the start winding an auxiliary winding once the motor reaches running speed. However, because the run capacitor must be designed for continuous use, it cannot provide the starting boost of a starting capacitor.

Typical starting torques for this type of motor are low, from 30 to 150 per cent of rated load, so these motors are not used in difficult starting applications. However, unlike the split-phase motor, PSC motors have low starting currents, usually less than 200 per cent of rated load current, making them excellent for applications with high cycle rates.

PSC motors have several advantages. They need no starting mechanism and so can be reversed easily, and designs can easily be altered for use with speed controllers. They can also be designed for optimum efficiency and high power factor at rated load.

Permanent split capacitor motors have a wide variety of applications depending on the design. These include fans, blowers with low starting torque and intermittent cycling uses such as adjusting mechanisms, gate operators and garage-door openers, many of which also need instant reversing.

Capacitor start–capacitor run motors

In appearance we can distinguish this motor because of the two capacitors that are mounted on the motor case.

This type of motor is widely held to be the most efficient single-phase induction motor, as it combines the best of the capacitor-start and the permanent split capacitor designs. It is also able to handle applications too demanding for any other kind of single-phase motor.

As shown in Figure 6.16 (page 243), it has a start capacitor in series with the auxiliary winding (like the capacitor-start motor) and this allows for high starting torque. However, like the PSC motor it also has a run capacitor that remains in series with the auxiliary winding after the start capacitor is switched out of the circuit.

Capacitor start–capacitor run motor

Another advantage of this type of motor is that it can be designed for lower full-load currents and higher efficiency, which means that it operates at a lower temperature than other single-phase motor types of comparable horsepower. Typical uses include woodworking machinery, air compressors, high-pressure water pumps, vacuum pumps and other high-torque applications.

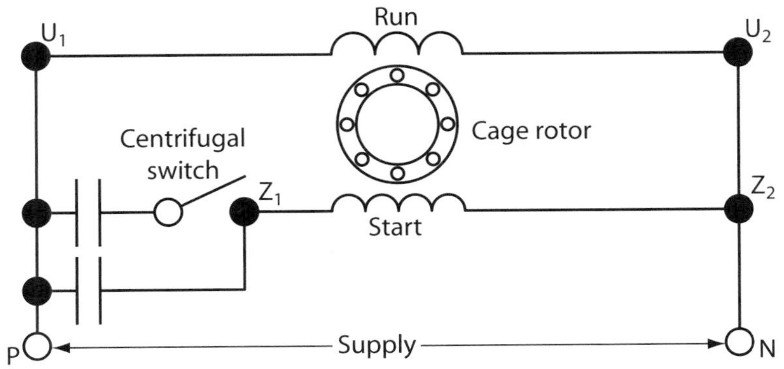

Figure 6.16 Capacitor start-capacitor run split-phase motor

Shaded-pole motors

One final type of single-phase induction motor that is worthy of mention is the shaded-pole type. We cover this last as, unlike all of the previous single-phase motors discussed, shaded-pole motors have only one main winding and no start winding.

Starting is by means of a continuous copper loop wound around a small section of each motor pole. This 'shades' that portion of the pole, causing the magnetic field in the ringed area to lag the field in the non-ringed section. The reaction of the two fields then starts the shaft rotating.

Because the shaded pole motor lacks a start winding, starting switch or capacitor, it is electrically simple and inexpensive. In addition, speed can be controlled merely by varying voltage or through a multi-tap winding.

The shaded-pole motor has many positive features but it also has several disadvantages. As the phase displacement is very small it has a low starting torque, typically in the region of 25–75 per cent of full-load torque. Also, it is very inefficient, usually below 20 per cent.

Low initial costs make shaded-pole motors suitable for light-duty applications such as multi-speed fans for household use and record turntables.

Motor construction

On completion of this topic area the candidate will be able to identify the basic principles of motor construction.

To help you understand how motors are put together, this section includes photographs and diagrams of the components used in motor construction.

Motor designs
The three-phase squirrel-cage induction motor

Components of a three-phase squirrel-cage induction motor

Field winding

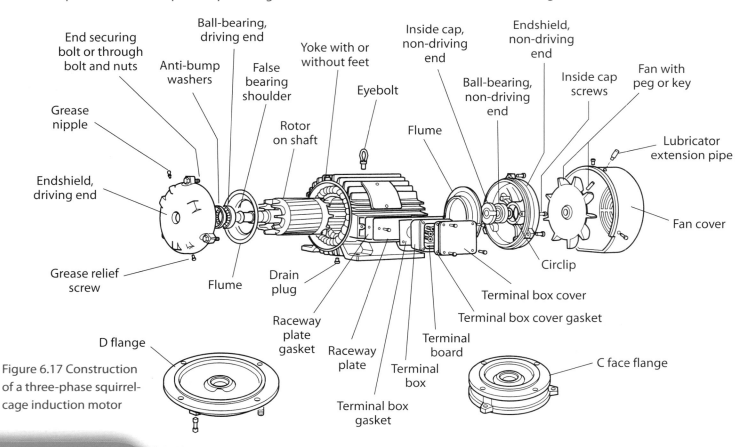

Figure 6.17 Construction of a three-phase squirrel-cage induction motor

The single-phase induction motor

This motor consists of a laminated stator wound with single-phase windings arranged for split-phase starting and a cage rotor. The cage rotor is a laminated framework with lightly insulated longitudinal slots into which copper or aluminium bars are fitted. The bars are then connected together at their ends by metal end-rings. No electrical connections are made to the rotor.

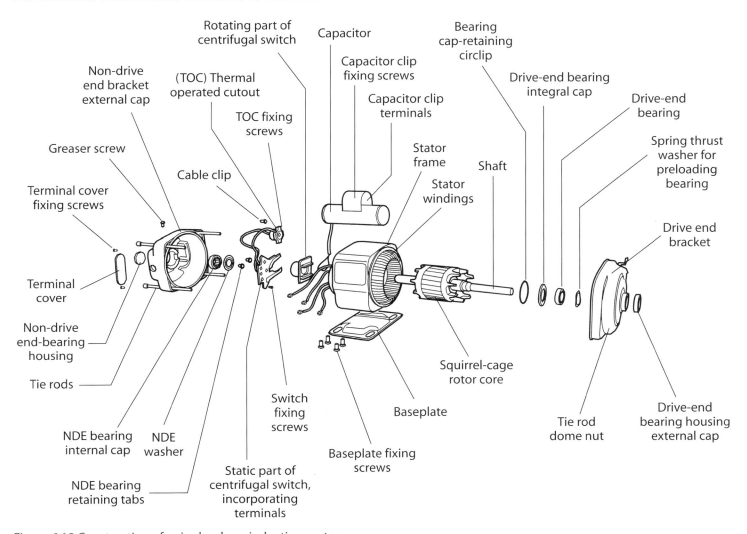

Figure 6.18 Construction of a single-phase induction motor

Motor speed and slip calculations
Speed of a motor

There are essentially two ways to express the speed of a motor.

- **Synchronous speed.** For an a.c. motor this is the speed of rotation of the stator's magnetic field. Consequently, this is really only a theoretical speed as the rotor will always turn at a slightly slower rate.

- **Actual speed.** This is the speed at which the shaft rotates. The nameplate on most a.c. motors will give the actual motor speed rather than the synchronous speed.

The difference between the speed of the rotor and the synchronous speed of the rotating magnetic field is known as the **slip**, which can be expressed either as a unit or in percentage terms. Because of this, we often refer to the induction motor as being an asynchronous motor.

Remember, the speed of the rotating magnetic field is known as the synchronous speed, and this will be determined by the frequency of the supply and the number of pairs of poles within the machine. The speed at which the rotor turns will be between 2–5 per cent slower, with an average of 4 per cent being common.

We already know that when a conductor passes at right angles through a magnetic field, current is induced into the conductor. The direction of the induced current is dependent on the direction of movement of the conductor, and the strength of the current is determined by the speed at which the conductor moves. If the rotating magnetic field and the rotor are now revolving at the same speed, there will be no lines of magnetic flux cutting through the rotor bars, no induced e.m.f. and consequently no resultant magnetic field around the rotor bars to interact with the rotating magnetic field of the stator. The motor will immediately slow down and, having slowed down, will then start to speed up as the lines of magnetic flux start to cut through the rotor bars again – and so the process would continue.

Standard a.c. induction motors therefore depend on the rotor trying, but never quite managing, to catch up with the stator's magnetic field. The rotor speed is just slow enough to cause the proper amount of rotor current to flow, so that the resulting torque is sufficient to overcome windage and friction losses and drive the load.

Determining synchronous speed and slip

All a.c. motors are designed with various numbers of magnetic poles. Standard motors have two, four, six or eight poles, and these play an important role in determining the synchronous speed of an a.c. motor.

As we said before, the synchronous speed can be determined by the frequency of the supply and the number of pairs of poles within the machine. We can express this relationship with the following formula:

$$\text{Synchronous speed } (n_s) \text{ in revolutions per second} = \frac{\text{frequency (f) in Hz}}{\text{number of pole pairs (p)}}$$

Remember

When a conductor passes at right angles through a magnetic field, current is induced into the conductor. The direction of the induced current will depend on the direction of movement of the conductor, and the strength of the current will be determined by the speed at which the conductor moves

Example 1

Calculate the synchronous speed of a four-pole machine connected to a 50 Hz supply.

$$n_s = \frac{f}{p}$$

As we know the motor has four poles. This means it has two pole pairs. We can therefore complete the calculation as:

$$n_s = \frac{50}{2} \quad \text{therefore } n_s = 25 \text{ revolutions per second (rps)}$$

To convert revolutions per second into the more commonly used revolutions per minute (rpm), simply multiply n_s by 60. This new value is referred to as N_s, which in this example will become 25×60, giving 1500 rpm.

We also said that we refer to the difference between the speed of the rotor and the synchronous speed of the rotating magnetic field as the slip, which can be expressed either as a unit (S) or in percentage terms (S per cent). We express this relationship with the following formula:

$$\text{per cent slip} = \frac{\text{synchronous speed } (n_s) - \text{rotor speed } (n_r)}{\text{synchronous speed } (n_s)} \times 100$$

Example 2

Note: In this example numbers have been rounded up for ease.

A six-pole cage-induction motor runs at 4 per cent slip. Calculate the motor speed if the supply frequency is 50 Hz.

$$S \text{ (per cent)} = \frac{(n_s - n_r)}{n_s} \times 100$$

We therefore need first to establish the synchronous speed. As the motor has six poles it will have three pole pairs. Consequently:

$$\text{Synchronous speed } n_s = \frac{f}{p} \text{ giving us 50 and therefore } n_s = 16.7 \text{ revs/sec}$$

We can now put this value into our formula and then, by transposition, rearrange the formula to make n_r the subject. Consequently:

$$S \text{ (per cent)} = \frac{(n_s - n_r)}{n_s} \times 100 \quad \text{giving us} \quad 4 = \frac{16.7 - n_r}{16.7} \times 100$$

Therefore by transposition:

$$4 = \frac{16.7 - n_r}{16.7} \times 100 \quad \text{gives us} \quad (16.7 - n_r) = \frac{4 \times 16.7}{100} = 0.668$$

Therefore by further transposition:

$$(16.7 - n_r) = 0.668 \qquad \text{becomes} \qquad 16.7 - 0.668 = n_r$$

Therefore $n_r =$ **16.032 rps** or $N_r =$ **962 rpm**

Synchronous a.c. induction motors

A synchronous motor, as the name suggests, runs at synchronous speed. Because of the problems discussed earlier, this type of motor is not self-starting and instead must be brought up to almost synchronous speed by some other means.

Three-phase a.c. synchronous motors

To understand how the synchronous motor works, assume that we have supplied three-phase a.c. power to the stator, which in turn causes a rotating magnetic field to be set up around the rotor. The rotor is then supplied via a field winding with d.c. and consequently acts a bit like a bar magnet, having north and south poles. The rotating magnetic field now attracts the rotor field that was activated by the d.c. This results in a strong turning force on the rotor shaft, and the rotor is therefore able to turn a load as it rotates in step with the rotating magnetic field.

It works this way once it has started. However, one of the disadvantages of a synchronous motor is that it cannot be started from standstill by just applying a three-phase a.c. supply to the stator. When a.c. is applied to the stator, a high-speed rotating magnetic field appears immediately. This rotating field rushes past the rotor poles so quickly that the rotor does not have a chance to get started. In effect, the rotor is repelled first in one direction and then the other.

An induction winding (squirrel-cage type) is therefore added to the rotor of a synchronous motor to cause it to start, effectively meaning that the motor is started as an induction motor. Once the motor reaches synchronous speed, no current is induced in the squirrel-cage winding, so it has little effect on the synchronous operation of the motor.

Synchronous motors are commonly driven by transistorised variable-frequency drives.

Single-phase a.c. synchronous motors

Small single-phase a.c. motors can be designed with magnetised rotors. The rotors in these motors do not require any induced current so they do not slip backwards against the mains frequency. Instead, they rotate synchronously with the mains frequency. Because of their highly accurate speed, such motors are usually used to power mechanical clocks, audio turntables and tape drives; formerly they were also widely used in accurate timing instruments such as strip-chart recorders or telescope drive mechanisms. The shaded-pole synchronous motor is one version.

As with the three-phase version, inertia makes it difficult to accelerate the rotor instantly from stopped to synchronous speed, and the motors normally require some sort of special feature to get started. Various designs use a small induction motor (which may share the same field coils and rotor as the synchronous motor) or a very light rotor with a one-way mechanism (to ensure that the rotor starts in the 'forward' direction).

Motor windings

A motor can be manufactured with the windings internally connected. If this is the case and there are three terminal connections in the terminal block labelled U, V and W, you would expect the motor windings to be connected in a delta configuration. This is shown in Figure 6.19.

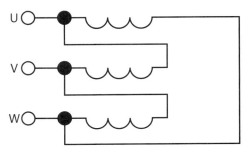

Figure 6.19 Motor windings with delta connection

However, there may be four connections in the terminal box labelled U, V, W and N. If this is the case, the windings would be arranged to give a star configuration, as shown in Figure 6.20.

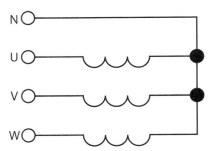

Figure 6.20 Motor windings with four connections

Alternatively, the terminal block may contain six connections: U1, U2, V1, V2, W1 and W2. This is used where both star and delta configurations are required. The terminal connections can then be reconfigured for either star or delta, starting within the terminal block. Figure 6.21 illustrates the connections that would come out to the terminal block.

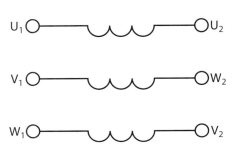

Figure 6.21 Motor windings with six connections

Remember

Some older motors may have different markings in the terminal block and you should therefore refer to the manufacturer's data

Motor starters

A practical motor starter has to do more than just switch on the supply. It also has to have provision for automatically disconnecting the motor in the event of overloads or other faults. The overload protective device monitors the current consumption of the motor and is set to a predetermined value that the motor can safely handle. When a condition occurs that exceeds the set value, the overload device opens the motor starter control circuit and the motor is turned off. The overload protection can come in a variety of types, including solid-state electronic devices.

The starter should also prevent automatic restarting should the motor stop because of supply failure. This is called **no-volts protection** and will be discussed later in this section.

The starter must also provide for the efficient stopping of the motor by the user. Provision for this is made by the connection of remote stop buttons and safety interlock switches where required.

The direct-on-line (DOL) starter

This is the simplest and cheapest method of starting squirrel-cage (induction) motors.

The expression 'direct-on-line' starting means that the full supply voltage is directly connected to the stator of the motor by means of a contactor-starter, as shown in Figure 6.22.

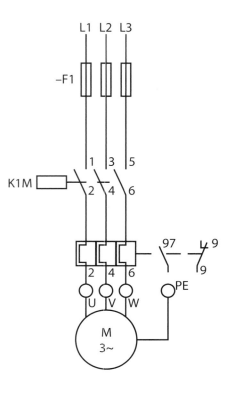

DOL starter

Figure 6.22 DOL starter

Since the motor is at a standstill when the supply is first switched on, the initial starting current is heavy. This 'inrush' of current can be as high as six to ten times the full load current, i.e. a motor rated at 10 A could have a starting current inrush as high as 60 A, and the initial starting torque can be about 150 per cent of the full-load torque. Thus you may observe motors 'jumping' on starting if their mountings are not secure. As a result, direct-on-line starting is usually restricted to comparatively small motors with outputs of up to about 5 kW.

DOL starters should also incorporate a means of overload protection, which can be operated by either magnetic or thermal overload trips. These activate when there is a sustained increase in current flow.

To reverse the motor you need to interchange any two of the incoming three-phase supply leads. If a further two leads are interchanged then the motor will rotate in the original direction.

Operating principle of a DOL starter

A three-pole switch controls the three-phase supply to the starter. This switch normally includes fuses, which provide a means of electrical isolation and also short-circuit protection. We shall look at the operation of the DOL starter in stages.

Let us start by looking at the one-way switch again. In this circuit (Figure 6.23), the switch is operated by finger and the contacts are then held in place mechanically.

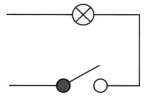

Figure 6.23 One-way switch

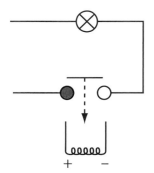

Figure 6.24 Switch relay

We could decide that we do not want to operate the switch this way and instead use a relay. In this system (Figure 6.24), when the coil is energised it creates a magnetic field. Everything in the magnetic field will be pulled in the direction of the arrow and the metal strip will be pulled onto the contacts. As long as the coil remains energised and is a magnet, the light will stay on.

Looking at Figures 6.23 and 6.24, we can see that the first option works well enough for its intended purpose. However, it could not be used in a three-phase system as we would need one for each phase and would have to trust to luck each time we tried to hit all three switches at the same time. However, the second option does provide us with an effective method of controlling more than one thing from one switch, as long as they are all in the same magnetic field.

Let us apply this knowledge to a DOL starter. We know that we cannot have three one-way switches in the starter. But it helps to try and think of the contacts; where the switches are not operated by finger, but are pulled in the direction of the arrow by the magnetic effect of the coil (see Figure 6.25), items affected by the same magnetic effect are normally shown linked by a dotted line. For ease of explanation we will use a 230 V coil.

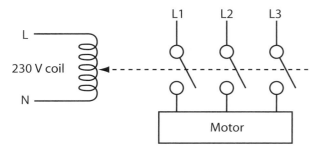

Figure 6.25 Three-switch relay

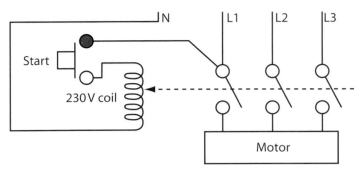

Figure 6.26 Three-switch relay with 'normally open' start button

From this we can see that as soon as we put a supply on to the coil it energises, becomes a magnet and pulls the contacts in. Obviously, this would be no good for our starter, as every time the power is put on the starter will become active and start operating whatever is connected to it. In the case of machinery this could be very dangerous. The starter design therefore goes one step further to include no-volts protection. The simple addition of a 'normally open' start button provides this facility, as shown in Figure 6.26.

So far so good. We now have to make a conscious decision to start the motor.

Our next problem though, is that every time we take our finger off the start button, the button springs back out, the supply to the coil is lost and the motor stops. This is the 'no volt' protection element in operation.

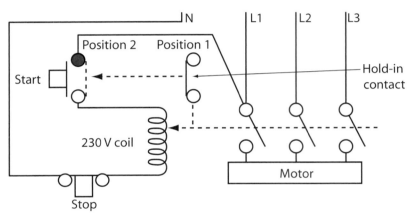

Figure 6.27 Restarting the motor

What we need for normal operation is a device called the 'hold in' or 'retaining' contact, as shown in Figure 6.27. This is a 'normally closed' contact (position 1) which is also placed in the magnetic field of the coil. Consequently, when the coil is energised it is also pulled in the direction shown, and in this case, across and on to the terminals of the start button (position 2).

We can now take our finger off the start button, as the supply will continue to feed the coil by running through the 'hold in' contact that is linking out the start button terminals (position 2).

This now means that we can only break the coil supply and stop the motor by fitting a stop button. In the case of starters fitted with a 230 V coil, this will be a normally closed button placed in the neutral wire.

Now, for the fraction of a second that we push the stop button, the supply to the coil is broken, the coil ceases to be a magnet and the 'hold in' contact returns to its original position (position 1). Since the start button had already returned to its original open position when we released it, when we take our finger off the Stop button everything will be as when we first started. Therefore any loss of supply will immediately break the supply to the coil and stop the motor; if a supply failure was restored, the equipment could not restart itself – someone would have to take the conscious decision to do so.

This system is basically the same as that of a contactor. In fact, many people refer to this item as the contactor starter. Such starters are also available with a 400 V coil, which is therefore connected across two of the phases.

Remote stop/start control

In the DOL starter as described, we have the means of stopping and starting the motor from the buttons provided on the starter enclosure. However, there are situations where control of the motor needs to take place from some remote location. This could be, for example, in a college workshop where, in the case of an emergency, emergency stop buttons located throughout the workshop can be activated to break the supply to a motor. Equally, because of the immediate environment around the motor, it may be necessary to operate it from a different location.

Commonly known as a remote stop/start station, the enclosure usually houses a start and a stop button connected in series. However, depending on the circumstances it is also possible to have an additional button included to give 'inch' control of a motor.

If we use the example of our DOL starter, as described in the previous diagrams, but now include the remote stop/start station, the circuit would look as in Figure 6.28 below, where for ease the additional circuitry has been shown in red.

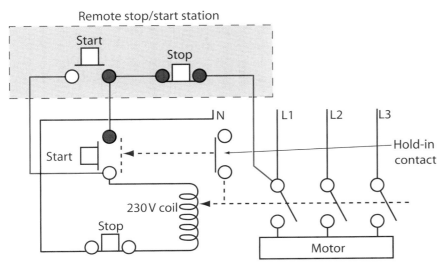

Figure 6.28 Remote stop/start control

As can be seen, the remote start button is effectively in parallel with the start button on the main enclosure, with the supply to both of these buttons being routed via the stop button of the remote station.

If the intention is to provide only emergency stops, omit the remote station shown so that these are all connected in series with the stop button on the main enclosure.

Hand-operated star-delta starter

This is a two-position method of starting a three-phase squirrel-cage motor, in which the windings are connected firstly in star for acceleration of the rotor from standstill, and then secondly in delta for normal running.

The connections of the motor must be arranged for star-delta, starting with both ends of each phase winding – six in all – brought out to the motor terminal block. The starter itself, in its simplest form, is in effect a changeover switch. Figure 6.29 gives the elementary connections for both star and delta.

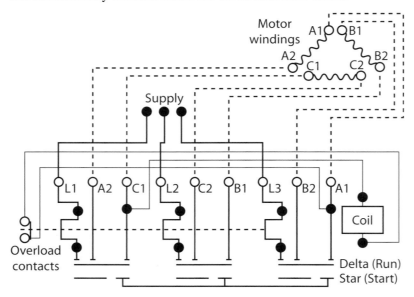

Figure 6.29 Hand-operated star-delta connections

When the motor windings are connected in star, the voltage applied to each phase winding is reduced to 58 per cent of the line voltage, thus the current in the winding is correspondingly reduced to 58 per cent of the normal starting value.

Applying these reduced values to the typical three-phase squirrel-cage induction motor, we would have: initial starting current of about two to three-and-a-half times full-load current and initial starting torque of about 50 per cent full-load value.

The changeover from star to delta should be made when the motor has reached a steady speed on star connection, at which point the windings will now receive full line voltage and draw full-rated current.

If the operator does not move the handle quickly from the start to run position, the motor may be disconnected from the supply long enough for the motor speed to fall considerably. When the handle is eventually put into the run position, the motor will therefore take a large current before accelerating up to speed again. This surge current could be large enough to cause a noticeable voltage dip. To prevent this, a mechanical interlock is fitted to the operating handle. However, in reality the handle must be moved quickly from start to run position, otherwise the interlock jams the handle in the start position.

The advantage of this type of starter is that it is relatively cheap. It is best suited for motors against no load or light loads, and it also incorporates no-volts protection and overload protection.

Automatic star-delta starter

Bearing in mind the user actions required of the previous hand-operated starter, the fully automatic star-delta contactor starter (as shown in Figure 6.30) is the most satisfactory method of starting a three-phase cage-induction motor. The starter consists of three triple-pole contactors, one employing thermal overload protection, the second having a built-in time-delay device, and the third providing the star points.

Star-delta starter

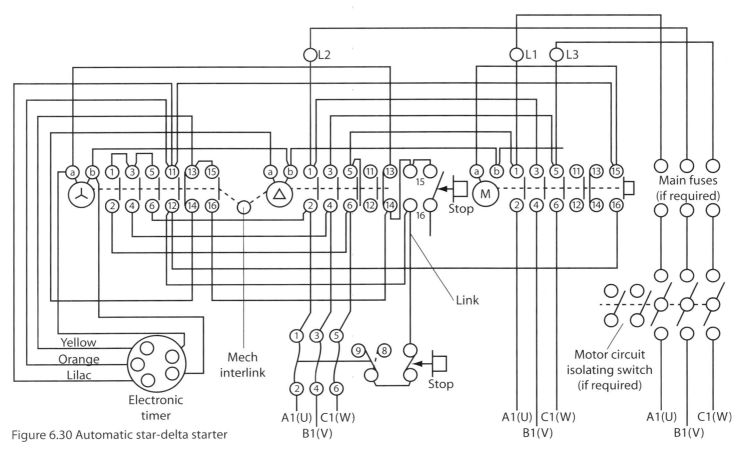

Figure 6.30 Automatic star-delta starter

The changeover from star to delta is carried out automatically by the timing device, which can be adjusted to achieve the best results for a particular situation.

Soft starters

A soft starter is a type of reduced-voltage starter that reduces the starting torque for a.c. induction motors. The soft starter is in series with the supply to the motor, and uses solid-state devices to control the current flow and therefore the voltage applied to the motor. In theory, soft starters can be connected in series with the line voltage applied to the motor, or can be connected inside the delta loop of a delta-connected motor, thereby controlling the voltage applied to each winding. Soft starters can also have a soft-stop function, which is the exact opposite to soft start. This sees the voltage gradually reduced and thus a reduction in the torque capacity of the motor.

The auto-transformer starter

This method of starting is used when star-delta starting is unsuitable, either because the starting torque would not be sufficiently high using that type of starter or because only three terminals have been provided at the motor terminal box – a practice commonly found within the UK water industry.

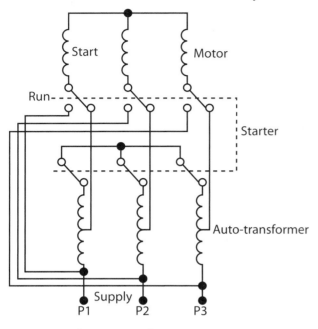

Figure 6.31 Connections for an auto-transformer starter

This again is a two-stage method of starting three-phase squirrel-cage induction motors, in which a reduced voltage is applied to the stator windings to give a reduced current at start. The reduced voltage is obtained by means of a three-phase auto transformer, the tapped windings of which are designed to give 40 per cent, 60 per cent and 75 per cent of the line voltage respectively. Although there are a number of tappings, only one tapping is used for the initial starting, as the reduced voltage will also result in reduced torque. Once this has been selected for the particular situation in which the motor is operating, it is left at that position and the motor is started in stages. This is much like the star-delta starter in that, once the motor has reached sufficient speed, the changeover switch moves on to the run connections, thus connecting the motor directly to the three-phase supply. Figure 6.31 illustrates the connections for an auto-transformer starter.

The rotor-resistance starter

This type of starter is used with a slip-ring wound-rotor motor. These motors and starters are primarily used where the motor will start against full load as an external resistance is connected to the rotor windings through slip-rings and brushes, which serves to increase the starting torque.

When the motor is first switched on, the external rotor resistance is at a maximum. As the motor speed increases, the resistance is reduced until at full speed, when the external resistance is completely eliminated, the machine runs as a squirrel-cage induction motor.

The starter is provided with no-volts and overload protection and an interlock to prevent the machine being switched on with no rotor resistance connected. (For clarity these are not shown in Figure 6.32, since the purpose of the diagram is to show the principle of operation.)

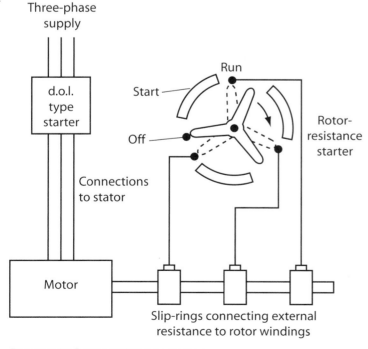

Figure 6.32 Rotor-resistance starter

Speed control of d.c. machines

We said at the beginning of this chapter that there are three types of d.c. motor: series, shunt and compound. One of the advantages of the d.c. machine is the ease with which the speed may be controlled.

Some of the more common methods used to achieve speed control on a d.c. machine are described below.

Armature resistance control

With this system of control we are effectively reducing the voltage across the armature terminals by inserting a variable resistor into the armature circuit of the motor. This creates the illusion of applying a lower-than-rated voltage across the armature terminals.

The disadvantages of this method of control are that we see much of the input energy dissipated in the variable resistor, a loss of efficiency in the motor and poor speed regulation in shunt and compound motors.

Although not discussed in this book, the principle of applying a lower-than-rated voltage across the armature terminals forms the basis of the Ward-Leonard system of speed control. In essence this provides a variable voltage to the armature terminals by controlling the field winding of a separate generator. Although expensive, this method gives excellent speed control and therefore finds use in situations such as passenger lifts.

Field control

This method works on the principle of controlling the magnetic flux in the field winding. This can be controlled by the field current and, as a result, controls the motor speed. As the field current is small, the power dissipated by the variable resistor is reasonably small. We can control the field current in the various types of motor as follows.

- **Series motor** – place a variable resistor in parallel with the series-field winding.
- **Shunt motor** – place a variable resistor in series with the field-shunt winding.

This method of speed control is not felt to be suitable for compound machines, as any reduction in the flux of the shunt is offset by an increase in flux from the series field because of an increase in armature current.

Pulse width modulation (PWM)

We know all about the problem with the variable resistor: although it works well, it generates heat and hence wastes power. PWM d.c. motor control uses electronics to eliminate this problem. It controls the motor speed by driving the motor with short pulses. These vary in duration to change the speed of the motor. The longer the pulses the faster the motor turns, and vice versa. The main disadvantages of PWM circuits are their added complexity and the possibility of generating radio frequency interference (RFI), although this can be minimised by the use of short leads or additional filtering on the power supply leads.

Speed control of a.c. induction motors

We have already established that synchronous speed is directly proportional to the frequency of the supply and inversely proportional to the number of pole pairs. Therefore the speed of an induction motor can be changed by varying the frequency and/or the number of poles. We can also control the speed by changing the applied voltage and the armature resistance.

For adjustable speed applications, variable speed drives (VSD) use these principles by controlling the voltage and frequency delivered to the motor. This gives control over motor torque and reduces the current level during starting. Such drives can control the speed of the motor at any time during operation.

Variable-frequency drive (VFD)

The phrase 'a.c. drive' has different meanings to different people. To some it means a collection of mechanical and electro-mechanical components (the variable frequency inverter and motor combination) which, when connected together, will move a load. More commonly – and also for our purposes – an a.c. drive should be considered as being a variable-frequency inverter unit (the drive), with the motor as a separate component. Manufacturers of variable-frequency inverters will normally refer to these units as being a variable-frequency drive (VFD).

A variable-frequency drive is a piece of equipment that has an input section – the converter – which contains diodes arranged in a bridge configuration. This converts the a.c. supply to d.c. The next section of the VFD, known as the constant-voltage d.c. bus, takes the d.c. voltage and filters and smoothes out the wave form. The smoother the d.c. wave form, the cleaner the output wave form from the drive.

The d.c. bus then feeds the final section of the drive, the inverter. As we know from reading earlier chapters of this book, this will invert the d.c. back to a.c. using Insulated Gate Bipolar Transistors (IGBT), which create a variable a.c. voltage and frequency output.

On the job: Motor noise

You are asked to investigate a problem in the plant room of a hospital, as the estates maintenance manager has reported a noisy motor on the air-conditioning system. When you look at the motor, you find that as well as it being noisy there is a vibration.

Do you think this is a real problem? If so, what actions would you recommend?

Activity

Have a look at some motors in and around your workplace. Decide what type of motors they are (a.c., d.c., single-phase, three-phase, induction,, shaded-pole etc.) and compare their physical sizes with their power rating.

FAQ

Q Can I reverse the direction of any motor?

A No, not all motors are capable of being reversed. Also, if you do reverse the direction of a motor you have to be careful that the load will not be damaged or create a harmful effect. For example, a boiler flue fan that was rotating in the wrong direction would cause a serious safety hazard.

Q How do I reverse the direction of a d.c. motor?

A Simply reverse the polarity of the supply connections.

Q How do I reverse the direction of a three-phase motor?

A Swap over any two of the phases and the motor will run the other way.

Q How do I reverse the direction of a single-phase motor?

A It depends on the type of motor. You would need to reverse either the start or run winding connections but often these are fixed (especially for smaller motors) so you wouldn't be able to reverse it.

Q Can I just connect up a three-phase motor and 'trust to luck' that it will spin in the right direction when it is switched on?

A No, you should always check that the phase rotation of the supply is correct before switching the motor on, otherwise serious damage could occur.

Q My single-phase motor starts sometimes and not others. What's the problem?

A With capacitor start motors it probably means that the capacitor has failed; try replacing it. With shaded pole motors, such as those found in boiler flues, it is usually that the bearings are worn or sticky and there simply isn't enough starting torque to start the motor. Replacing the motor is usually the best solution.

Knowledge check

1. What will be the percentage slip of an induction motor having a synchronous speed of 1500 rpm and a rotor speed of 1425 rpm?

2. What is the type of motor most likely to be used in a domestic vacuum cleaner?

3. Describe the principle of operation of a series-wound (universal) motor.

4. How would you change the direction of rotation of a three-phase induction motor?

5. What are the component parts of a squirrel-cage rotor?

6. Describe how to reverse the direction of a single-phase split-phase motor.

7. What is the purpose of the centrifugal switch in a capacitor start motor?

8. What type of motor uses the rotor-resistance starter?

chapter 7

Safe, effective and efficient completion of installations

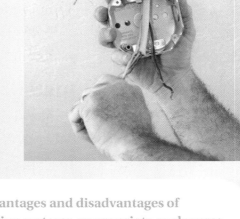

Unit 2 outcome 1

As we have already seen in chapter 2, the improvement of working practices is crucial to enable best practice to be used in the workplace. We will now explore this topic in further depth.

In addition to safety, it is important that working practices are also effective and efficient. After all we need to be able to achieve the things we promise contractors on time and on budget!

This chapter builds on topics already covered at Level 2 and in chapter 2 of this book.

On completion of this chapter the candidate will be able to:

- describe types of work and factors to be included in safe installation activities and parties involved and the role of documentation and communication
- state the factors and parties that need to be considered when planning installations
- describe the procedures involved with producing a formal contract, the role of the contract and the importance of co-ordinating contracts
- state standard voltages for transmission and distribution and the importance of load balancing
- describe safety arrangements and the applications/limitations of protective equipment

- describe advantages and disadvantages of common wiring systems, appropriate enclosures for installations and conditions of effective termination
- state factors and calculations affecting selection of conductors
- use BS 7671 and other data to determine disconnection times
- select appropriate cable sizes for conduits
- describe earthing and bonding and earthing arrangements for TT, TNS, TNC-S and the installation of consumer's switchgear.

Safe working procedures

On completion of this topic area the candidate will be able to describe the methods and procedures necessary to make an area safe before starting work and describe the general types of work that need to be included in installation activities.

Health and safety at work

This area has been largely covered at Level 2 and chapters 1 and 2 of this book.

The Health and Safety at Work Act 1974 places a duty on employers and employees to ensure that their work activities do not endanger anyone. To this end risk assessments are undertaken, with precautions implemented to reduce the risk from hazards identified. Useful precautions (safety measures) include:

- barriers and warning tapes – these establish a safe perimeter
- warning signs
- communication – inform anyone who may be affected by an activity; always advise people of precautions they may need to take and any restrictions that may arise as a result of your activities
- isolation of supplies – this is covered by the Electricity at Work Regulations 1989; in most cases it will be necessary to isolate the supply to circuit (a more detailed look at safe isolation procedures was undertaken at Level 2
- permit to work – this was covered in depth in chapter 2.

Installation activities

In the Level 2 book we looked in detail at the sectors within the construction industry. As such we will give only a brief summary of that material here.

The types of installations that electricians are likely to make are as follows.

- Lighting and power installations – lighting can be anything from normal office lighting to floodlights in stadiums. Power installations cover anything from sockets in a household to machines in factories.
- Alarm, security and emergency installations – ranging from fire alarm systems to complex anti-burglar devices.
- Building management and control installations – for example 'intelligent' lighting, air conditioning, energy management, machine control systems, computer systems.
- Communication, data and computer installations – includes networking of computers and cabling for voice, data, video and control signals.

Specialist work may include:

- cable jointing – connecting main power cables
- electrical machine drive installations – energy-efficient motors
- panel building – control panels used to control electricity distribution
- instrumentation – installing process control and measurement systems
- electrical maintenance – electronics, safety issues and preventative maintenance.

Planning an installation

On completion of this topic area the candidate will be able to state the factors that need to be considered when planning installations.

Installation factors

For any installation project to be successful, many factors have to be taken into consideration. First the selection of appropriate wiring systems, cable management systems, distribution systems, equipment, plant and accessories has to be made for the installation requirements and customer's needs. The installation plan also has to take into account the type of building, its location and the access available, as well as storage facility and security.

Structure of the building

The structure of the building and the order in which it is being erected will affect the choice of components. For example, switch cubicles, switchgear and standby generators (UPS) will be used in different building structures.

This will also determine the delivery requirements to site. For example, if a large switch cubicle was to be sited in the basement main switch room, then delivery would have to be arranged while access to the basement was possible. Similarly, if the standby generator was to be located on the roof, then delivery of that component before the roof was constructed would mean making provision to store it safely and securely until it could be installed.

The building fabric

The **building fabric** is also a major factor in the planning and design of an installation. This will, in most cases, dictate the type of fixing requirements, whether an installation is flush or surface and, therefore, the wiring systems.

The building fabric can also dictate the installation sequence. For example, if internal walls are constructed of brick or concrete then any provision for cables within those has to be made before they are constructed. Conversely, installations into walls made of stud partitioning will mean this can be done later.

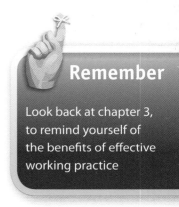

Remember

Look back at chapter 3, to remind yourself of the benefits of effective working practice

Definition

Building fabric – the materials from which a building is made. Prefabricated buildings are assembled from a series of pre-prepared components

External influences

External influences affect the choice of equipment to be installed. Any equipment installed outdoors must be able to cope with the weather conditions such as rain, extreme cold and sunshine.

BS 7671 Regulation 522 requires that every item of equipment shall either be of a design appropriate to the situation in which it is to be used or its mode of installation shall take account the conditions likely to be encountered. Chapter 52 of BS 7671, Selection and Erection of Wiring Systems, refers to the external influences to which an installation may be exposed. These are:

- ambient temperature
- external heat sources
- presence of water or high humidity
- presence of foreign bodies
- presence of corrosive or polluting substances
- impact
- vibration
- other mechanical stresses
- presence of flora and/or mould growth
- presence of fauna
- solar radiation and ultraviolet radiation
- building design.

Regulation 522.5 requires that the presence of any corrosive or polluting substances likely to give rise to corrosion or deterioration of the wiring system shall be taken into account and materials resistant to such substances should be used. This would affect the choice of conduits to be buried in walls with a cement or plaster finish, as these can be acidic or corrosive.

Regulation 522.11 requires that wiring systems exposed to solar or ultraviolet (UV) radiation shall be suitable for the conditions in which they are to be installed. Protection is provided to cables clipped to the surface of walls through the colour of the sheath of the cable (black) as this is able to absorb the UV rays without degradation.

Storage of parts and materials

Storage of parts and materials is another essential factor in the planning process for any installation. The longer materials are left on site, the greater the risk of loss or breakage. Good planning, such as the staged delivery of materials throughout the job, will minimise losses.

Storage can be facilitated by the main contractor in areas within compounds or temporary buildings specially set aside for the purpose. However, this provision would need to be confirmed at the tendering stage, as an allowance may have to be made for the cost of hiring or supplying temporary storage or accommodation.

If the installation is over a large area or many floors, then provision will need to be made for safe storage of materials and tools localised to the work area. Lockable site boxes are frequently used for this purpose. Tools and equipment have to be locked away safely during breaks, not just at the end of the day.

Tools and equipment

Access to equipment and specialist tools that may be required for the completion of the job has to be identified during the planning process, with the cost of the supply and delivery of equipment accounted for at the appropriate stage. Many items will be needed during the course of an installation. Some of these will be very specific, for example for work on oil refineries and large power stations it is possible the contractor would also have to make allowance for the transport of staff and equipment around the site.

Minimising disruption

The key to any successful project is good project management. Always plan the installation process and ensure that start and finish times are met. One of the major causes of delay to a project are clashes that occur between trades trying to work in the same area.

To avoid this, it is essential that all parties follow the same programme. Of course, even on the best managed sites there will still be problems. However, with a good programme that everyone follows, these should be kept to a minimum. Figure 7.01 shows the programme of work and timescale for the construction of a house.

Did you know?

The customer only pays for the materials once – any replacement of lost or damaged materials will be your responsibility (and at your expense!) until installed

Remember

Any plant or equipment left lying around unused is dead money – what would be the point of hiring a portable scaffold for the duration of the job if it is only required for the first few days?

Activity	Number of weeks						
	11	12	13	14	15	16	17
Electrical 1st fix ground	■						
Plumbing 1st fix ground		■					
Plaster ground floor			■				
Painters 1st floor				■			
Plumbers 1st fix 2nd floor			■				
Electricians 1st fix 2nd floor		■					
Plaster 2nd floor				■			
Painters 2nd floor					■		
Elec test & commission						■	
Plum test & commission							■

Figure 7.01 Programme of work and timescale for construction of a house

Using a system like this can make some of the planning seem very obvious and straightforward. Clearly, the electrical testing and commissioning cannot take place before the installation is complete. Neither can the plumber test and commission his heating system until the electrician has proved it is safe to put the supply on!

By arranging different times for the plumber and electrician to be on site, this effectively minimises disruption to the work areas of each of the trades, making it a safer working environment.

Estimating installation time

Estimating the time needed for an installation can be quite simple. In many cases you will follow four simple steps.

1. Break the job down into various tasks.

2. Put them in the correct sequence.

3. Allot a time to each task.

4. Total up the time for the job.

This will give a total time for the job to be completed in hours or days. However, where the installation will have to link in with other workers – such as builders, plumbers, painters and decorators – the task of estimating the actual time needed to complete the installation becomes more complicated.

How much time to allot to the task?

Over a period of time, a record can be kept that shows how long it has taken to install a particular type of material in a particular installation situation. This type of document is often referred to as a **book of rates**. The type of information contained within such a document will refer to the time and cost of completing a certain type of work. For example, it could record the hourly rates for installing steel conduit per metre, either fixed on the surface, the ceiling, the floor or laid in concrete.

In practice, we would use the book of rates to reach decisions regarding estimations for the customers. If our book of rates records that we can fix 10 metres of steel conduit every hour, then for 120 metres of steel we would estimate 12 hours (120 divided by 10).

Books of rates have been compiled by individual companies over a period of time. These rates reflect what companies know they are able to achieve and thereby serve as an accurate estimate of time. The differences in installation times that may exist between one company and another can be the major factor in why one company's tender price for a project may be less than another's. For that reason, the book of rates is generally kept very secure within the organisation.

There are now many versions of books of rates in the marketplace, as well as CDs and estimating package. If all companies used these, then prices for projects would all be the same, but in reality companies use these products as a basis for the tendering process and from their experience adjust the rates accordingly.

Parties concerned with installations and their relationships

On completion of this topic area the candidate will be able to identify the parties concerned with installations and the relationships between them.

Various people are involved in the successful completion of an installation project. The most important person is the client, since it is their money that will be spent! If the client is going to be made happy, then they must believe that they have obtained value for money and that the installation/project meets their requirements fully.

Most clients are largely non-technical and they therefore require the services of people to help and guide them through the process – hence the need for professional help. Table 7.01 (below and page 268) identifies the parties involved with a project and shows the relationships between them.

Client	• Person or organisation that wants the work done and is paying for it. • Specifies the purpose of the building. • Usually gives an idea of number of rooms, size, design etc. and anything specific. • May also give an idea of the price they are willing to pay for the work.
Architect	• Designs the appearance and construction of the building so that it fulfils its proper function. • Advises the client on the practicality of their wishes. • Ideally provides a design solution that satisfies the client and also complies with the appropriate rules and regulations for the type of building. • For small projects, an architect may draw up a complete plan. For larger, more complex buildings they will consult specialist design engineers about technical details. • Engages on behalf of the client, the main contractor, to complete the work.
Main contractor	• Usually the builders, because they have the bulk of the work to carry out. • Has the contract for the whole project. • Employs subcontractors to carry out different parts of the work, generally through a competitive tendering process. • In refurbishment projects, where the amount of building work is small, the electrical contractor could be the main contractor. • Pay and co-ordinate subcontractors.

(continued over)

Nominated subcontractors	• Named (nominated) specifically in the contract by the client or architect to carry out certain work. • Must be used by the main contractor. • Normally have to prepare a competitive tender. • Subcontractors will include electrical installation companies.
Non-nominated subcontractors	• Companies chosen by the main contractor (i.e. not specified by the client). • Their contract is with the main contractor.
Nominated supplier	• Supplier chosen by the architect or consulting engineer to supply specific equipment required for the project. • Main contractor must use these suppliers.
Non-nominated supplier	• Selected by main contractor or subcontractors. • For electrical supplies, this will be a wholesaler who can provide the materials needed for the project at the best competitive price.
Consulting engineers (Design engineers)	• Act on behalf of the architect, advising on and designing specific services such as electrical installation, heating and ventilation etc. • Create a design that satisfies the client and architect, the supply company and regulations. • Ensure that cable sizes have been calculated properly, that the capacities of any cable trunking and conduit are adequate, and that protective devices are rated correctly. • Produce drawings, schedules and specifications for the project that will be sent out to the companies tendering for the contract. • Answer any questions that may arise from this. • Once the contract has been placed, they will produce additional drawings to show any amendments. • Act as a link between the client, the main contractor and the electrical contractor.

Table 7.01 Parties concerned with installations

Regulatory requirements

On completion of this topic area the candidate will be able to state the role of the relevant parties in respect of the regulatory requirements.

Building Regulations

Communities and local government are responsible for the Building Regulations. These exist principally to ensure the health and safety of people in and around buildings. The regulations apply to most new buildings and many alterations of existing buildings in England and Wales including domestic, commercial and industrial installations. Building Regulations apply to the construction, access and energy conservation of buildings.

Failure to comply with Building Regulations may result in the local authority taking the builder to a magistrates' court, where they can be fined up to £5000 with a further fine of up to £50 for each day that contravention continues. Alternatively, if the owner failed to comply, the local authority may serve an enforcement notice on the owner of the construction requiring them to alter or remove the work that contravenes regulations. In this event the local authority also has the power to undertake the work itself and recover the costs from the owner.

Building Regulations protect those people in and around buildings

All parties have duties to comply with Building Regulations.

Did you know?

In July 2004, part P of the Building Regulations were published, bringing domestic electrical installations in England and Wales under building regulation control. An installation that complies with BS 7671 will comply with part P of the Building Regulations.

What has changed is the need to register with the local authority building control officer the work to be done. You will also need, on completion of the work, to provide the appropriate certification. This can now only be done by someone who is registered as a competent person on the self-certification scheme. Such registration schemes are offered by NICEIC, BRE Certification Ltd, BSI and ELECSA Ltd. To be registered on to these schemes, individuals will have to demonstrate their competence and have at least one year of work inspection experience

The client

Should the client or any of his or her agents, fail to comply with the requirements of the Building Regulations, then the client may have to meet the costs of correction or removal in its entirety of the project as well as the prospect of a fine imposed by the magistrates court.

The main contractor

The main contractor, subcontractors, suppliers and consultants all have a responsibility to design, install and supply materials that are compliant with the Building Regulations. Any one of the above parties may find themselves in court as the client attempts to minimise the losses imposed on him or her by the local authority building control office.

Environmental regulations

Remember

The implications of environmental legislation affect everybody involved in a design and build project

All the parties associated with a project, from the client through to the suppliers and installers, have duties to comply with environmental regulations. We covered these in chapter 1 (pages 27–29). However, the main regulations are:

- Pollution Prevention and Control Regulations 2000
- Environmental Protection Act 1990
- Clean Air Act 1993
- Controlled Waste Regulations 1998
- Radioactive Substances Act 1993
- Dangerous Substances and Preparations and Chemicals Regulations 2000.

Ultimately, any person or organisation that is in contravention of the legislation can expect to be prosecuted. This can be either as an offence in common law, where nuisance, negligence and/or damage to property results in a claim, or a statutory offence, where the crime can result in a fine or imprisonment.

Health and safety

Safety tip

The planning supervisor – prepares a health and safety plan at the pre-tender stage.

The principal contractor – develops the plan and keeps it up to date throughout the construction process

As well as the Health and Safety at Work Act 1974, the Construction (Design and Management) Regulations 1994 places a duty on all those who can contribute to the health and safety of a construction project. The person ordering the construction work (the client) has to:

- appoint a planning supervisor to co-ordinate the management and health and safety issues during the design and early stages of preparation
- appoint a principal contractor to co-ordinate and manage health and safety issues during the construction work (main contractor).

More information on relevant health and safety legislation can be found in chapter 1 pages 2–12. Failure to comply with the requirements of these acts and regulations can result in a criminal prosecution.

BS 7671 Wiring Regulations

The Wiring Regulations relate to the design, selection, erection, inspection and testing on electrical installations. Although they are not a statutory document, the regulations were written with existing legislation in mind. Therefore, an installation that failed to comply with the requirements of BS 7671 would be in contravention of the Electricity at Work Regulations 1989. The Electricity Safety, Quality and Continuity Regulations 2002 empowers the local electricity distributor to refuse to connect a supply to an installation if it fails to meet the requirements of BS 7671.

It can also be argued that failure to meet BS 7671 may be in contravention of the Health and Safety at Work Act 1974. If, as a result, someone was injured, it may be argued that the Health and Safety at Work Act 1974 was contravened. However, this view has yet to be tested in court.

Did you know?

Installations failing to meet the requirements of BS 7671 will almost certainly contravene the requirements of the Building Regulations, in particular part P of those Regulations

Working with documentation, drawings and specifications

On completion of this topic area the candidate will be able to read and interpret drawings and specifications to prepare materials requisitions.

It should be clear by now that documents, especially drawings and specifications, are very important for the success of a project. Without them, everyone would have a different idea about how things should be done and the result would be chaotic. Documents provide the technical information that everyone needs to do their job properly. However, they are only useful if the information they provide is clear and users can understand them. Unfortunately, this does not always happen.

In the construction industry the estimator uses the drawings and specification provided by the consulting engineer during the tendering stage. The electrician will also need the drawings during the installation stage. In the project specifications (which usually come in two parts: general and particular) there might be hundreds of different drawings, particularly if it is a large project like a school or hospital.

There are several types of drawing that will be used in a project; each one has a specific purpose and no single document will contain all the information needed, except possibly on the simplest level. Different people will need different documents depending on their task. You will often have to use several documents, reading them alongside each other to get all the information you need.

At Level 2 we learned about scaled drawings and preparing materials lists. But there are other drawings that could help in preparing a materials requisition, as described in the text that follows.

Site plans

Site plans give an overall layout of the site, for example indicating the position of the construction in relation to roads. Site plans are usually drawn to scale, perhaps 1:250 or higher. If it was proposed to run a cable from one building to another, then this type of drawing would be used to get a measurement of the length required and the route to be taken.

Block diagrams

These are used to show the order or sequence of control of an installation. They are not drawn to scale. A distribution diagram would have detail of the switchgear, contactors, isolators and busbar chambers as well as the size and types of cables used between the control gear.

Location diagrams

As with block diagrams, these show the position of the equipment around the installation. From this the measurements, for example of cable runs, can be established.

Circuit diagrams

Like block diagrams these are not drawn to scale. Circuit diagrams are used to show how circuits work. The drawing can be used to get the detail of specific items of equipment that may be needed to construct the circuit.

Contracts

On completion of this topic area the candidate will be able to state the procedures involved with the production of a formal contract, the role of contracts in the installation activity and methods used to monitor contract progress.

Agreeing contracts

Before any work starts, **contracts** must be agreed and entered into by all the firms involved. Any failure to comply with the details of the contract by either party could result in a court action and heavy financial damages. This might happen if the contractor does not complete the work or uses substandard materials, or if the client does not pay.

Contracts do not have to be made in writing. A verbal agreement, even one made on the telephone, can constitute a legal contract. However, most companies use written contracts that cover all aspects of the terms and conditions of the work to be carried out.

Did you know?

A fixed-price contract is exactly what it says: the contractor agrees to complete the job for the price quoted in the tender, even if material or labour costs go up before the project is completed

Several conditions must be met for a contract to be binding.

1. An offer must be made which is clear, concise and understandable to the customer.

2. The customer must accept the offer and this acceptance must be received by the contractor. The acceptance must be unqualified (i.e. with no additional conditions). Up to this point there is no agreement or obligation binding either side. The contractor is free to withdraw the offer and the customer can reject it.

3. There must be a 'consideration' on both sides. This is what each party is agreeing to do for the other.

There are several reasons why a contract may not be made. Here are four of them.

1. **Withdrawal.** The contractor can withdraw the offer at any time until the offer is accepted. The contractor must notify the customer of the withdrawal.

2. **Lapse due to time.** Most offers put a time limit on acceptance. After this time, the offer expires and the contractor is under no obligation, even if the customer later accepts the offer.

3. **Rejection.** The customer can reject the contractor's offer; no reason has to be given. If the customer asks the contractor to submit a second offer (for additional work or simply to lower the price), the contractor is under no legal obligation to quote again. Each quote is self-contained; the terms or conditions for a previous quote do not automatically apply.

4. **Death of contractor.** If the contractor dies before the offer is accepted the customer must be notified; otherwise the customer could agree (within the offer time limit) and the contract would become valid.

Breach of contract occurs when one of the parties does not fulfil the terms of the contract, for example, if the contractor does not perform the work to the specification in the offer (it is the contractor's responsibility to ensure that the installation is in complete compliance with the specification) or if the customer refuses to pay for the work.

Definition

Breach of contract – when one party does not fulfil the contract terms

Contract law is very complex, so cannot be covered fully on this course. To minimise the risks you can use a standard form of contract. The Joint Contracts Tribunal (JCT) 'Standard Form of Contract' (normally for projects of a non-complex nature or in excess of 12 months' duration) or the 'Intermediate Form of Contract' are typical of contracts used in the industry.

If you are involved in any kind of contract it is always advisable to seek professional legal assistance before making or accepting an offer. Once contracts are agreed, construction and installation work can begin and many more people become involved.

Monitoring contract progress

Once a contract has been agreed with a client and signed, it becomes equally important to monitor the progress made on the contract. Effectively this means being kept aware of the progress made towards completion of the job. There are several areas that will need to be monitored and recorded:

- labour requirements during the project
- scheduling of deliveries of materials to site to prevent delays
- expenditure on materials
- expenditure on labour
- additional work done outside of the original contract.

At Level 2 we looked at how charts and schedules can help monitor progress through making information easy to understand and allowing the user to clearly see what they need to know. The key examples of this is a critical path analysis, which can be done either as a network (Critical Path Network, or CPN), an activity sequence or a bar chart.

Bar charts and CPNs will help you to see which activities affect your work and how they fit in with everyone else's working schedule. In this way you can:

- plan which areas to work in
- see when you can start each activity
- make sure you have the correct materials and equipment ready at the right time
- avoid causing delays to others and the overall contract.

Site documentation

We covered the documentation found on site at Level 2. Remember, there are a great deal of documents and forms you will need to use during your work. These will be invaluable sources of information to monitor the progress of contracts, as they will contain a detailed record of all the activity on the site. Common examples include:

- job sheets
- variation order
- day worksheets
- time sheets
- delivery notes
- manufacturers' data
- site reports and memos
- site diaries.

Programming

On completion of this topic area the candidate will be able to explain the importance of co-ordinating the electrical installation with other trades and recognise the implication of variations.

As we have seen, the larger a project becomes the greater the number of people involved in its day-to-day operation. For many projects there will be representatives in more than one trade working on site. When working on an installation of this nature, it is essential that the work is co-ordinated. As we have seen, the work of one trade may need to be completed before another can begin. Without central direction, time and money will be wasted.

For a successfully completed project the criteria are:

1. completed on time

2. completed within budget

3. completed without sustaining accidents or injuries.

For this to be achieved it is essential that the project is managed effectively. To this end a programme of work has to be created before the job is started. This sets down overall start and finish dates for the project, with interim targets being set for individual tasks/specialisms. With all parties working from the same programme, everyone understands exactly what they need to do and what the implications of their actions will be on everyone else.

Failure to keep to the programme may result in:

- clashes between trades
- too many people trying to work in one area
- delay
- materials not being delivered to site at the appropriate time
- safety risks
- overspend
- implementation of penalty clauses
- loss of future work
- prosecution under health and safety legislation.

However, it is important to remember that for a programme of work to be successful it needs to be flexible. Delays at the beginning of the project can cause variations to the programme throughout the entire project. In this scenario, we must adjust the plan – otherwise we will be working to a programme that is already impossible to achieve! Once a delay has been noticed the work has to be reprogrammed, with new interim dates set.

Remember

Delays can happen – but you don't want them to! If a delay does happen, look back at it and work out how to avoid it happening in the future

Communication

On completion of this topic area the candidate will be able to explain the importance of effective communications in maintaining good relationships.

Effective communication

Communication is about much more than speaking or writing. You communicate an enormous amount by how you look, the gestures and facial expressions you use, and the way you behave. Even something as simple as a smile can make a big difference to your communication.

The benefits of good communication are huge. Good attitude, appearance and behaviour will immediately put your customers at ease and earn you respect from others. Clearly conveyed thoughts and ideas, when backed up with good working practices and procedures, will invariably improve productivity, increase profitability and produce satisfied customers.

Speaking, writing, appearance, attitude and behaviour all combine to make up your personal 'communication package'. It affects your relationships with everyone you deal with, such as:

- clients
- client representatives
- architects
- main contractors
- surveyors.

- local authorities
- other site trades
- supervisors
- work colleagues

When dealing with any of those listed above, always attempt to answer any questions they may ask of you as clearly and accurately as possible. If you are unable to answer their question, tell them so and offer to find out the answer. Or, if you are unable to do this, refer them to your supervisor or someone who you believe may be able to help them.

If you promise to help then find out the answer to their question, follow through on your promise. Do not let them down by forgetting either to find out the answer or failing to report the answer to them. Worst of all, never promise to do something and then not do it!

Often it will be necessary to communicate in writing. After all, it is common practice to confirm the details of any verbal conversation in writing. We do this to:

- ensure that a record of the conversation exists
- confirm understanding of what may have been agreed or said.

Letter and report writing

We covered techniques for report writing in chapter 1 pages 25–26. Good written communication skills are vital for avoiding confusion, especially in an industry that relies heavily on documentation. Writing good letters and reports is a part of the communication process. Letters can be used to:

- request information, e.g. delivery dates from a wholesaler or progress updates from a main contractor
- inform people of situations.

Reports can be used to:

- summarise investigations into the causes and effects of problems or trends and recommend solutions, e.g. why there has been an increase in absenteeism; why the work is below standard
- provide statistical or financial summaries, e.g. end-of-week materials costs, annual accounts
- record decisions made at meetings
- supply information for legal purposes, e.g. accident reports
- monitor progress, e.g. negotiations, construction work, implementation of a new system
- look into the feasibility of introducing new procedures, processes or products, or changing company policy.

Make your writing interesting

Do not let your document become boring and repetitive. Use a **thesaurus** to find alternative words to use. A thesaurus contains lists of words with similar meanings grouped together. For example, instead of the word 'wherewithal' in the sentence 'I have the wherewithal to pay the bill', the thesaurus gives 'means', 'resources' or 'ability' (among others) as possible alternatives that you could use.

Definition

Thesaurus – a type of dictionary that lists words with similar meanings

Standard voltages

On completion of this topic area the candidate will be able to state the standard supply voltages for transmission and distribution, the systems of distribution for single- and three-phase four-wire systems and the importance of load balancing.

Generation

We covered the generation of electricity in chapter 5 page 168. You will remember that electricity is generated at a power station and produced by alternators at voltages of 25 kV. This voltage is then stepped up by a transformer for transmission.

Transmission

Transmission is the process of carrying electrical power from one point to another, often over long distances. The principles of transmission are covered in chapter 5 (page 168).

In order to reduce power losses and keep cable sizes to the minimum, voltages are transmitted at very high values, which results in small values of current in the conductors. By keeping the values of current low the power lost over the transmission lines is kept to a minimum.

- **275 kV to 400 kV** is supplied for the supergrid. The supergrid links power stations together up and down the country and enables supply to be maintained.

- **132 kV** is supplied to the original national grid. This is a network of lines across the country, which enables supplies to be shared and spread regionally between power stations.

- **33 kV to 66 kV** is supplied for secondary transmission from the grid to the remote ends.

Distribution

Distribution refers to the supply of electricity in a local area.

- 33 kV to 11 kV is supplied to transformers. Transformers within an area are generally connected in a ring – hence the term ring main. The use of the ring connection guarantees supply. If one leg of the ring fails, the supply can be continued and maintained while repairs are made. Some large industrial installations, will take supplies direct at 33/11 kV.

- 11 kV to 400 V is supplied for primary distribution to consumers from substations. Industrial users will take the supply direct to their own substations and distribute it within their own installation.

- 230 V is supplied to domestic consumers from the local 11 kV/400 V substation.

Single- and three-phase four-wire systems of distribution

Figure 7.02 shows the secondary winding of an 11 kV to 400 V substation transformer. The secondary winding of the transformer is connected in star. This gives a common connection to the transformer coils. This common point establishes the neutral position and, by connecting this point to earth, limits the potential that can be developed to earth in the event of a fault to 230 volts.

Did you know?

The supergrid enables power to continue to be distributed around the country, even if one power station needs to be shut down

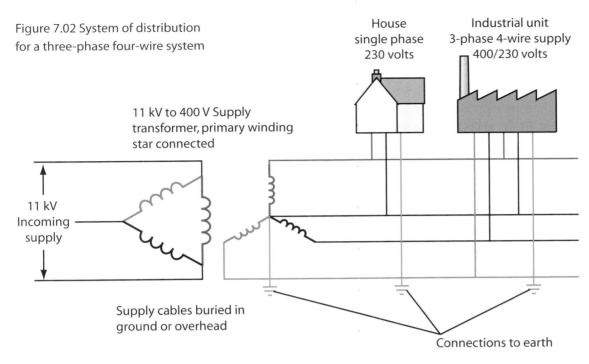

Figure 7.02 System of distribution for a three-phase four-wire system

The output of the supply transformer is 400 volts when measured across any two phases. The voltage when measured between any of the phases and the neutral point is 230 volts. We are therefore able to derive four types of supply from this system.

- Three-phase four-wire supply at 400/230 volts.
- Three-phase three-wire supply at 400 volts.
- Single-phase two-wire supply at 400 volts, used on remote farms where 400 volt equipment is needed but the supply authority considers it too costly to run four cables.
- Single-phase two-wire supply at 230 volts.

Load balancing

Load balancing was introduced at Level 2 and as a concept is fairly straightforward. When a three-phase induction motor is connected to a three-phase supply, then we describe the load as balanced. This is because each of the three coils in the motor are identical in terms of electrical resistance. Therefore the current drawn from each phase is identical. This seems simple enough. However, load balancing is not always so easy. What happens when we connect a single-phase load to a three-phase supply?

In this scenario we must try to spread the loads equally across the three phases. If the load distribution is widely uneven, this would result in more current being present in one phase conductor than in the others, which can lead to overloading. If we are to design and install economic installations by trying to spread the load as evenly as possible across the three phases of the supply, we must use cables and switchgear.

Dangers from electricity

On completion of this topic area the candidate will be able to identify the dangers of using electricity and describe methods of controlling the risk.

Electric shock

We covered electric shock comprehensively at Level 2. BS 7671 states the precautions we need to provide both basic and fault protection. We will now focus on the devices used to protect against electric shock.

PEBADS (formerly known as EEBADS)

Protection by Protective Equipotential Bonding and Automatic Disconnection of the Supply (PEBADS) is the main method of fault protection. This is achieved by joining all exposed extraneous conductive parts within the installation together. This prevents a difference of potential existing between any two items of extraneous metalwork, even under fault conditions.

Having joined all the extraneous conductive parts together, these are then connected, along with all the exposed conductive parts, to earth, providing a low resistance (impedance) path for current to flow around. A fault will bring about an automatic disconnection of the supply and activate the protective device.

For the device to work correctly, disconnection of the supply not only needs to be automatic but within the time constraints of BS 7671.

Protection by class II equipment

This method of protection relies on the electrical equipment having two layers of insulation, which prevents contact with any live parts. If you were to look in a kitchen, equipment such as food processors, mixers and blenders are all double-insulated and have the square within a square symbol stamped on the side.

Class II equipment has no circuit protective conductor; therefore it is supplied by two core cables.

Protection by non-conducting location

This type of protection is used in specialised production areas. To put it simply, if the walls and floors are made of non-conducting materials, then anyone coming into indirect contact with a live conductor will not be able to receive an electric shock as there will be no path for the current to flow along.

Figure 7.03 The double-insulated symbol

Protection by a free local equipotential bonding

This is used in specialist areas, such as television repair workshops. All the metalwork within an installation is bonded together but not connected to earth, preventing anyone receiving an electric shock to earth.

Protection by electrical separation

In areas where there is an increased risk of electric shock, all supply is derived from a safety isolating transformer. In this there are no electrical connections between the primary and secondary sides of the transformer. On the secondary side there is no connection to earth. This means that anyone using the equipment is unable to receive an electric shock to earth.

Fire

We covered fire safety at Level 2. The measures taken above would normally meet the requirements of protecting an installation from a fire arising from an earth fault.

However, there are some situations where additional protection is required to specifically protect people from the possibility of a fire. A thatched roof on a property is a perfect example of this. Obviously the thatch increases the possibility of fire risk. This means, in addition to EEBADS, circuits passing through the roof space would need to be protected from vermin, such as rats, by a 500 mA residual current device (RCD).

Fire and burns

Electrical equipment generates heat; some generate heat as part of their function, for example fan heaters and lights. When deciding on the location of equipment, we must take into account the heat given off by the equipment and the impact this will have.

The installation of down lighters into a ceiling void requires the shielding of cables or wooden structures to help prevent fire. Manufacturers for these light fittings produce 'top hats' made from flameproof material that fit over the light fitting, reducing fire risk.

Did you know?

A good example of electric separation is an electric shaver socket in a bathroom

Remember

Fires needs heat, fuel and oxygen to burn – if we break the circuit we remove the heat and prevent the fire

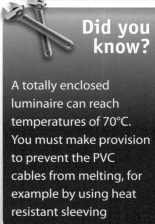

Did you know?

A totally enclosed luminaire can reach temperatures of 70°C. You must make provision to prevent the PVC cables from melting, for example by using heat resistant sleeving

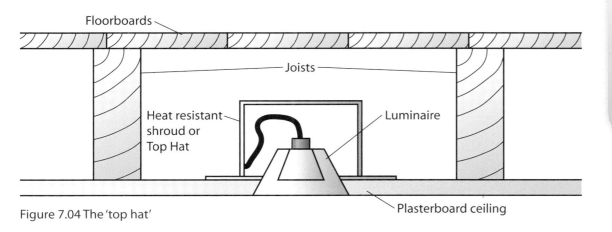

Figure 7.04 The 'top hat'

Floorboards

Joists

Heat resistant shroud or Top Hat

Luminaire

Plasterboard ceiling

BS 7671 requires that all people, fixed equipment and fixed materials adjacent to electrical equipment are protected against the harmful effects of heat. It is important not to have a fire hazard in the vicinity of electrical equipment. Table 42.1 from BS 7671 is reproduced below (Table 7.02). This gives the temperature limit under normal load conditions for an accessible part of equipment (meaning one within arms reach). Therefore, any equipment exceeding this temperature limit must be placed out of reach.

The temperature limit under normal load conditions for an accessible part of equipment, within arm's reach		
Part	**Material of accessible surface**	**Maximum temperature (°C)**
A hand-held means of operation	Metallic Non-metallic	55 65
A part intended to be touched but not hand-held	Metallic Non-metallic	70 80
A part which need not be touched for normal operation	Metallic Non-metallic	80 90

Table 7.02 BS 7671 table 42.1

Isolation and switchgear

It is essential that when people work on electrical circuits/equipment they:

- safely isolate supplies
- ensure that the supply cannot be restored, while work continues.

Safe isolation

BS 7671 requires that all circuits are controlled by an isolator that disconnects all live conductors from the supply. Before working on any circuit or equipment it should be safely isolated from the supply and a locking system should be used to prevent the isolator being turned back on. The key should be placed in a secure place and remain under your control at all times.

Before you start work, you should check that safe isolation has been proved. To do this, use an appropriate voltage tester. Isolation is covered fully at Level 2.

Injury from electrically actuated machinery

Before working on any rotating machine or electrical motor, the same isolation procedures should be used. There should be provision for switching off any electrical device that requires maintenance. BS 7671 requires electrical motors above 0.37 kW to be provided with control equipment, which incorporates means of protection against overload and the means of preventing automatic restarting after stoppage.

A starter, used to control an electric motor, is essentially an electrically operated switch (contactor) through which the supply to the motor is controlled by a set of contacts operated by a coil.

When the start button of a motor is pressed the coil is energised. As a result the contacts close, allowing the supply into the motor. In the event of a power failure the coil de-energises and the contacts re-open. However, in the event that the supply returns, the motor cannot automatically restart as it requires the start button to be pressed in order to energise the coil. This technology provides no-volt or under-voltage protection and can prevent injury to machine operators who may be making general maintenance on the machine during its 'downtime'.

Wiring systems and enclosures

On completion of this topic area the candidate will be able to identify appropriate wiring systems and enclosures for use in installations.

Part 1 of BS 7671 (the fundamental principles of protection for safety) requires that installations are designed to provide for the safety of persons, livestock and property against dangers and damage which may arise in the reasonable use of electrical installations. In order to meet these specifications, BS 7671 requires that installations have to be designed to match the requirements of their location.

The choice and type of wiring systems and methods of installation are determined by considering:

- use of the structure
- environmental conditions
- current demand
- overcurrent protection.

Use of the structure

It is important to make allowances for what the building will be used for – in other words, the work activities that will be undertaken there. In a domestic installation, PVC sheaths on cables provide minimal mechanical protection; however, when combined with flush fitting the cable into the fabric of the building and running cable underneath the flooring, it is ideal for that situation. For a warehouse, an installation will need a far more robust wiring system, capable of taking knocks without sustaining damage. Here we would choose to use steel conduit wired with PVC single-core cables.

Environmental conditions

Equally important is an understanding of the environmental conditions in which the installation will be used. Clearly material installed inside will encounter very different conditions to material installed outside. For example, PVC/SWA cable may be supported by a plain mild steel tray inside but a galvanised cable tray outside in order to protect from corrosion (not that you would always use a plain mild steel tray inside – where corrosive fumes may be given off it would be more appropriate to use a plastic-coated cable tray).

The risk from fire or explosion is also an environmental issue. In the petrochemical industry the electrical accessories used are from a heavy metal construction. As such the wiring systems are either MICC cable or PVC/SWA cable. It is vital in this environment that equipment is unable to generate any spark that could lead to ignition.

All electrical equipment is given an IP rating. This refers to the index of protection, a numerical reference system used to indicate the degree to which an item of equipment can be penetrated by solid objects or liquids. BS 7671 requires enclosures to reach the minimum standard of IP2X. This means that the live parts within the enclosure cannot be touched when a test finger 12mm in diameter and 80mm in length is inserted into it. Equipment outside has to meet at least IP44 – protected from objects 1mm in diameter and against water.

The final major condition to be considered is the effect of direct sunlight. UV rays can cause degradation of the cable sheath. Cables with a black oversheaf are suited to places where there will be direct exposure to the sun, as black absorbs the harmful UV rays.

Current demand

Current demand determines the size of cable used. When current passes along the cable it gives off heat. If the cable carries its full rated current then it will reach a working temperature of 70°C. As long as the cable is able to dissipate this heat, this is not a problem. However, if it is not able to loose heat to the atmosphere, the cable temperature will continue to rise, causing damage. A cable enclosed in trunking will especially struggle with this, while a larger cable will carry proportionally less than its full current-carrying capacity.

To counter voltage drop a larger cable may also be required, in order to reduce the voltage dropped in the supply cable.

Current-carrying capacities for differing size and types of cable can be found in appendix 4 of BS 7671.

Overcurrent protection

Protective devices should operate at predetermined values of current and voltage and at times which are appropriate to the characteristics of the circuit and the possibilities of danger. Some protective devices will, therefore, work quicker than others. It is during this period that harm can be done.

A rewirable fuse has a fusing factor of approximately two. This means that it will require twice its rated current to cause the device to operate instantaneously; anything less than this will cause the device to operate over a greater time period. Appendix 3 of BS 7671 has tables of time and current characteristics of overcurrent protective devices. From these tables it is possible to determine how quickly, or slowly, a protective device will operate for a given fault current.

For example, a fault current of 55 amps will cause a 30 amp rewirable fuse to operate in approximately 2000 seconds. Therefore, the wiring system must be able to withstand, without sustaining damage, this overcurrent and the heat associated with it for approximately 30 minutes. This should be taken into account at the design stage, and influence the size of cable to be used. Increasing the size of the cable at this stage will reduce the overheating effect under fault conditions.

The choice of protective device can be affected by the prospective short-circuit current that it may need to handle. A rewirable fuse meeting BS 3036 will handle fault current levels up to around 4 kA. Above this point the device has problems extinguishing the arc created across its terminals. Heat damage will result in the cables and distribution board being damaged beyond repair. The choice of protective device is not only down to its size, but also its type and its ability to clear high levels of fault current quickly.

The following list indicates the differing levels of fault current that protective devices are able to clear safely:

- BS 3036 – 4 kA
- BS 1361 – 30 kA
- BSEN 60898 – 6/10 kA.
- BS 3871 – 6/16 kA
- BS 88 – 80 kA

Remember

An increase in current will result in an increase of heat. For the period that this increased current is allowed to flow, the heat will damage the cables and the connections

Advantages and limitations of wiring systems and enclosures

On completion of this topic area the candidate will be able to describe the advantages and limitations of various wiring systems and enclosures.

Wiring systems

In chapter 5 we looked at details connected to certain types of wiring system. We will now look at the advantages and limitations of the types of cable mentioned.

MICC

Mineral-insulated metal sheathed cables is a generic term for cables that have conductors, protected by mineral insulation (magnesium oxide), enclosed within a metal casing. These cables can be made of aluminium or copper. Widely used throughout the industry is Mineral Insulated Copper Cable (MICC).

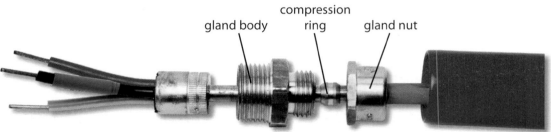

MICC cable

Advantages of this type of cable include the following.

- Fire resisting properties – copper can withstand 1000°C and magnesium oxide 2800°C. As such they are often used for emergency systems such as fire alarms and emergency lighting, or in areas with high ambient temperature. This type of cable also does not emit smoke or toxic gas.

- Robustness – the cable does not require any additional mechanical protection.

- Higher current-carrying capacity is achieved than for PVC-insulated cables, resulting in the ability to use smaller size cables.

- The relative space between the conductors and sheath is maintained when the cable is flattened, not affecting the cable's insulation values.

- The cable is non-ageing, with many cables installed in the 1930s still in operation today.

- The cable is totally waterproof.

- Low smoke and fume (LSF) sheathing is available on request from the manufacturer.

- The copper sheath can be used for earth continuity, saving the need for a separate protective conductor in the cable.

The disadvantages of this cable are as follows.

- Cost – this is very high compared with FP 200 and other wiring systems.

- Handling – removing bends for surface work is necessary but difficult.

- Weight – a 100m coil of 2.5mm² is difficult to handle, anything above this size may require a two-handed operation.

- When installing bare MICC, precautions must be taken to prevent the sheath from being scratched or damaged as this is the CPC.

- MICC cable with a PVC sheath cannot be installed when the temperature falls below 0°C as the sheath may crack and break.

- PVC sheath is required outside to prevent oxidisation and the cable turning green.

PVC/SWA/PVC cable

Shape stranded copper conductor — PVC insulation — Extruded bedding — Galvanised steel wire armour — PVC oversheath

PVC/SWA/PVC cable

Remember

This type of cable is used extensively for mains and sub-mains and wiring within industrial installations

The advantages of this type of cable are:

- robust construction, which makes it ideal for use in heavy industry as the wire armouring provides excellent mechanical protection

- very flexible and easy to handle

- it can be laid on a tray, clipped direct to the surface or on ladder racking; it can also be laid directly in the ground.

The disadvantages of this cable are:

- cost, when compared with other wiring systems

- no PVC sheath, meaning it cannot be installed when the temperature falls below 0°C as the sheath may crack and break

- the larger size cables lose their flexibility and can be more difficult to manoeuvre

- the larger cables will require cable rollers, cable jacks and a drum bar

- the larger cables are purchased on cable drums, which can be 2m or more in diameter. In addition, 1 metre of four-core 300mm² weighs approximately 50kg.

Single-core PVC-insulated sheathed and unsheathed cable

We looked at the features of this type of cable at Level 2.

Advantages of this cable are it is:

- relatively cheap in comparison to other wiring systems
- flexible and easy to handle in the smaller sizes
- readily drawn into conduit or laid in trunking.

Disadvantages of this cable are:

- it requires additional mechanical protection
- when used in a metal enclosure, live conductors associated with a circuit must pass through the same openings to prevent eddy currents being induced
- the smaller solid, stranded conductors should not be used where vibration can be experienced, such as termination to electric motors
- it cannot be installed when the temperature drops below 0°C, as the insulation may crack and break.

PVC-insulated and sheathed flat wiring cables

We looked at the features of this type of cable at Level 2.

Advantages of this cable are:

- flexibility
- it is easy to handle and feed through holes in joists
- it can be used on the surface or in flush work
- it can be laid on cable tray above false ceilings
- it is relatively cheap when compared to other wiring systems.

Disadvantages of this cable are:

- the PVC sheath offers very little mechanical protection
- capping or oval conduit needs to be used when there is a risk of damage to the cable
- it cannot be installed when the temperature drops below 0°C.

Steel conduit

Steel conduit is annealed mild steel tubing. It is used as a cable management system to contain PVC-insulated cables and has three types of finish: black enamel, galvanised and stainless steel. It is used widely in commercial and industrial installations where a high degree of mechanical protection is required.

Black enamelled steel conduit is used inside installations where the conduit will not be exposed to a damp environment. Galvanised conduit is used in external situations and anywhere where the conduit will be exposed to moisture. Stainless steel conduit is the more expensive and is generally used in food production areas.

The advantages of conduit are:

- robust and will withstand most impacts
- provides a rewirable system
- additions to the system are relatively easy to make
- provides an excellent cpc.

The disadvantages of conduits are:

- expensive to install as it has to be handled twice (first installation, then the wiring)
- expensive to purchase compared to twin and earth cable
- difficult to install and requires specialist tools and equipment, such as pipe vices, bending machines and stocks and dies to thread the conduit, as well as electric drills and drill bits for fixings
- joints need to be mechanically and electrically sound (tight)
- needs to have joints and exposed threads painted to avoid rust
- requires skill and training
- has sharp edges which must be removed if damage to the cable, as well as cuts to hands and fingers, are to be avoided.

Plastic conduit

Plastic conduit provides a fairly robust cable management system, although not as robust as steel conduit. It is used mainly in the commercial environment, either in complete systems or to provide protection to twin and earth cable where it drops from a false ceiling to switches and socket outlets.

The advantages of plastic conduit are it:

- is cheaper to buy and install than steel conduit
- is much lighter to handle than steel and therefore helps provide a higher production rate through the day
- is simple to bend

- is a rewirable system and easy to make additions
- is easy to cap off
- is available in two colours (black or white) and does not require painting
- produces minimum condensation due to low thermal conductivity
- can be used indoors or outdoors, laid in the ground, installed on the surface or flush within the building fabric
- will not corrode.

The disadvantages of plastic conduit are it:

- is expensive to use in comparison with twin and earth cable
- has to be handled twice – once to install and again when wired
- needs adequate allowance for expansion; plastic conduit expands five times greater than steel conduit
- requires an additional cpc to be installed
- cannot be installed when the temperature drops below 0°C
- joints in the conduit need to be glued with a PVC adhesive
- requires bending springs to prevent conduit collapsing
- frequently bends 'spring out' and need to be done again.

Safety tip

When using the PVC adhesive, adequate ventilation is essential. PVC adhesive can also burn your skin and eyes

Trunking

We covered trunking fully on pages 189–200 in chapter 5 of this book.

The advantages of trunking systems are they:

- provide a rewirable system and additions can be made at a later date
- can be installed before the final position of equipment is known
- can be used to contain many circuits and therefore can be cost effective
- have socket outlets and switches fitted in the lid
- provide an excellent cpc.

The disadvantages of trunking systems are:

- the removal of an individual circuit can be difficult, as the cables often intertwine with each other
- earth continuity has to be maintained across each joint for metal trunking
- a separate cpc has to be installed for each circuit, contained within the plastic trunking
- sharp edges on the metal trunking can damage cables and fingers

- segregation for multi-compartment trunking must be maintained at joints, crossovers and bends etc.

- if proprietary bends and sets are not available then these will have to be manufactured on site

- they can be over filled with cables

- special tools may be required for cutting the trunking when working with trunking larger than 200 × 200mm.

Cable tray and cable ladder

We looked at cable tray on page 201 of chapter 5.

Cable tray types include the following.

- **Standard cable tray** – suitable for light and medium duty installation work, available from 50mm wide up to 915mm.

- **Heavy duty tray** – as its name implies, suitable for carrying larger cables and is available from 152mm to 610mm wide.

- **Return-flange cable tray** – the return edge provides this tray with its strength, and therefore can be used to carry much bigger cables.

- **Heavy duty return-flange cable tray** – this tray is of a much heavier gauge and is designed for use in circumstances where high loading or adverse conditions can be experienced.

- **Cable ladder** – sometimes referred to as a ladder rack. This is an effective method of transporting cables across long unsupported spans or places where the number of supports are reduced. Often used in adverse site conditions, it is formed into what appears to be a ladder, with the 'rungs' slotted to allow cable cliques and ties etc. It is possible to get cable ladders up to 2m wide.

Standard cable tray

Figure 7.05 Return-flange cable tray

Figure 7.06 Heavy-duty return-flange cable tray

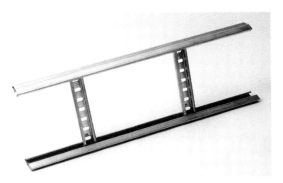
Cable ladder

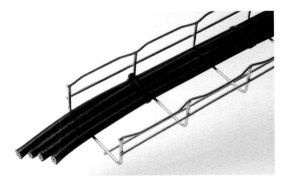
Cable basket

- **Cable basket** – the steel wire is formed into a basket shape, which makes for a very strong and lightweight support system on which cables can be laid.

Cable tray finishes include the following.

- **Plain, mild steel** – used in areas that will not be exposed to moisture and will be painted to the customer's requirements.

- **Painted finish** – used in areas that will not be exposed to moisture.

- **Galvanised** – to prevent corrosion and used in areas where exposure to moisture can be expected, such as outdoors.

- **Plastic coating** – used in industrial areas where corrosive fumes may be given off. Different finishes are available, such as PVC and polyethylene, with the required finish determined by the fumes given off.

- **Stainless steel** – used in food processing industries.

The advantages of cable tray systems are:

- they are easy and quick to assemble

- cables are laid on to the tray, as opposed to having to be drawn through

- additional cables can be installed easily at a later date

- the tray allows maximum airflow around the cables

- they are able to support many cables

- they can be located above false ceilings

- the slots in the tray make it easy to dress and fix in place with cable ties

- accessories such as crossovers, tees, bends and offsets are all available.

The disadvantages of cable tray systems are:

- as with all metal objects, sharp edges can cut and damage cables and fingers

- larger sizes of tray and cable ladder require more than one person to handle them

- planning the route and making fixings can take longer

- they can be overloaded with cables

- require bonding (earth)

- if accessories such as crossovers and tees etc, are not available, these will have to be manufactured on site, requiring extra training and skill

- trays over 300mm will require more than a hacksaw to cut them; tools such as jigsaws or large flat trunking saws will be needed.

Effective termination

On completion of this topic area the candidate will be able to state how effective terminations of cables and conductors are made.

Terminating cables and flexible cords

The entry of the cable end into an accessory is known as a termination. In the case of a stranded conductor the strands should be lightly twisted together with pliers before terminating. Care must be taken not to damage the wires. BS 7671 requires that a cable termination of any kind should securely anchor all the wires of the conductor that may impose any appreciable mechanical stress on the terminal or socket. A termination under mechanical stress is liable to disconnection.

When current is flowing in a conductor a certain amount of heat is developed. The consequent expansion and contraction may be sufficient to allow a conductor under stress, particularly tension, to be pulled out of the terminal or socket. One or more strands or wires left out of the terminal or socket will reduce the effective cross-sectional area of the conductor at that point. This may result in increased resistance and probably overheating. When terminating flexible cords into a conduit box, the flex should be gripped with a flex clamp. Any cord grips that are used to secure flex or cable should be clamped down on to the protective outer sheathing.

Remember

Despite the implication, trailing cables are not indestructable and care must be taken to regularly inspect them for damage

Connecting to terminals

Types of terminals

There is a wide variety of conductor terminations. Typical methods of securing conductors in accessories are pillar terminals and screwheads with nuts and washers.

Pillar terminals 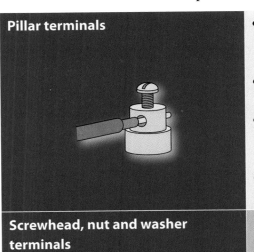	• A pillar terminal is a brass pillar with a hole through its side into which the conductor is inserted and secured with a setscrew. • If the conductor is small in relation to the hole it should be doubled back. • When two or more conductors are to go into the same terminal they should be twisted together. • Care should be taken not to damage the conductor by excessive tightening.
Screwhead, nut and washer terminals	• Using round-nosed pliers form the conductor end into an eye. This should be slightly larger than the screw shank but smaller than the out-side diameter of the screwhead, nut or washer. • The eye should be placed in such a way that rotation of the screwhead or nut tends to close the joint in the eye.

(continued over)

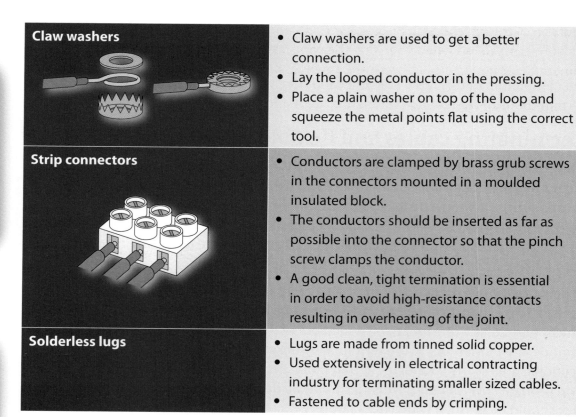

Claw washers	• Claw washers are used to get a better connection. • Lay the looped conductor in the pressing. • Place a plain washer on top of the loop and squeeze the metal points flat using the correct tool.
Strip connectors	• Conductors are clamped by brass grub screws in the connectors mounted in a moulded insulated block. • The conductors should be inserted as far as possible into the connector so that the pinch screw clamps the conductor. • A good clean, tight termination is essential in order to avoid high-resistance contacts resulting in overheating of the joint.
Solderless lugs	• Lugs are made from tinned solid copper. • Used extensively in electrical contracting industry for terminating smaller sized cables. • Fastened to cable ends by crimping.

Figure 7.07 Types of terminal

Terminating cable ends to crimp terminals

In order to terminate conductors effectively it is sometimes necessary to use crimp terminals, for instance in the termination of bonding conductors to earth clamps and sink tops. The terminals are usually made of tinned sheet copper with silver braised seams. The crimping tool is made with special adjustable steel jaws so that a range of cable end terminals can be crimped.

Cable joints and connections

Joints and connections should be avoided wherever possible. Where they must be used, they should be mechanically and electrically suitable. They must also be accessible for inspection and testing, unless exempted by Regulation 526.3 as follows:

- a joint designed to be buried in the ground
- a compound filled or an encapsulated joint
- a connection between a cold tail and a heating element (e.g. a ceiling and floor system, a pipe trace-heating system)
- a joint made by welding, soldering, and brazing or compression tool
- a joint forming part of equipment.

The joints in non-flexible cables should be made by soldering, brazing or welding, or be of a compression type and be insulated. The devices used should relate to the size of cable and be insulated to the voltage of the system being used. Connectors used must be of the appropriate British Standard, and the temperature of the environment must be considered when choosing the connector. EAW Regulations state that every joint and connection in a system should be installed so as to prevent corrosion.

Plastic connector

Among the most common types of connector, plastic connectors often come in a block of 10 or 12, sometimes nicknamed a 'chocolate block'. The size required is very important and it should relate to the current rating of the circuit being used. They are available in 5, 15, 20, 30 and 50 amp ratings.

Porcelain connectors

These are used where a high temperature may be expected. They are often found inside appliances such as water heaters, space heaters and luminaries. They should also be used where fixed wiring has to be connected into a totally enclosed fitting.

Screwits

These are porcelain connectors with internal porcelain threads that twist onto the cable conductors. They are now obsolete, but you may come across them in older installations.

Compression joints

These include many types of connector that are fastened on to the conductors, usually by a crimping tool. The connectors may be used for straight through joints or special end configurations. If the conductors are not clean when making the joint with a crimping tool this may result in a high-resistance joint, which could cause a build up of heat and eventually lead to fire risk.

Uninsulated connectors

These are often required inside wiring panels and fuse boards etc. and are used to connect earth cables and protective conductors.

Junction boxes

The two important factors to consider when choosing junction boxes are the current rating and the number of terminals. Junction boxes are usually either for lighting or socket outlet circuits.

Soldered joints

Although soldering of armoured and larger cables is still common, the soldering of small-circuit cables is rarely carried out for through joints. This joint was often referred to as a marriage joint, and involved stripping back the insulation, carefully twisting together conductors, soldering and then taping up.

A four-terminal junction box

A modern RB4 junction box

> **Remember**
>
> When a cable is not long enough or alterations are necessary, connectors may have to be used. These are often used in mineral-insulated systems. Where actual switching circuits are wired or where cables are taken from ring final circuits, junction boxes will normally be used

Selection of conductor size

On completion of this topic area the candidate will be able to state the factors affecting the selection of conductor size. The candidate will be able to calculate the size of the cable after: determining the design current (applying correction factors and diversity); ensuring compliance with volt drop requirements; ensuring that the cable complies with thermal constraints and shock protection.

Selection

Before calculating the size of the cable required, we first need to decide on the type of cable we are going to install. In order to do this we need to have information about the installation, specifically:

- where the cable is to be routed
- any external influences which may affect our choice.

If, for example, we need a cable to clip to the surface, should we use PVC/PVC cable or a more robust cable offering a higher level of mechanical protection, such as a PVC/SWA/PVC cable? The choice of cable affects not only the type of cable but also the current-carrying capacity for a given size.

The four areas we will primarily consider when selecting size of cable are:

- overcurrent protection
- voltage drop constraint
- shock protection
- thermal constraint.

Overcurrent protection

When calculating overcurrent protection we use the following formula:

$$I_b \leq I_n \leq I_t$$

where:

- I_b = the design current of the load
- I_n = the rating of the protective device
- I_t = the tabulated value of current-carrying capacity of the cable chosen.

The above statement may look complicated but in fact it is a statement of the obvious.

If the protective device is smaller than the design current of the circuit then, as soon as the circuit is switched on, the protective device would operate. Equally, if the size of the protective device is greater than the current-carrying capacity of the cable, then the cable would be damaged without the protective device operating. In this situation the cable would be protecting the fuse!

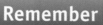

Remember

$\leq$ is a mathematical symbol meaning less than or equal to

Co-ordinating the selection of the size of the protective device in relation to the size of the load, and then selecting the cable in relation to the size of the protective device, will mean that in the event of an overcurrent situation, the protective device will operate, protecting the cable and the load.

Voltage drop constraint

We need to ensure that sufficient voltage arrives at the terminals of the load, in order to ensure that the system can function safely. The safe operating voltage of an item of equipment can usually be found on an information plate attached to the equipment or in the operating instructions supplied from the manufacturer.

Voltage drop constraint can be expressed as:

Actual volt drop (Vd) ≤ Maximum permitted volt drop

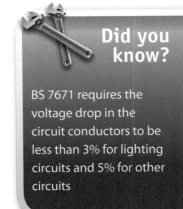

Did you know?

BS 7671 requires the voltage drop in the circuit conductors to be less than 3% for lighting circuits and 5% for other circuits

Shock protection

We need to be sure that automatic disconnection of the supply will take place in the event of an earth fault. This needs to happen within the constraints set out by BS 7671: within 0.4 seconds for circuits of 32A or less or within 5 seconds for circuits of greater than 32A.

Shock protection requirements can be expressed as:

Actual Z_s ≤ Maximum Z_s

where:

- actual Z_s is the calculated value of earth fault loop impedance of the design circuit
- maximum Z_s is the value of earth fault loop impedance related to the size and type of protective device used, the value of which is obtained from tables in Part 4 of BS 7671.

Thermal constraint

In fault conditions a large value of current can be present in the conductors. We need the conductors to withstand the large rise of temperature this will cause, without sustaining damage.

Generally, the smallest cable that forms a part of the earth fault loop path is the cpc (circuit productive conductor). Therefore the calculation concentrates on confirming the minimum size of cpc required that will withstand the increased current and heat. The logic behind this is that, if the smallest cable can withstand the increase in temperature without sustaining damage, the largest cables in the circuit should also be undamaged.

Thermal constraint can be expressed as:

Actual size of cpc chosen ≥ Minimum calculated size of cpc

Calculating the appropriate size

Design current (I_b)

You will have to calculate this; it is the normal resistive load current in the circuit. The following formulae apply to single- and three-phase supplies:

Single-phase supplies:

$$U_o = 230 \text{ V}$$

$$I_b = \frac{\text{Power}}{U_o}$$

where U_o is the phase voltage to earth (supply voltage).

Three-phase supplies:

$$U_o = 400 \text{ V}$$

$$I_b = \frac{\text{Power}}{\sqrt{3} \times U_o}$$

In a.c. circuits, the effects of either highly inductive or highly capacitive loads can produce a poor power factor (PF). You will have to allow for this. To find the design current you may need to use the following equations:

Single-phase circuits:

$$I_b = \frac{\text{Power}}{U_o \times \text{PF}}$$

Three-phase circuits:

$$I_b = \frac{\text{Power}}{\sqrt{3} \times U_o \times \text{PF}}$$

where PF is the power factor of the circuit concerned.

Diversity

Did you know?

Information on applying diversity can be found in appendix 1 of the *IEE On Site Guide*

You will also need to take diversity into account and make a professional judgement on the load. Will the cable supply an individual load that will be either on or off, or will the cable supply a group of loads, some on and some off? In the latter situation, some diversity may be allowed when calculating the design current. The effect of applying diversity is to reduce the design current of the load, which in turn will result in smaller, more economic cables being required.

An example of applying diversity can be seen when calculating the size of cable required to supply an electric cooker. If the supply voltage is 230 volts and the total load of the cooker is 6 kW in the design current, then I_b would be calculated by dividing the power by the voltage:

$$I_b = \frac{P}{V}$$

$$I_b = \frac{6000}{230} = \textbf{26 amps}$$

Looking at appendix 1 of the *IEE On-Site Guide*, Table 1B, it can be seen that the supply to a cooker in a domestic installation can have diversity applied to it. In this case the diversity is as follows: 10 amperes plus 30 per cent of the full load of the connected cooking appliance in excess of the first 10 amperes.

Therefore $I_b = 10 + (16 \times 0.3) = 10 + 4.8 = $ **14.8 amperes**

Now instead of sizing the cable to supply 26 amps we need only supply a cable that can carry 14.8 amps. More information on diversity can be found on pages 310–311.

Rating of the protective device (I_n)

When you have worked out the design current of the circuit (I_b), you must next work out the current rating or setting (I_n) of the protective device.

Regulation 433.1.1, on page 73 in the *IEE Wiring Regulations*, states that current rating (I_n) must be no less than the design current (I_b) of the circuit. The reason for this is that the protective device must be able to pass enough current for the circuit to operate at full load, but without the protective device operating and disconnecting the circuit.

Protective devices come in standard values, for example 13 A for plug tops. You can find these values in the *IEE Wiring Regulations*, Tables 41.2, 41.3 and 41.4.

The table you use to select the value of your protective device will depend on the type of equipment or circuit to be supplied and the requirements for disconnection times.

Reference methods

Table 4A2 from the *IEE Wiring Regulations* lists all the methods that can be used to install a cable.

You need to decide at this stage in the cable selection process which method of installation to use. This will ensure that the correct cable column is chosen in the later stages of cable selection. This choice is also important when you calculate correction factors for thermal insulation.

Correction factors

As the 'designer' of the installation, you need to know the five correction factors, know where they are required and then apply these to the nominal rating of the protection (I_n).

Correction factor	Tables for correction factor values	Symbol
Ambient temperature	Tables 4B1, 4B2 and 4B3	Ca
Grouping factors	Tables 4C1, 4C2, 4C3, 4C4 and 4C5	Cg
Thermal insulation	Regulation 523.7	Ci
BS 3036 fuse	0.725	Cr
Mineral-insulated cable	0.9 Table 4G1A	N/A

Table 7.03 Tables for correction factor values

Ambient temperature

This is the temperature of the surroundings of the cable, often the temperature of the air in a room or building in which the cable is installed. When a cable carries current it gives off heat. Therefore, the hotter the surroundings of the cable, the more difficult it is for the cable to get rid of this heat. But if the surrounding temperature is low, then the heat given off could be easily let out and the cable could carry more current. Cables must give off this heat safely or they could be damaged and there would be a risk of a fire.

You can find the correction factor for **ambient temperature** in Table 4B1 of the *IEE Wiring Regulations*. These tables are based on an ambient temperature of 30°C. This means that any cables installed in an ambient temperature above this will need the correction factor applying to them. This is because the cable will not be able to get rid of the heat it gives off safely when carrying current.

Note that an ambient temperature below 30°C (e.g. 25°C) will give a correction factor that will allow your choice of cable to be improved. Because a lower ambient temperature will allow the cable to 'give off' heat more easily and safely, therefore you could reduce the size of cable.

When a cable runs through areas experiencing different ambient temperatures, correction factors should be applied to the highest temperature only. You should note that Table 4B2 is used for cables buried in the ground or in underground conduits at an ambient temperature other than 20°C. Table 4B3 is for cables buried in soils with different thermal properties.

Grouping factors

If a number of cables are run together and touching each other, they will all produce heat when carrying current. The effect of this is that they are less able to cool down, so when installing cables we should keep them separated to reduce overheating.

If this is impossible, then you must apply a grouping factor to the cables. Grouping factors are in Tables 4C1, 4C2, 4C3, 4C4 and 4C5. You will see from the notes with these tables that the method of installation can affect the correction factor to be used.

- If each cable is separated by a clearance to the next surface of at least one cable diameter (De), they are classified as 'spaced'. You must take care in choosing the correction factor.

- Where the horizontal clearance to the next cable is more than 2De, then no correction factor is needed.

It is important that you apply these correction factors to the number of circuits or multicore cables that are grouped and not the number of conductors. You will find that you are normally using Table 4C1.

Thermal insulation

Thermal insulation has the effect of wrapping a cable in a fur coat on a hot summer's day. The heat produced when the cable carries current cannot escape. There are two ways that thermal insulation is relevant in electrical installation.

- Where flat twin and earth cable is in contact with thermal insulation, you must choose the appropriate reference method from Table 4A2 and select the cable from table 4D5.

- Where the cable is totally enclosed in thermal insulation for any length from 50mm up to 500mm, the cable must have a correction factor applied as stated in Regulation 523.7 Table 52.2 (see Table 7.04). If more than 500mm of the cable is to be totally enclosed, then the correction factor to be applied must be 0.5. The cable rating is then effectively halved.

Length in insulation (mm)	De-rating factor
50	0.89
100	0.81
200	0.68
400	0.55

Table 7.04 Cable rating

BS 3036 fuse

The only device that does not disconnect in sufficient time is the BS 3036 semi-enclosed rewireable fuse. If this particular protective device is used, then we must always use a factor of 0.725 when calculating current-carrying capacity.

Mineral-insulated cable

Table 4G1A (BS 7671) states that for bare cables (i.e. no PVC outer covering) exposed to touch, the tabulated values should be multiplied by 0.9.

Table 4G2A states that no correction factor for grouping need be applied.

Application of correction factors

You will have seen that nearly all of these factors have a value of less than 1. We also know that the cable must be able to carry current of an equal or greater value than the rating of the fuse that protects the cable.

If the cable's ability to give off heat is reduced by external conditions, then the only solution is to increase the size of the cable.

The use and application of the correction factors will require you to increase the physical cross-sectional area of the cable, which in turn increases the cable's capacity to carry current and give off heat.

Firstly we need to calculate the value of the design current (I_b) to allow us to determine the rating of the protective device (I_n). We then need to apply the correction factors to I_n to produce what is known as the tabulated value of current (I_t). Hence:

$$I_t > = \frac{I_n}{C_a \times C_g \times C_i \times C_r}$$

Now you can use the value of I_z together with the correct cable table from BS 7671, Table 4D1A to 4J4B, to find the size of cable that you need to use given these correction factors.

Example

A single-phase power circuit has a design current of 30 A. It is to be wired in flat two-core 70°C PVC-insulated and sheathed cable with cpc to BS 6004. It will have copper conductors and be grouped with four other similar cables, all clipped direct. If the ambient temperature is 30°C and the circuit is to be protected by a BS 3036 fuse, what should be the nominal current rating of the fuse and the minimum csa of the cable conductor?

When answering these types of questions it is a good idea to construct a table of information. This will help you understand and retain the process of cable selection. It also makes cable selection easier to understand.

Installation method no. is 20 (clipped direct), therefore reference method = C

From Table 4C1, as grouped, C_g = 0.6 (there being five circuits)

From Table 4B1, as 70° PVC, C_a = 1.0 (as 30°C ambient temperature)

Design current (I_b) = 30 A

As the protective device rating (I_n) must be ≥ I_b, we can take I_n = 30 A

$$I_z = \frac{30}{0.6 \times 1.0 \times 0.725} = 69A$$

$I_t \geq I_z$, and from Table 4D5 column 6, it is found that the minimum conductor csa that can be used is 16mm² cable with a tabulated value (I_t) of 85 A.

Over the following pages we will develop this example further to include compliance with voltage drop, shock protection and thermal protection.

Voltage drop

Regulation 525.3 states; 'The voltage drop between the origin of the installation (usually the supply terminals) and a socket outlet or the terminals of the current-using equipment should not exceed the values in Appendix 12.'

This means that, for lighting circuits the maximum voltage drop allowed is 3% and for other circuits it is 5%.

The voltage drops either because the resistance of the conductor becomes greater as the length of the cable increases or the cross-sectional area of the cable is reduced. This means that, on long cable runs, the cable cross-sectional area may have to be increased. This reduces the resistance, allowing current to 'flow' more easily and reducing the voltage drop across the circuit.

You can calculate the maximum permissible voltage drop as follows.

1. For 230 volt lighting circuits:

$$3\% = \frac{230 \times 3}{100} = 6.9 \text{ volts}$$

2. For other 230 volt circuits:

$$5\% = \frac{400 \times 5}{100} = 11.5 \text{ volts}$$

Therefore, when calculating cable runs we are allowed to have a system that gives as little as 218.5 volts if supplied at 230V. It is better to keep the voltage drops as low as possible because low voltages can reduce the efficiency of the equipment being supplied. Lamps will not be as bright and heaters may not give off full heat. The values for cable voltage drop are given in the accompanying tables of current-carrying capacity in appendix 4 of the *IEE Wiring Regulations*. The values are given in millivolts per ampere per metre (mV/A/m).

You should use the formula below to calculate the actual voltage drop.

$$\text{Voltage drop (Vd)} = \frac{m/V/A/m \times I_b \times L}{1000}$$

Where: mV/A/m   is the value given in the Regulation Tables

I_b is the circuit's design current

L   is the length of cable used in the circuit measured in metres.

Example

Using the same question from the previous example (page 302) you are going to work out the voltage drop on the cable. Assume the cable run is 27m.

Installation method no. is 20 (clipped direct), therefore reference method = C

From Table 4C1, as grouped, C_g = 0.6 (there being five circuits)

From Table 4B1, as 70° PVC, C_a = 1.0 (as 30°C ambient temperature)

Design current (I_b) = 30 A

As the protective device rating (I_n) must be $\geq I_b$, we can take $I_n = 30$ A

$$I_z = \frac{30}{0.6 \times 1.0 \times 0.725} = 69 \text{ A}$$

$I_t \geq I_z$, and from Table 4D5 column 6, it is found that the minimum conductor csa that can be used is 16mm² cable with a tabulated value (I_t) of 85 A.

From Table 4D5 the voltage drop (Vd) is 2.8 mV/A/m

$$\text{Therefore VD} = \frac{2.8 \times 30 \ (I_b) \times 27 \ (\text{length})}{1000} = 2.27 \text{ volts}$$

Actual Vd must not exceed 5 per cent of the supply voltage. 5 per cent of 230 V = 11.5 V. Therefore, for this example the volt drop of 2.27 V is acceptable.

Shock protection

The next stage in cable selection is to check that the protective device will work inside the specified time limits if there is a fault.

The speed of operation of the protective device is extremely important. It will depend on the size of the fault current. This in turn will depend on the impedance of the earth fault loop path.

Table 7.05 gives the maximum disconnection times for various voltages.

System	50V < U_0 ≤ 120V seconds		120V < U_0 ≤ 230V seconds		230V < U_0 ≤ 400V seconds		U_0 > 400V seconds	
	a.c.	d.c	a.c.	d.c.	a.c.	d.c.	a.c.	d.c
TN	0.8	NOTE 1	0.4	5	0.2	0.4	0.1	0.1
TT	0.3	NOTE 1	0.2	0.4	0.07	0.2	0.04	0.1

NOTE 1: Disconnection is not required for protection against electric shock but may be required for other reasons, such as protection against thermal effects.

Table 7.05 Maximum allowed disconnection times

You need to do a calculation to check that the circuit protective device will operate within the required time. You do this by checking that the actual Z_s is lower than the maximum Z_s given in the relevant table for the protective device you have chosen.

Maximum Z_s can also be found from *IEE Wiring Regulations* or manufacturers' data. Use the following formula to calculate the actual Z_s:

$$\text{Actual } Z_s = \frac{Z_e + (\{R_1 + R_2\} \times \text{multiplier (mf)} \times \text{length})}{1000}$$

where:

- Z_e = the external impedance on the supply authority's side of the earth fault loop. You can get this value from the supply authority. Typical maximum values are: TN-C-S (PME) system 0.35 ohms; TN-S (cable sheath) 0.8 ohms; TT system 21 ohms.

- $R_1 + R_2$ = the resistance of the line conductor plus the cpc resistance. You can find values of resistance per metre for $R_1 + R_2$, for various combinations of line and cpc conductors up to and including 50mm in Table 9A of the *IEE On-Site Guide* or Table G1–G2 in the *Amicus Guide* (page 51).

- Tables 9B and 9C of the *IEE On-Site Guide* give you multipliers (mf) to apply to the values given in Table 9A to calculate the resistance under fault conditions. If the conductor temperature rises, resistance in the conductors will increase. Tables 9B and 9C are based on the type of insulation used and whether the cpc is: incorporated in a cable; bunched with the cables; bare and in contact with the cable covering (54.2); a core in a cable; bunched with cables (54.3).

This information, together with the length of the circuit, can now be applied to the formula. Then you can check to see that the actual Z_s is less than the maximum Z_s given in the appropriate tables.

When you have done this, you can prove that the protective device will disconnect the circuit in the time that is specified in the appropriate table. In other words, if the actual Z_s is less than the maximum Z_s you have compliance for shock protection.

Example

This is the same question that was used in the previous two examples. This time you are also required to work out the requirements for shock protection. Assume the cable supplies a domestic cooker and the Z_e is 0.8.

Installation method no. is 20 (clipped direct), therefore reference method = C

From Table 4C1, as grouped, $C_g = 0.6$ (there being five circuits)

From Table 4B1, as 70° PVC, $C_a = 1.0$ (as 30°C ambient temperature)

Design current $(I_b) = 30$ A

As the protective device rating (I_n) must be $\geq I_b$, we can take $I_n = 30$ A

$$I_z = \frac{30}{0.6 \times 1.0 \times 0.725} = 69 \text{ A}$$

$I_t \geq I_z$, and from Table 4D5 column 6, it is found that the minimum conductor csa that can be used is 16mm² cable with a tabulated value (I_t) of 85 A.

From Table 4D5 the voltage drop (Vd) is 2.8 mV/A/m

Therefore Vd = 2.8 × I_b × 27 (length)/1000 = 2.27 volts

Actual Vd must not exceed 5 per cent of the supply voltage. 5% of 230 V = 11.5 V. Therefore, for this example the volt drop of 2.27 V is acceptable.

From Table 41.2, the maximum Z_s for a 30 A BS 3036 fuse is 1.09 Ω.

$$\text{EFLI } (Z_s) = \frac{Z_e + ([R_1 + R_2] \times \text{any multipliers} \times \text{length})}{1000}$$

$Z_e = 0.8$

From Table 9A of the *IEE On-Site Guide*, 16mm² phase + 6mm² cpc

$$R_1 + R_2 = 4.23 \text{ m}\Omega/\text{m}$$

From Table 9C of the *IEE On-Site Guide*, the multiplier for a grouped cable is 1.20

Therefore, $Z_s = 0.8 + \dfrac{(4.23 \times 1.2 \times 27)}{1000}$ giving $Z_s = 0.8 + \dfrac{(137.05)}{1000} = 0.937 \ \Omega$

As the actual value of Z_s is less than the permitted value (1.09 Ω), our cable of 16mm² is acceptable.

Thermal constraints

Now that you have chosen the type and size of cable to suit the conditions of the installation, you must look at **thermal constraints**. This is a check to make sure that the size of the cpc, 'the earth conductor', complies with the *IEE Wiring Regulations*.

If there is a fault on the circuit, which could be a short circuit or earth fault, a fault current of hundreds or thousands of amperes could flow. Imagine that this is a 1mm² or 2.5mm² cable; if this large amount of current was allowed to flow for a short period of time, i.e. a few seconds, the cable would melt and a fire would start.

We need to check that the cpc will be large enough to be able to carry this fault current without causing any heat/fire damage. The formula that is used to check this situation is the **adiabatic equation**. The cpc will only need to carry the fault current for a short period of time, until the protective device operates.

Regulation 543.1.1 states: 'The cross-sectional area of every protective conductor shall be calculated in accordance with Regulation 543.1.3 (adiabatic equation) or selected in accordance with Regulation 543.1.4 (Table 54.7).' You will see that Regulation 543.1.4 asks that reference should be made to Table 54.7. This table shows that, for cables 16mm² and below with the cpc made from the same material as the line conductor, the cpc should be the same size as the line conductor. A line conductor between 16mm² and 35mm² requires a cpc to be 16mm². A line conductor above 35mm² requires a cpc to be at least half the cross-sectional area.

Multicore cables have cpcs smaller than their respective line conductors, except for 1mm² which has the same-sized cpc. Regulation 543.1.2 of BS 7671 gives two options, calculation or selection. If we were to apply option (ii) of this Regulation then selection would make these cables contravene the Regulations. Clearly, it is not intended that composite cables should have their cpcs increased in accordance with the table, therefore calculation by the adiabatic equation required in option (i) of the same Regulation should be applied.

On the job: Intermittent fault

You have been asked by your employer to investigate an 'earthing' problem that has been reported by a customer. The customer concerned has just moved into a large 80 year-old house that will allow him to work from home as a computer software designer. The house has large gardens, a separate garage, three outside security lights and a gas-fired central-heating system. The customer will be working from one of the bedrooms on his computers.

The customer has reported that he is getting intermittent RCD trips on the main consumer unit. The customer has noticed that the trips seem to occur only at night and only when it has been raining. Rain during the day does not cause the same problems. There is also a circuit out to the garage that has its own RCD and this does not trip.

What do you think could be causing the problem?

The adiabatic equation referred to in the introduction enables a designer to check the suitability of the cpc in a composite cable. If the cable does not incorporate a cpc, a cpc installed as a separate conductor may also be checked.

The equation is as follows:

$$S = \frac{\sqrt{I^2 \times t}}{k}$$

where:

S = the cross-sectional area of the cpc in mm²

I = the value of the earth fault current

t = the operating time of the disconnecting device in seconds

k = a factor depending on the conductor and its insulating material.

In order to apply the adiabatic equation, we first need to calculate the value of I (fault current) from the following equation:

$$I = \frac{U_o}{Z_s}$$

where:

U_o is the nominal supply voltage

Z_s is the earth fault loop impedance.

If you are using method (i) from Reg 543.1.2 and applying the adiabatic equation, you must find out the time/current characteristics of the protective device. A selection of time/current characteristics for standard overcurrent protective devices is given in Appendix 3 of *IEE Wiring Regulations*. You can get the time (t) for disconnection to the corresponding earth fault current from these graphs.

If you look at the time/current curve you will find that the scales on both the time (seconds) scale and the prospective current (amperes) scale are logarithmic and the value of each subdivision on the graph depends on the major division boundaries into which it falls.

For example, on the current scale all the subdivisions between 10 and 100 are in quantities of 10, while the subdivisions between 100 and 1000 are in quantities of 100, and so on. This also occurs with the time scale subdivisions, thus between 0.01 and 0.1 they are in hundredths and between 0.1 and 1 are in tenths etc.

As an example, if you look at the graph and table shown in Figure 7.08 you will see that for a BS 88 fuse with a rating of 32 A, a fault current of 200 A will cause the fuse to clear the fault in 0.5 seconds.

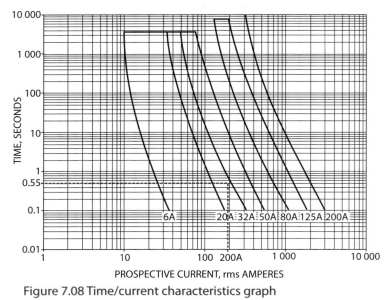

Figure 7.08 Time/current characteristics graph

TIME/CURRENT CHARACTERISTICS FOR FUSES TO BS 88: PART 2 AND PART 6					
Fuse rating	Current for time (seconds)				
	0.1	0.2	0.4	1	5
6 A	36 A	31 A	27 A	23 A	17 A
20 A	175 A	150 A	130 A	110 A	79 A
32 A	320 A	260 A	220 A	120 A	125 A
50 A	540 A	450 A	380 A	310 A	220 A
80 A	1100 A	890 A	740 A	580 A	400 A
125 A	1800 A	1500 A	1300 A	1050 A	690 A
200 A	3000 A	2500 A	2200 A	1700 A	1200 A

In addition to this graph, the IEE has produced a small table showing some of the more common sizes of protective devices and the fault currents for a given disconnection time.

Next, you need to select the k factor using Tables 54.2 to 54.6. The values in these tables are based on the initial and final temperatures shown in each table. This is where you may need to refer to the cable's operating temperature shown in the cable tables.

Now you must substitute the values for I, t and k into the adiabatic equation. This will give you the minimum cross-sectional area for the cpc. If your calculation produces a non-standard size, you must use the next largest standard size.

From a designer's point of view, it is advantageous to use the calculation method as this may lead to savings in the size of cpc.

Example

This example is similar to those used in the previous three examples, but this time you also need to complete calculations for thermal constraints.

Installation method no. is 1 (clipped direct), therefore reference method = 1

From Table 4C1, as grouped, $C_g = 0.6$ (there being five circuits)

From Table 4B1, as 70° PVC, $C_a = 1.0$ (as 30°C ambient temperature)

Design current $(I_b) = 30$ A

As the protective device rating (I_n) must be $\geq I_b$, we can take $I_n = 30$ A

$$I_z = \frac{30}{0.6 \times 1.0 \times 0.725} = 69 \text{ A}$$

$I_t \geq I_z$, and from Table 4D5 column 6, it is found that the minimum conductor csa that can be used is 16mm^2 cable with a tabulated value (I_t) of 85 A.

From Table 4D5 the voltage drop (Vd) is 2.8 mV/A/m

$$\text{Therefore Vd} = \frac{2.8 \times 30 \ (I_b) \times 27 \ (\text{length})}{1000} = 2.27 \text{ volts}$$

Actual Vd must not exceed 5 per cent of the supply voltage. 5% of 230 V = 11.5 V. Therefore, for this example the volt drop of 2.27 V is acceptable.

From Table 41.2, the maximum Z_s for a 30 A BS 3036 fuse is 1.09 Ω.

$$\text{EFLI } (Z_s) = Z_e + \frac{([R_1 + R_2] \times \text{any multipliers} \times \text{length})}{1000}$$

$$Z_e = 0.8$$

From Table 9A of the *IEE On-Site Guide*, 16mm^2 phase + 6mm^2 cpc

$$R_1 + R_2 = 4.23 \text{ m}\Omega/\text{m}$$

From Table 9C of the *IEE On-Site Guide*, the multiplier for a grouped cable is 1.20

$$\text{Therefore, } Z_s = 0.8 + \frac{([4.23] \times 1.2 \times 27)}{1000} \text{ giving } Z_s = 0.8 + \frac{137.05}{1000} = 0.937 \ \Omega$$

As the actual value of Z_s is less than the permitted value (1.09 Ω), our cable of 16 mm² is acceptable.

To apply the adiabatic equation, $I = \dfrac{U_o}{Z_s} = \dfrac{230}{0.937} = 245.46$ A

From Fig. 3.2A of Appendix 3 of BS 7671, t = 0.3 second. From Table 54.3 for 70° PVC, k = 115

$$S = \frac{\sqrt{I^2 \times t}}{k} = \frac{\sqrt{245.46^2 \times 0.3}}{115} = \frac{77.62}{115} = 1.17 \text{ mm}^2$$

Therefore our 6mm² cpc is acceptable.

Diversity

Appendix 1, Table 1B, of the *IEE On-Site Guide* contains a method allowing diversity to be applied depending upon the type of load and installation premises. The individual circuit and load figures are added together to determine the total 'assumed current demand' for the installation. This value can then be used as the starting point to determine the rating of a suitable protective device and the size of cable, considering any influencing factors in a similar manner to that applied to final circuits.

Note: Remember that the calculation of maximum demand is not an exact science and a suitably qualified electrical engineer may use other methods of calculating maximum demand.

Example

A 230 volt domestic installation consists of the following loads:

- 15 × filament lighting points
- 6 × fluorescent lighting points, each rated at 40 watts
- 4 × fluorescent lighting points each rated at 85 watts
- 3 × ring final circuits supplying 13 A socket outlets
- 1 × A3 radial circuit supplying 13 A socket outlets for adjoining garage
- 1 × 3 kW immersion heater with thermostatic control
- 1 × 13.6 kW cooker with a 13 A socket outlet incorporated in the control unit.

Determine the maximum current demand for determining the size of the sub-main cable required to feed this domestic installation. The circuit protection is by the use of BS 1361 fuses.

Lighting

Tungsten light points (see Table 1A of the IEE On-Site Guide)

15 × 100 W minimum = 1500 W

Fluorescent light points (see Note 2 below Table 1A of the *IEE On-Site Guide*)

6 × 40 W with multiplier of 1.8 (40 × 1.8 = 72) = 432 W

4 × 85 W with multiplier of 1.8 (85 × 1.8 = 153) = 612 W

Total = 2544 W

Using Item 1 in Table 1B of the *IEE On-Site Guide*, we can apply diversity as being 66 per cent of the total current demand. Therefore:

66% of 2544 = 1679 W   and since $I = \dfrac{P}{V}$ this gives us $\dfrac{1679}{230} = 7.3$ A

Power (See Item 9 in Table 1B of the *IEE On-Site Guide*)

3 × ring final circuits

1 × ring at 100% rating (30 A) = 30 A

2 × ring at 40% rating (40% of 30 A) = 24 A

20 A A3 radial circuit
1 × radial at 40% rating (40% of 20 A) = 8 A

3 kW immersion heater (see Item 6 in Table 1B of the *IEE On-Site Guide*)
3 kW heater with no diversity   I = P/V = 3000/230 which gives us = 13 A

13.6 kW cooker with socket outlet (see Table 1A and Item 3 in Table 1B of the *IEE On-Site Guide*)

$I = \dfrac{P}{V}$ giving $\dfrac{13600}{230}$ which gives us a total rated current of 59A

The first 10 A, plus 30% of the remainder of the overall rated current, plus 5A for the socket.

59 A − 10 A = 49 A   and therefore 30% of 49 A = 14.7 A

The allowable total cooker rating is therefore   10A + 14.7 A + 5 A = 29.7 A

Our total assumed current demand is therefore:

7.3 + 30 + 24 + 8 + 13 + 29.7   giving a total of **112 A**

Determining the size of conduit

On completion of this topic area the candidate will be able to determine the size of conduit required relative to the size and number of cables to be installed.

BS 7671 requirements

The only requirements BS 7671 has for the installation of cables into conduit, and for that matter trunking, is that the wiring system should be selected and erected so as to minimise damage to the sheath, insulation of cables, insulated conductors and their terminations during installation, use and maintenance.

Allowances for grouping of cables within conduit are made at the design stage when the cable size is calculated, so no additional allowance for grouping needs to be made at this point.

Tables in appendix 5 of the *IEE On-Site Guide* give information and data to be used in the sizing of conduit. Cable factors are given for various sizes and types of cable. From this information a total cable factor for the cables to be contained within conduit can be found. Using this table of conduit factors, it is simply a case of selecting the conduit size that matches.

However straightforward this seems, before we can calculate the size of the conduit required we need to determine the conduit route. We also need to know the number of bends and draw-in boxes being established.

Example

Calculate the size of plastic conduit to be used if the total length of the run is 10m, which comprises two 3m lengths each with one bend in and one 4m straight length. The PVC cables to be installed are two 10mm² line and neutral and one 6mm² cpc.

Figure 7.09 Diagram to help calculate the size of plastic conduit

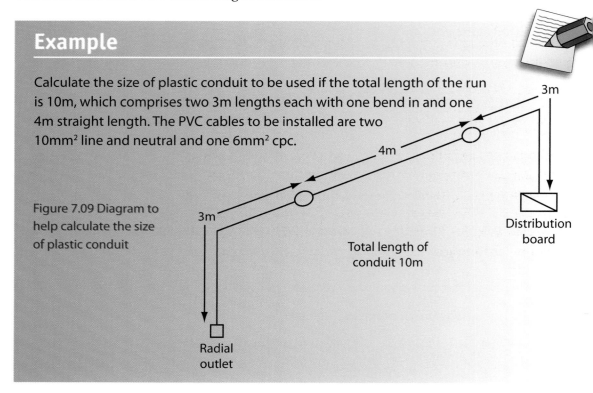

From the *IEE On-Site Guide*, appendix 5, Table 5C, we can obtain cable factors for use in conduit in long straight runs over 3m or runs of any length incorporating bends.

The cable factor for 10mm² is 105, and for 6mm² is 58

Therefore: Total cable factor = (2 × 105) + 58 = **268**

Using Table 5D, we are able to find the size of conduit required. This is done one section at a time. First we have to consider a 3m run of conduit with one bend in it, then a 4m straight run of conduit. On the left-hand side of the table you will find the length of run; check this against the conduit factor equal to or larger than the total cable factor.

For the 3m run with one bend we find a conduit factor for 20mm conduit at 270. 16mm at 167 would be too small and 25mm conduit at a factor of 487 would be unnecessary. If we now do the same for a 4m straight run of conduit we find that **20mm** conduit with a factor of **286** is the size required.

However, with different conduit runs and cable sizes, the requirement for the conduit with the bends may have resulted in a larger size conduit being needed. That is why it is essential to size the conduit in different sections to establish the overall requirement.

Did you know?

Trunking is sized in exactly the same way as conduit, using the relevant tables in appendix 5 of the *IEE On-Site Guide*. However, for sizes of trunking not included in appendix 5, a space factor of 45 per cent should not be exceeded when calculating the size of the trunking

Earthing and bonding

On completion of this topic area the candidate will be able to explain the terms 'earthing' and 'bonding' and describe earthing systems.

We covered definitions of Earthing at Level 2. BS 7671 describes it as a connection of the exposed conductive parts of installation to the main earthing terminal of the installation.

Bonding is not defined, but a bonding conductor is a protective conductor providing equipotential bonding, thereby maintaining various exposed conductive parts and extraneous conductive parts at substantially the same potential. Protective equipotential bonding and automatic disconnection of the supply, previously referred to as EEBADS, is the main method of fault protection. At Level 2 we covered the main earthing systems, their principles and operations.

Protective devices

On completion of this topic area the candidate will be able to describe the applications and limitations of overcurrent protective devices, including fuses, circuit breakers and residual current circuit breakers (RCB).

Overcurrent protection

We covered overcurrent protection devices, such as circuit breakers and fuses, at Level 2. These are designed to operate within specific limits, disconnecting the supply automatically in the event of a current overload or fault current (such as short circuits or earth faults).

Fuse systems

We covered fuses at Level 2. Fuses are split into several different types. These are:

- BS 3036 rewirable fuses
- BS 1361/1362 cartridge fuses
- BS 88 high breaking capacity (HBC) fuse
- Type 'D' and Neozed fuses.

Miniature Circuit Breakers (MCBs)

Miniature Circuit Breakers were introduced in chapter 5 on page 226. That section explained the purpose and uses of circuit breakers. In this section we will look at the principles behind the operation of circuit breakers.

Definition

Overload – an overcurrent occuring in a circuit which is electrically sound

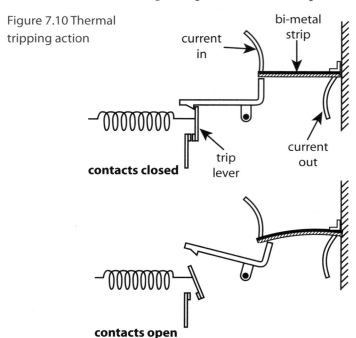

Figure 7.10 Thermal tripping action

The circuit breaker is so arranged that normal currents will not affect the latch, whereas excessive currents will move it to operate the breaker. There are two basic methods by which overcurrent can operate or 'trip' the latch.

Thermal tripping

The load current is passed through a small heater coil wrapped around a bi-metal strip inside the MCB housing; the heat created depends on the current it carries. This heater is designed to warm the bi-metal strip either directly (the current passes through the bi-metal strip which in effect is part of the electrical circuit) or indirectly (where a coil of current-carrying conductor is wound around the bi-metal strip) and the excess current warms the bi-metal strip.

The bi-metal strip is made of two different metals, normally brass and steel (brass expanding more than steel). These two dissimilar metals are securely riveted or welded together along their length. The rate of expansion of the two metals is different so that when the strip is warmed, it will bend and will trip the latch.

The bi-metal strips are arranged so that normal currents will not heat the strip to tripping point.

Remember

If current increases beyond the rated value, the heater increases in temperature and thus the bi-metal strip is raised in temperature, trips the latch and opens the contacts

Magnetic tripping

The principle used here is the force of attraction, which can be set up by the magnetic field of a coil carrying the load current. At normal current the magnetic field is not strong enough to attract the latch, but overload currents operate the latch and trip the main contacts.

The magnetic field is set up by a current in the flexible strip that attracts the strip to the iron. This releases the latch. This is often used in miniature circuit breakers and combined with a thermal trip.

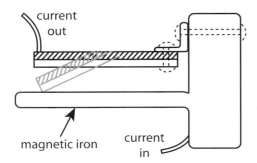

Figure 7.11 A simple attraction type of magnetic trip

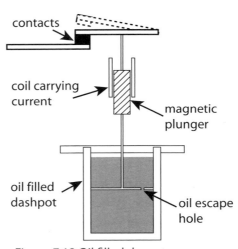

Figure 7.12 Oil filled daspot

The **oil dashpot** solenoid type is used on larger circuit breakers. The time lag is adjustable by varying the size of the oil-escape hole in the dashpot position. Current rating is adjustable by vertical movement of the plunger.

Combined tripping

As we have discussed, circuit breakers use two basic methods in their operation. There is always some time delay in the operation of a thermal trip, as the heat produced by the load current must be transferred to the bi-metal strip. Thermal tripping is therefore best suited to small overloads of comparatively long duration.

Magnetic trips are fast acting for large overloads or short circuits. The two methods are often combined to take advantage of the best characteristics of each.

They are then classified according to their instantaneous tripping current, i.e. the current at which they will operate within 100 ms. Types 1, 2, 3 and 4 are circuit breakers to BS 3871, and Types B, C and D are used with BS EN 60898.

If you were to take, for example, a 20 A Type B MCB this would have an instantaneous tripping current of three to five times I_n, i.e. between $3 \times 20 = 60$ A and $5 \times 20 = 100$ A. Whereas a 20 A type D MCB could have an instantaneous tripping as high as 20 times, I_n (400A). They are clearly very different devices.

Guidance on the selection of circuit breakers to BS 3871 or BS EN 60898 is given in Table 7.2B of the *IEE On-Site Guide* as shown below, where I_n is the value of current that can be carried indefinitely by the device.

MCB type	Instantaneous trip current	Application
1 B	2.7 to $4 \times I_n$ 3 to $5 \times I_n$	Used in domestic and commercial installations with little or no switching surge
2 C 3	4.0 to $7.0 \times I_n$ 5 to $10 \times I_n$ 7 to $10 \times I_n$	General use in commercial/industrial installations, where the use of fluorescent lighting, small motors etc. can produce switching surges that would operate a Type 1 or B circuit-breaker. Type C or 3 may be necessary in highly inductive circuits such as banks of fluorescent lighting.
4 D	10 to $50 \times I_n$ 10 to $20 \times I_n$	Suitable for transformers, X-ray machines and industrial welding equipment etc., where high in-rush currents may occur.

Table 7.06 MCB selection

Type 1 and B are the more sensitive devices. Please note that, as part of the harmonisation of electrical standards, BS EN 60898 Types B, C and D will eventually replace the current BS 3871 types.

Disadvantages of MCBs	Advantages of MCBs
• They have mechanical moving parts. • They are expensive. • They must be regularly tested. • Ambient temperature can change performance.	• They have factory-set operating characteristics which cannot be altered. • They will maintain transient overloads and trip on sustained overloads. • Easily identified when they have tripped. • The supply can be quickly restored.

Table 7.07 Advantages and disadvantages of MCBs

Fusing factor

It is evident that each of the protective devices discussed in the previous section provide different levels of protection, e.g. rewirable fuses are slower to operate and less accurate than MCBs. In order to classify these devices it is important to have some means of knowing their circuit breaking and 'fusing' performance. This is achieved for fuses by the use of a fusing factor.

$$\text{Fusing factor} = \frac{\text{Fusing current}}{\text{Current rating}}$$

This is the ratio of the fusing current. It is the minimum current that will cause the fuse to blow and the stated current rating of the fuse or MCB (which is the maximum current that the fuse can sustain without blowing). Fusing currents can be found in appendix 3 of BS 7671. These tables are logarithmic and the scales increase by factors of 10, not uniformly as may be expected, and therefore the interpretation of these scales will require some practice. The rating of the fuse is the current it will carry continuously without deterioration.

Fusing factors for the above devices can generally be grouped as follows:

- BS 3036: 1.8–2.0
- BS 1361: 1.6–1.9
- BS 88: 1.25–1.7
- MCBs: up to 1.5.

The higher the fusing factor, the less accurate – and therefore the less reliable – the device selected will be.

You may, while looking at fuses, have noticed a number followed by the letters kA stamped on to the end cap of an HBC fuse or printed on to the body of a BS 1361 fuse. This is known as the breaking capacity of fuses and circuit breakers. When a short circuit occurs, the current may, for a fraction of a second, reach hundreds or even thousands of amperes. The protective device must be able to break or make such a current without damage to its surroundings by arcing, overheating or the scattering of hot particles. The breaking capacities of MCBs are indicated by an 'M' number, e.g. M6 or a figure within a rectangular box like this 6000. This means that the breaking capacity is 6 kA or 6000 A. The breaking capacity will be related to the prospective short circuit current.

Fuse and circuit breaker operation

In this part the following symbols will be referred to.

I_z	The maximum current-carrying capacity of a cable for continuous service, under the particular installation conditions concerned (appendix 4 of BS 7671)
I_n	The nominal current or current setting of the device protecting the circuit against overcurrent (appendix 3 of BS 7671)
I_2	The operating current, i.e. the fusing current or tripping current for the conventional operating time (appendix 3 of BS 7671)
I_b	The design current of the circuit, i.e. the current intended to be carried by the circuit in normal service
I_{ef}	Earth-fault current
I_{sc}	Short-circuit current
t	Time in seconds
k	Material factor found from Tables 43.1 and 54.2 to 54.6 of BS 7671
s	Conductor cross-sectional area in mm^2

Table 7.08

Consider a protective device (fuse or MCB) rated at 20 A. This value of current can be carried indefinitely by the device and is known as its nominal current setting, I_n. The value of current that will cause the device to operate, I_2, will be larger than I_n and will be dependent on the device's fusing factor.

This fusing factor figure, when multiplied by the nominal setting I_n, will give the value of operating current I_2.

For fuses to BS 1361 and BS 88 and circuit breakers to BS 3871 and BS/EN 60898, the fusing factor has been approximated to 1.45. Therefore our 20 A device would operate when the current reached 29 A (1.45 × 20).

BS 7671 Regulation 433.1 requires co-ordination between conductor and protective device when an overload occurs such that:

(i)   The nominal current or setting (I_n) of the protective device is not less than the design current (I_b) of the circuit; therefore $I_b \leq I_n$ (the symbol $\leq$ means less than or equal to). For example, if the circuit design current was 20 A then the protective device would need to have a rating of at least 20 A.

(ii) The nominal current or current setting (I_n) of the protective device does not exceed the lowest value of the current-carrying capacities (I_z) of any of the conductors in the circuit. Therefore $I_n \leq I_z$.

The formulae can be combined so that $I_b \leq I_n \leq I_z$.

In simple terms this means that the design current of the circuit must be less than or equal to the protective device rating, which in turn must be less than or equal to the current-carrying capacity of the cable.

(iii) The current (I_2) causing effective operation of the protective device does not exceed 1.45 times the lowest of the current-carrying capacities (I_z) of any of the conductors of the circuit.

Therefore:

$$I_2 \le I_z \times 1.45$$

We now have:

$$I_2 = I_n \times 1.45 \text{ and } I_2 \le I_z \times 1.45$$

We can now say:

$$I_n \times 1.45 \le I_z \times 1.45$$

We can now remove the 1.45 from the equation, as it is common to both the left-hand side (LHS) and the right-hand side (RHS) of the equation.

Therefore, this leaves:

$$I_n \le I_z.$$

This means that the current rating of the protective device must be less than or equal to the current-carrying capacity of the cable.

This is only true for situations where BS 1361 and BS 88 fuses and circuit breakers complying with BS 3871 and BS/EN 60898 are in use.

Now considering our 20 A device, if the cable is rated at 20 A, then condition (ii) is satisfied. The fusing factor is 1.45; therefore condition (iii) is also satisfied, because:

$$I_2 = I_n \times 1.45 \text{ which is } 20 \text{ A} \times 1.45 = 29 \text{ A}$$

This is the same as 1.45×20 A (I_z) cable rating because:

$$1.45 \times 20 \text{ A} = 29 \text{ A (rating of the cable)}.$$

Problems with BS 3036 rewirable fuses

This type of fuse may have a fusing factor as high as 2. Using the previous example we would see that:

$$I_n \times 2 \le I_z \times 1.45 \qquad \text{would not be true.}$$

In order to comply with condition (iii), I_n should be less than or equal to 0.725 times I_z. (This figure is derived from $1.45 \div 2 = 0.725$.)

For example, if a cable is rated at 20 A, then I_n for a BS 3036 device should be less than or equal to:

$$0.725 \times 20 = 14.5 \text{ A}$$

As the fusing factor is 2, the operating current I_z would now be:

$$2 \times 14.5 = 29 \text{ A}$$

This conforms to condition (iii) if the fuse size was 14.5 A. However, if the design current required was the same as in the previous example, and hence, the fuse size (20 A) was the same, then this condition would not apply – the cable size (and hence its rating) would need to be increased. Be wary of using this type of fuse!

If all these requirements are met then this will ensure that the conductor insulation is undamaged when an overload occurs.

Discrimination

Discrimination was covered at Level 2. Discrimination is said to have taken place when a smaller rated local device operates before the larger device in a fault. This is generally only a problem when a system uses a mixture of devices. Obviously a particular type of fuse will discriminate against a similar type of fuse if it is of larger rating.

Activity

Imagine you are going to install a new shower circuit into a typical domestic property. Produce a written estimate of the expected installation time. This can be done by:

- breaking down the job into various tasks
- putting them in the correct sequence
- allotting a time to each task
- totalling-up the time for the job.

FAQ

Q Do I have to be Part P registered to carry out electrical installation work?

A Strictly speaking no, you do not have to be Part P registered to work as an electrician. However, if your work comes within the scope of Part P (for example, work in domestic kitchens, bathrooms and outdoors, although there are others), then it would probably be more cost-effective and efficient to be Part P registered. If not, you would have to pay someone else to certificate your work for you.

Q Now that Part P has been introduced, does that mean I have to inspect, test and certificate my work?

A Actually, under BS 7671 it has been a requirement to do this for many years, so you and your colleagues should have been doing this anyway. In practical terms the only real difference is that work within the scope of Part P has to be reported to the local authority.

Q Is BS 7671 law?

A In England and Wales, no. However, failure to comply with BS 7671 means that it is likely that you have failed to comply with the Electricity at Work Regulations 1989, which is law.

Q My boss insists that our installations exceed the standards laid down in BS 7671. Why is this?

A It's always worth remembering that the standards laid down in BS 7671 are considered to be minimum standards. Individual company and client requirements may demand higher standards and as long as they produce an installation which is no less safe than BS 7671 requires (and they're prepared to pay for it) that's OK.

Q I've been told that joint boxes are a 'non-preferred' method of terminating cables. Why is that?

A BS 7671 states that every joint has to be 'accessible for inspection, testing and maintenance' (unless it is designed to be buried, i.e. mechanically strong such as a soldered or compression joint). In this case the problem is not with the joint box itself, but the location you intend to place it. You cannot bury a joint box in the fabric of a building as it will then be impossible to inspect.

Knowledge check

1. List four types of installation activities that may be carried out by an electrician.

2. What is meant by the term 'external influences' in relation to an electrical installation?

3. Why is it important to estimate the anticipated time taken to do a job?

4. List the types of installation work that would come within the scope of Part P of the Building Regulations.

5. Which type of diagram (e.g. site plan, block diagram, location diagram, circuit diagram) would be most useful when carrying out the following activities:

    (a) fault finding on an item of electrical equipment

    (b) running a cable between two buildings on a site

    (c) isolating part of the building from a source of supply

    (d) installing a cable to supply a machine.

6. List the standard voltages for generation, transmission, distribution and utilisation of electricity in the UK.

7. Describe how the EEBADS system of fault protection works.

8. List the relative advantages and disadvantages of steel and plastic conduit.

9. When terminating the sheathing of a conductor at an accessory, what must be done?

10. List different methods by which you can connect to terminals.

11. Explain the meaning of the term 'diversity'.

12. State the four main correction factors which affect the current-carrying capacity of a conductor.

13. Draw a simple, labelled diagram of the earth fault loop path.

14. What problem could you have when using a type B MCB to protect an X-ray machine?

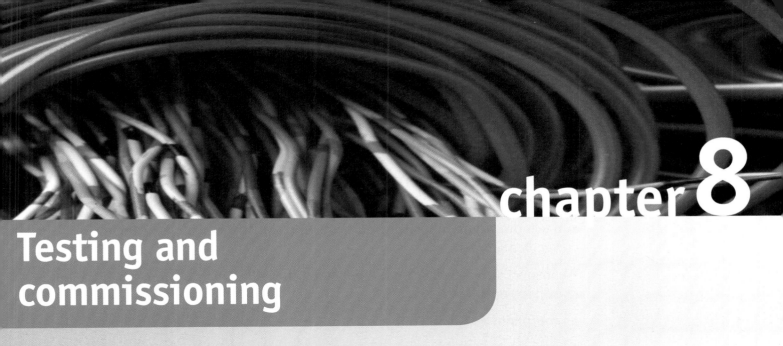

chapter 8

Testing and commissioning

Unit 2 outcome 2

It is important that electricians are not just able to construct; they should also be able to recognise faults and take action to help prevent them. As such, using the correct means to test and inspect material is vital. Not all faults will be easily visible. Some will be concealed and only take effect over a long period of time. Regular inspection, tests and maintenance will limit such faults. Testing should take place not only at the completion of work but also during the installation process.

On completion of work, you will need to have the work commissioned and certified. This is to make sure that it matches the specifications and meets all requirements of safety. In many cases you will not be the one carrying out these tasks. However, it is important that you know what they involve.

On completion of this outcome the candidate will be able to:

- state the purpose of inspection and commissioning and the factors to be considered
- identify the purpose and conditions of periodic inspection
- list relevant sources of information to facilitate testing and inspection
- list items associated with visual inspection prior to commissioning and identify instruments suitable for use
- explain the importance of test instruments, calibration and documentary evidence
- describe how to carry out common tests
- explain the need to comply with test values
- describe the certification process
- describe the requirements of testing and procedures for dealing with documentation and clients.

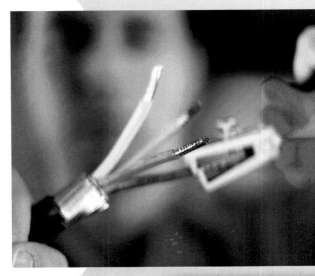

The purpose of inspection and commissioning

On completion of this topic area the candidate will be able to state the purpose of inspection and commissioning and the factors to be considered.

Initial verification procedures

In order to make sure that this work is carried out satisfactorily the inspection and test procedure must be carefully planned and carried out and the results correctly documented.

We inspect and commission material after the completion of work for three key reasons to ensure:

- compliance with BS 7671
- compliance with the project specification (commissioning)
- that it is safe to use.

Compliance with BS 7671

BS 7671 Part 6 states that every electrical installation shall, either during construction, on completion, or both, be inspected and tested to verify, so far as is reasonably practicable, that the requirements of the Regulations have been met. In carrying out such inspection and test procedures, precautions must be taken to ensure no danger is caused to any person or livestock and to avoid damage to property and installed equipment.

BS 7671 requires that the following information be provided to the person carrying out the inspection and test of an installation:

- the maximum demand of the installation expressed in amperes per phase
- the number and type of live conductors at the point of supply
- the type of earthing arrangements used by the installation, including details of equipotential bonding arrangements
- the type and composition of circuits, including points of utilisation, number and size of conductors and types of cable installed (this should also include details of the 'reference installation method' used)
- the location and description of protective devices (fuses, circuit breakers etc.)
- details of the method selected to prevent danger from shock in the event of an earth fault, e.g. earthed equipotential bonding and automatic disconnection of supply
- the presence of any sensitive electronic devices.

It is important to remember that periodic inspection and testing must be carried out on installations to ensure that the installation has not deteriorated and still meets all requirements. Tests will also need to be carried out in the event of minor alterations or additions being made to existing installations.

Did you know?

The requirements for periodic inspection and testing are given in BS 7671 Chapter 62

Remember

All electrical items *must* be tested before finally being put into service

Did you know?

For commercial or industrial installations, the requirements of the Electricity Supply Regulations 1988 and the Electricity at Work Regulations 1989, both of which are statutory instruments, should also be taken into account

Compliance with the project specification

Once the installation is complete, we need to test it against the original specification for the work. This is in order to check that the finished installation matches the requirements laid out by the customer and is fit for use in the environment where it will be used. The tasks involved in checking compliance with project specification are as follows.

- **Pre-commissioning** – this involves a full inspection of the installation and the carrying out of all tests required before the installation is energised (continuity, polarity and insulation resistance).

- **Commissioning** – includes all tests which require power to be available, e.g. the measurement of the earth-fault loop impedance and functional testing of residual current devices (RCDs).

As commissioning involves the initial energising of an installation, this task has to be carried out in a controlled manner with the knowledge of everyone involved. This means that all other persons working on site at the time of the energising must be informed that power will be applied to the installation, so that all precautions can be taken to prevent danger.

The commissioning process is intended to confirm that the installation complies with the designer's requirements. As such, commissioning includes the functional testing of all equipment, isolation, switching, protective devices and circuit arrangements.

Safe to use

The final act of the commissioning process is to ensure the safe and correct operation of all circuits and equipment which have been installed, and that the customer's requirements have been met. This will also confirm that the installation works and, more importantly, will work under fault conditions; after all, it is under fault conditions that lives and property will be at risk.

The testing of electrical installations can cause some degree of danger; it is the responsibility of the person carrying out the tests to ensure the safety of themselves and others. *Health and Safety Executive Guidance Note GS38 (Electrical test equipment for use by electricians)* details relevant safety procedures and should be observed in full.

When using test instruments the following points will help to achieve a safe working environment.

- The person carrying out the tests must have a thorough understanding of the equipment being used and its rating.

- The person carrying out the tests must ensure that all safety procedures are being followed, e.g. erection of warning notices and barriers where appropriate.

- The person carrying out the tests should ensure that all instruments being used conform to the appropriate British Standard, i.e. BS EN 61010 or older instruments manufactured to BS 5458 (provided these are in good condition and have been recently calibrated).

Remember

Many industrial processes have very complicated control systems. This may require the supplier of the control panel as well as the client to be present when commissioning is carried out

Safety tip

Particular attention should be paid when using instruments capable of generating a test voltage in excess of 50 volts, e.g. insulation resistance testers. If the live terminals of such an instrument are touched a shock will be received.

- The person carrying out the tests should check that test leads including probes and clips are in good condition, are clean and have no cracked or broken insulation. Where appropriate the requirements of GS38 should be observed, including the use of fused test leads.

Competence of the inspector

A final consideration when carrying out inspection and tests is the competence of the inspector. Any person undertaking these duties must be skilled and experienced and have sufficient knowledge of the type of installation. It is the responsibility of the inspector to:

- ensure no danger occurs to people, property and livestock
- confirm that test and inspection results comply with the requirements of BS 7671 and the designer's requirements
- express an opinion as to the condition of the installation and recommend remedial works
- make immediate recommendations, in the event of a dangerous situation, to the client to isolate the defective part.

Factors to be considered
Isolation

Isolation is covered in detail at Level 2.

Test instruments

All test equipment must be regularly checked to make sure it is in good and safe working order.

If you are checking the equipment yourself, the following points should be noted. If you have any doubt about an instrument or its accuracy, ask for assistance; test instruments are very expensive and any unnecessary damage caused by ignorance should be avoided.

Voltage indicating devices

Instruments used solely for detecting a voltage fall into two categories.

- Some detectors rely on an illuminated lamp (test lamp) or a meter scale (test meter). Test lamps are fitted with a 15 watt lamp and should not give rise to danger if the lamp is broken. A guard should also protect it.

Remember

You must ensure that your test equipment is calibrated; this indicates that the instrument is working properly and providing accurate readings. If you do not do this, test results could be void

- Other detectors use two or more independent indicating systems (one of which may be audible) and limit energy input to the detector by the circuitry used. An example is a two-pole voltage detector, i.e. a detector unit with an integral test probe, an interconnecting lead and a second test probe.

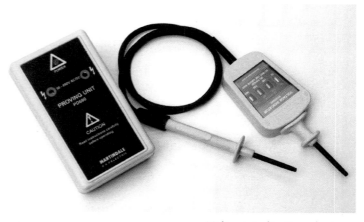

Voltage indicating device

Both these detectors are designed and constructed to limit the current and energy that can flow into the detector. This limitation is usually provided by a combination of the circuit design, using the concept of protective impedance, and current-limiting resistors built into the test probes. The detectors are also provided with in-built test features to check the functioning of the detector before and after use. The interconnecting lead and second test probes are not detachable components. These types of detector do not require additional current-limiting resistors or fuses to be fitted, provided that they are made to an acceptable standard and the contact electrodes are shrouded.

It is recommended that test lamps and voltage indicators are clearly marked with the maximum voltage which may be tested by the device and any short time rating for the device, if applicable. This rating is the recommended maximum current that should pass through the device for a few seconds, as these devices are generally not designed to be connected for more than a few seconds.

Operating devices

It is not an adequate precaution simply to isolate a circuit electrically before commencing work on it. **It is vital to ensure that once a circuit or item of electrical equipment has been isolated, it cannot inadvertently be switched back on.**

A good method of providing full electrical and mechanical isolation is to lock off the device (or distribution board containing the device) with a padlock (see Figure 8.01), with the person working on the isolated equipment keeping the key in his or her pocket. A skilled person should be the only one allowed to carry out this or similar responsible tasks concerning the electrical installation wiring.

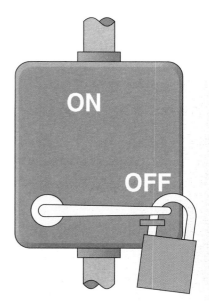

Figure 8.01 Device locked off with padlock

Systems and equipment

Prior to commencing work on a new electrical installation, a full specification is prepared. The specification should set out the detailed design and provide sufficient information to enable competent persons to construct the installation and to commission it.

Precise details of all equipment installed should be obtained from the manufacturers or suppliers to check that the required standards have been met, to ensure satisfactory methods of installation have been used and to provide the information necessary to confirm its correct operation.

The Health and Safety at Work Act 1974 and the Construction (Design and Management) Regulations 1994 both state the requirements for the provision of such information to be included in the operation and maintenance manual prepared for the project. This should also contain a description of how the installation is to operate and include copies of distribution board details, identifying the circuits, the protective devices, the wiring system and the installation methods as well as the technical data for all items installed, such as switchgear, luminaires and any special control systems that may have been incorporated.

Sources of information

Initial verification is intended to confirm that the installation complies with the designer's requirements and has been constructed, inspected and tested in accordance with BS 7671.

Sources of information will be covered in greater detail later in this chapter on pages 341–345.

Marking and labelling

We covered the importance of marking and labelling at Level 2. To recap quickly, labelling should show the origin of every installation and indicate:

- differences in voltages
- earthing and bonding connections
- residual current devices.

Contact with relevant parties

The testing and inspection procedure needs to be a planned activity as it will affect others living in, working in or passing through the installation. If disruption to others is to be kept to a minimum they will need to be advised of when, and in what areas, the activity will be taking place. They will also need to know the amount of time that the supply may be disrupted. For other contractors it may be necessary to provide them with temporary supplies, derived from other sources, to enable them to continue their work activities.

Following the inspection and testing of all new installations, alterations and additions to existing installations an Electrical Installation Certificate, together with a schedule of test results, should be given to the person ordering the work. Included on the test certificate is a recommendation as to when the installation should be re-tested; this will need to be shown to the client to draw it to his or her attention.

Periodic inspection and testing

On completion of this topic area the candidate will be able to identify the purpose and need for periodic inspection and testing.

Purpose of a periodic inspection

Initial inspection and testing is necessary on all newly completed installations. In addition, because all the electrical installations deteriorate due to a number of factors (such as damage, wear and tear, corrosion, excessive electrical loading, ageing and environmental influences) periodic inspection and testing must be carried out at regular intervals.

The purpose of a periodic inspection and test is to:

- confirm the safety of persons and livestock against the effects of electric shock or burns
- ensure protection against damage to property by fire or heat arising from an installation defect
- confirm that the installation has not been damaged and has not deteriorated to the extent that it may impair safety
- identify any defects in the installation or non-conformity with the current edition of the Regulations that may cause danger.

The intervals between tests is determined by the following.

- Legislation requires that all installations must be maintained in a safe condition and therefore must be periodically inspected and tested. Table 8.01 details the maximum period between inspections of various types of installation.
- Licensing authorities, public bodies, insurance companies and other authorities may require public inspection and testing of electrical installations.
- The installation must comply with BS 7671.
- It is also recommended that inspection and testing should occur when there is a change of use of the premises, any alterations or additions to the original installation, any significant change in the electrical loading of the installation, and where there is reason to believe that damage may have been caused to the installation.

Domestic Or if change of occupancy	10 years
Highway power supplies	6 years
Churches	5 years
Commercial Or if change of occupancy	5 years

(continued over)

Educational premises	5 years
Hospitals	5 years
Hotels	5 years
Laboratories	5 years
Offices	5 years
Public houses	5 years
Residential accommodation	5 years
Restaurants	5 years
Shops	5 years
Village halls/community centres	5 years
Agricultural/horticultural	3 years
Caravans	3 years
Cinemas	3 years
Emergency lighting	3 years
Industrial	3 years
Leisure complexes	3 years
Places of public entertainment	3 years
Theatres	3 years
Caravan parks	1 year
Fire alarms	1 year
Fish farms	1 year
Launderettes	1 year
Marinas	1 year
Petrol filling stations	1 year
Swimming pools	1 year
Construction sites	3 months

Table 8.01 Frequency of inspection

In the case of an installation that is under constant supervision while in normal use, such as a factory or other industrial premises, periodic inspection and testing may be replaced by a system of continuous monitoring and maintenance of the installation, provided that adequate records of such maintenance are kept.

When carrying out the design of an electrical installation, and particularly when specifying the type of equipment to be installed, the designer should take into account the likely quality of the maintenance programme and the periods between periodic inspection and testing to be specified on the Electrical Installation Certificate.

Both Section 6 of the Health and Safety at Work Act and the Construction (Design and Management) Regulations require information on the requirements for routine checks and periodic inspections to be provided. The advice of the Health and Safety Executive in their Memorandum of Guidance on the Electricity at Work Regulations indicates that practical experience of an installation's use may indicate the need for an adjustment to the frequency of checks and inspections, i.e. more often or less frequent depending on the likely deterioration of the installation during its normal use. This would be a matter of judgement for the duty holder.

Routine checks

Electrical installations should still be routinely checked in the intervening time between periodic inspection and testing. In domestic premises it is likely that the occupier will soon notice any damage or breakages to electrical equipment and will take steps to have repairs carried out. In commercial or industrial installations a suitable reporting system should be available for users of the installation to report any potential danger from deteriorating or damaged equipment.

In addition to this, a system of routine checks should be set up to take place between formal periodic inspections. The frequency of these checks will depend entirely on the nature of the premises and the usage of the installation. Routine checks are likely to include activities such as those listed in Table 8.02.

Activity	Check
Defect reports	Check that all reported defects have been rectified and that the installation is safe.
Inspection	Look for: • breakages • wear or deterioration • signs of overheating • missing parts (covers/screws) • switchgear still accessible • enclosure doors secure • labels still adequate (readable) • loose fittings.
Operation	Check operation of: • switchgear (where reasonable) • equipment (switch off and on) • RCD (using test button).

Table 8.02 Routine checks

The recommended period between both routine checks and formal inspections are given in Table 3.2 taken from *IEE Guidance Note 3*. The requirements for such inspections are stated in BS 7671 Chapter 62 and specify that all inspections should provide careful scrutiny of the installation without dismantling or with only partial dismantling where absolutely necessary. It is considered that the unnecessary dismantling of equipment or disconnection of cables could produce a risk of introducing faults that were not there in the first place.

In summary, the inspection should ensure that:

- the installation is safe
- the installation has not been damaged
- the installation has not deteriorated so as to impair safety
- any items that no longer comply with the Regulations or may cause danger are identified.

In practical terms the inspector is carrying out a general inspection to ensure that the installation is safe. However, the inspector is required to record and make recommendations with respect to any items that no longer comply with the current edition of the Regulations.

Statutory and non-statutory documentation

These include the following

1. The Electricity Supply Regulations (1988).
2. The Electricity at Work Regulations.
3. BS 5266 Pt. 1 Code of Practice for emergency lighting systems (other than cinemas). Other Regulations and intervals cover testing of batteries and generators.
4. BS 5839 Pt. 1 Code of Practice for the design, installation and servicing of fire alarm systems.
5. Local authority conditions of licence.
6. SI 1995 No 1129 (clause 27) The Cinematography (Safety) Regulations.

The inspection process

In new installations, inspection should be carried out progressively as the installation is installed and must be done before it is energised. As far as is reasonably practicable, an initial inspection should be carried out to verify that:

- all equipment and material is of the correct type and complies with applicable British Standards or acceptable equivalents
- all parts of the fixed installation are correctly selected and erected
- no part of the fixed installation is visibly damaged or otherwise defective
- the equipment and materials used are suitable for the installation relative to the environmental conditions.

The following items must be covered in an inspection.

1. **Connection of conductors:** Every connection between conductors and equipment or other conductors must provide durable electrical continuity and

adequate mechanical strength. Requirements for the enclosure and accessibility of connections must be considered.

2. **Identification of each conductor:** Table 51 of BS 7671 provides a schedule of colour identification of each core of a cable and its conductors. It should be checked that each core of a cable is identified as necessary. Where it is desired to indicate a phased rotation or a different function for cores of the same colour, numbered sleeves are permitted.

3. **Routing of cables:** Cable routes shall be selected with regard to the cable's suitability for the environment, i.e. ambient temperature, heat, water, foreign bodies, corrosion, impact, vibration, flora, fauna, radiation, building use and structure. Cables should be routed out of harm's way and protected against mechanical damage where necessary. Permitted cable routes are clearly defined in the *IEE On-Site Guide*; alternatively, cables should be installed in earthed metal conduit or trunking.

4. **Current-carrying capacity:** Where practicable, the cable size should be assessed against the protective device based upon information provided by the installation designer. Reference should be made as appropriate to appendix 4 of BS 7671.

5. **Verification of polarity:** It must be checked that no single pole switch or protective device is installed in any neutral conductor. A check must also be made that all protective devices and switches are connected in the line conductor only (unless the switch is a double pole device) and that the centre contact of Edison screw lamp holders are connected to the line conductor. No switches are permitted in the cpc.

6. **Accessories and equipment:** Correct connection is to be checked. Table 55.1 of BS 7671 is a schedule of types of plug and socket outlets available, the rating and the associated British Standards. Particular attention should be paid to the requirements for a cable coupler. Lamp holders should comply with BS 5042 and be of temperature rating T2.

7. **Selection and erection to minimise the spread of fire:** A fire barrier or protection against thermal effects should be provided if necessary to meet the requirements of BS 7671. The Regulations require that each ceiling arrangement be inspected to verify that it conforms with the manufacturer's erection instructions. This may be impossible without dismantling the system and it is essential, therefore, that inspection should be carried out at the appropriate stage of the work and that this is recorded at the time for incorporation in the inspection and test documents.

8. **Protection against direct contact:** Direct contact as defined in BS 7671 is the contact of persons or livestock with live parts. Live parts are conductors or conductive parts intended to be energised in normal use including a neutral conductor but by convention not a combined Protective Earthed Neutral (PEN) conductor. Protection is provided using the following methods.

   - **Insulation.** Is the insulation damaged or has too much been removed? Although protection by insulation is the usual method there are other methods of providing basic protection.

- **Barriers.** Where live parts are protected by barriers or enclosures, these should be checked for adequacy and security. Have all covers, lids and plates been securely fitted?

- **Obstacles.** Protection by obstacles provides protection only against an intentional contact. If this method is used, the area shall be accessible only to skilled persons or to instructed persons under supervision. Obstacles can include a fence around a transformer sub-station and barbed wire fencing on power pylons.

- **Out of reach.** Placing out of reach protects against direct contact. Increased distance is necessary where bulky conducting objects are likely to be handled in the vicinity. The requirements for this method are given more fully in appendix 3 of the *Memorandum of Guidance to the Electricity at Work Act*.

9. **Fault protection:** Fault protection as defined by BS 7671 is the contact of persons or livestock with exposed conductive parts which have become live under fault conditions. An exposed conductive part is a conductive part of equipment which can be touched but is not live although it can become live under fault conditions. Examples of exposed conductive parts could include metal trunking, metal conduit and the metal case of an electrical appliance, e.g. a classroom overhead projector. Earthing provides protection against this type of fault. We also need to check that extraneous conductive parts have been correctly bonded with protective conductors. An extraneous conductive part is a conductive part that is liable to introduce a potential, generally earth potential, and not form part of an electrical installation; examples of extraneous conductive parts are metal sink tops and metal water pipes. The purpose of the bonding is to ensure that all extraneous conductive parts which are simultaneously accessible are at the same potential. Methods of fault protection are given in BS 7671 as:

   - earthed equipotential bonding and automatic disconnection of supply (most common)
   - use of class II equipment
   - non-conducting location
   - earth-free local equipotential bonding
   - electrical separation.

10. **Protective devices:** Have they been set correctly for the load? If rewirable fuses have been fitted, has the correct size of fuse wire been used? If a socket is to be provided for outdoor equipment, has a 30 mA rated RCD been fitted?

11. **Checks on documentation:** Diagrams, schedules, charts, instructions and any other information must be available if inspection and testing is to be carried out in a satisfactory manner.

12. **Checks on warning notices:** These should be fixed to equipment operating in excess of 250 volts where this voltage would not normally be expected.

Did you know?

Typical earthing arrangements for domestic installations are shown diagrammatically in the *IEE On-Site Guide*. For other types of installation the appropriate size of earthing and bonding conductors should be determined in accordance with BS 7671 Chapter 54

Preparing for inspection

Where a diagram, chart or tables are not available, a degree of exploratory work may be necessary so that inspection and testing can be carried out safely and effectively. Notes should be made of any known changes in environmental conditions, building structure and alterations, which may have affected the suitability of the wiring for its present load and method of installation.

A careful check should be made of the type of equipment on site so that the necessary precautions can be taken, where conditions permit, to disconnect or short out electronic and other equipment which may be damaged by subsequent testing. Special care must be taken where control and protective devices contain electronic components. It is essential to determine the degree of these disconnections before planning the detailed inspection and testing.

For safety, it is necessary to carry out a visual inspection of the installation before beginning any tests or opening enclosures, removing covers etc. So far as is reasonably practicable; the visual inspection must verify that the safety of persons, livestock and property is not endangered. A thorough visual inspection should be made of all electrical equipment that is not concealed and should include the accessible internal condition of a sample of the equipment. External conditions should be noted and damage identified or, if the degree of protection has been impaired, the matter should be recorded on the schedule of the report. This inspection should be carried out without power supplies to the installation, wherever possible, in accordance with the Electricity at Work Regulations 1989.

The inspection should include a check on the condition of all electrical equipment and materials, taking into account any available manufacturer's information with regard to the following:

- safety
- wear and tear
- corrosion
- damage
- excessive loading (overloading)
- age
- external influences
- suitability.

The assessment of condition should take account of known changes in conditions influencing and affecting electrical safety, e.g. extraneous conductive parts, plumbing, structural changes etc. It would not be practicable to inspect all parts of an installation; thus a random sample should be inspected. This should include:

- checking that joints and connections are properly secured and that there is no sign of overheating
- checking switches for satisfactory electrical and mechanical conditions
- checking that protective devices are of the correct rating and type; check for accessibility and damage
- checking that conductors have suffered no mechanical damage and have no signs of overheating
- checking that the condition of enclosures remains satisfactory for the type of protection required.

Remember

With periodic inspection and testing, inspection is the vital initial operation and testing is subsequently carried out in support of that inspection

Did you know?

Guidance notes 3 3.8.1 and 3.8.2 give advice and information necessary to carry out safe inspection and testing

Periodic testing

Periodic testing is supplementary to periodic inspection and the same level of testing as for a new installation is not necessarily required.

When the building being tested is unoccupied, isolation of the supply for testing purposes will not cause a problem, but where the building is occupied, then testing should be carried out to cause as little inconvenience as possible to the user. Where isolation of the supply is necessary, then the disconnection should take place for as short a time as possible and by arrangement with the occupier. It may be necessary to complete the tests out of normal working hours in order to keep disruption to a minimum.

The following tests (where applicable) should be applied:

- continuity of all protective conductors (including equipotential bonding conductors and continuity of ring circuit conductors where required)
- insulation resistance
- polarity
- earth electrode resistance
- earth-fault loop impedance
- operation of devices for isolation and switching
- operation of residual current devices
- operation of circuit breakers.

Detailed test procedures

Continuity of protective conductors and equipotential bonding conductors

Where the installation can be safely isolated from the supply, then the circuit protective conductors and equipotential bonding conductors can be disconnected from the main earthing terminal in order to verify their continuity.

Where the installation cannot be isolated from the supply, the circuit protective conductors and the equipotential bonding conductors must not be disconnected from the main earthing terminal, as under fault conditions extraneous metalwork could become live. Under these circumstances a combination of inspection, continuity testing and earth loop impedance testing should establish the integrity of the circuit protective conductors.

When testing the effectiveness of the main bonding conductors or supplementary bonds, the resistance value between any service pipe or extraneous metalwork and the main earthing terminal should not exceed 0.05 ohms.

Insulation resistance

Insulation resistance tests can only be carried out where it is possible to safely isolate the supply. All electronic equipment susceptible to damage should be disconnected or alternatively the insulation resistance test should be made between line and neutral conductors connected together and earth. Where practicable the tests should be carried out on the whole of the installation with all switches closed and all fuse links in place. Where this is not possible, the installation may be subdivided by testing each separate distribution board one at a time.

BS 7671 Table 61 states a minimum acceptable resistance value of 1 megohms. However, if the measured value is less than 2 megohms then further investigation is required to determine the cause of the low reading.

Where individual items of equipment are disconnected for these tests and the equipment has exposed conductive parts, then the insulation resistance of each item of equipment should be checked. In the absence of any other requirements, the minimum value of insulation resistance between live components and the exposed metal frame of the equipment should be not less than 1 megohms.

Polarity

Polarity tests should be carried out to check that:

- polarity is correct at the intake position and the consumer unit or distribution board
- single pole switches or control devices are connected in the line conductor only
- socket outlets and other accessories are connected correctly
- centre contact bayonet and Edison screw type lamp holders have their outer or screwed contact connected to the neutral conductor
- all multi-pole devices are correctly installed.

Where it is known that no alterations or additions have been made to the installation since its last inspection and test, then the number of items to be tested can be reduced by sampling. It is recommended that at least 10 per cent of all switches and control devices should be tested together with any centre contact lamp holders and 100 per cent of socket outlets. However, if any cases of incorrect polarity are found then a full test should be made of all accessories in that part of the installation and the sample of the remainder should be increased to 25 per cent.

Earth-fault loop impedance

Earth-fault loop impedance tests should be carried out at:

- the origin of each installation and at each distribution board
- all socket outlets
- at the furthest point of each radial circuit.

Results obtained should be compared with the values documented during previous tests, and where an increase in values has occurred these must be investigated.

Operation of devices for isolation and switching

All main switches and isolators should be inspected for correct operation and clear labelling, and to check that access to them has not been obstructed. Where the operation of the contacts of such devices is not visible it may be necessary to connect a test lamp between each line and the neutral on the load side of the device to ensure that all supply conductors have been broken.

Operation of residual current devices

RCDs should be tested for correct operation by the use of an RCD tester to ensure they trip out in the time required by BS 7671 as well as by use of the integral test button. A check should also be made that the tripping current for the protection of a socket outlet to be used for equipment outdoors should not exceed 30 mA.

Operation of circuit breakers

All circuit breakers should be inspected for visible signs of damage or damage caused by overheating. Where isolation of the supply to individual sub-circuits will not cause inconvenience, each circuit breaker should be manually operated to ensure that the device opens and closes correctly.

Periodic inspection and test report

BS 7671 requires that the results of any periodic inspection and test should be recorded on a periodic inspection and test report of the type illustrated in Figure 8.02. The report should include the following:

- a description of the extent of the inspection and tests and what parts of the installation were covered
- any limitations (e.g. portable appliances not covered)
- details of any damage, deterioration or dangerous conditions which were found
- any non-compliance with BS 7671
- schedule of test results.

On the job: Periodic inspection

You are asked to carry out a periodic inspection at a small workshop. All goes well until you come to check the office that is attached to the factory, where you notice that about 20 staff are working on computers. No installation drawings exist for the office, as it turns out the factory was built as an extension to the offices many years ago.

1. Can you identify the problems you will encounter in this job?
2. How would you deal with the situation?

If any items are found which may cause immediate danger, these should be rectified immediately. If this is not possible then they must be reported to a responsible person without delay.

When inspecting older installations, which may have been installed in accordance with a previous edition of the *IEE Wiring Regulations*, provided that all items which do not conform to the present edition of BS 7671 are reported, the installation may still be acceptable, provided that no risk of shock, fire or burns exists.

Figure 8.02 Periodic inspection and test report

Details of the client

Name : . . . .
Address : . . . .
. . . .
. . . .

Purpose of report : . . . .

Details of the installation

Occupier : . . . .
Installation address (if different from above) : . . . .
. . . .
Description of premises: Domestic ☐   Commercial ☐   Industrial ☐   Other ☐
Estimated age of the installation : . . . . years . Evidence of alterations or additions : . . . . . . .Yes/No
Date of last inspection : . . . .   Records available : . . . . .Yes/No

Extent and limitations of the inspection

Extent of the installation covered by this report : . . . .
. . . .
. . . .
Limitations : . . . .
. . . .
. . . .
This inspection has been carried out in accordance with BS7671 (*IEE Wiring Regulations*). Cables concealed within trunking and conduit or cables concealed under floors, buried underground, installed in roof spaces or generally hidden within the fabric of the building have not been inspected.

Next inspection

I/We recommend that this installation should be further inspected and tested after an interval of not more than. . . . . months/year, providing that any observations requiring urgent attention are attended to without delay.

Declaration

Inspected and tested by

Name : . . . .   Signature : . . . .
For and on behalf of : . . . .   Position : . . . .
Address : . . . .
. . . .
. . . .   Date : . . . .

Summary of periodic testing

Type of test	Recommendations
Continuity of protective and ring conductors	Tests to be carried out between: • all main bonding connections • all supplementary bonding connections. Note: When an electrical installation cannot be isolated, protective conductors including bonding conductors must not be disconnected.
Insulation resistance	If this test is to be carried out, then test: • the whole installation with all protective devices in place and all switches closed • where electronic devices are present, the test should be carried out between phase and neutral conductors connected together and earth.
Polarity	Tests to be carried out: • origin of installation • all socket outlets • 10 per cent of control devices (including switches) • 10 per cent of centre contact lamp holders. Note: If incorrect polarity is found then a full test should be made on that part of the installation and testing on the remainder increased to 25 per cent. If further faults are found the complete installation must be tested.
Earth-fault loop impedance	Tests to be carried out at: • origin of installation • each distribution board • each socket outlet • extremity of every radial circuit.
Additional Protection	Test to verify the effectiveness of the RCDs
Phase sequence	Test that the phase sequence is correct
Functional	Activities to be carried out: • all isolation and switching devices to be operated • all labels to be checked • all interlocking mechanisms to be verified • all RCDs to be checked both by test instrument and by test button • all manually operated circuit breakers to be operated to verify they open and close correctly.
Voltage drop	This may be verified by measuring the circuit impedance or by calculation

Relevant sources of information

On completion of this topic area the candidate will be able to list relevant sources of information required to facilitate inspection and testing.

Statutory and non-statutory legislation

We covered many of the details of statutory legislation affecting inspection and testing, such as the Health and Safety at Work Act, at Level 2. You will also be familiar with the principles of *IEE Wiring Regulations* BS 7671. Here we will look in detail at how these Regulations relate to inspection and testing.

BS 7671 states that, as far as is reasonably practicable, an inspection shall be carried out to verify that:

- all equipment and materials used in the installation are of the correct type and comply with the appropriate British Standards or acceptable equivalent
- all parts of the installation have been correctly selected and installed
- no part of the installation is visibly damaged or otherwise defective
- the installation is suitable for the surrounding environmental conditions.

BS 7671

You will use BS 7671 as the document against which a professional judgement needs to be made. Along with the regulations and accompanying documents, you will also need the *IEE On-Site Guide* and the *IEE Guidance Note 3* on inspection and testing. These contain tables and charts with maximum values and checklists, against which comparisons have to be made.

Information

Before carrying out the inspection and test of an installation, BS 7671 requires the person carrying out the work to be provided with the following information:

- the maximum demand of the installation expressed in amperes per phase after diversity has been applied
- the number and type of live conductors, both for the source of energy and for each circuit to be used within the installation (e.g. single-phase two-wire a.c. or three-phase four-wire a.c. etc.)
- the type of earthing arrangements e.g. TNS, TNCS, TT etc.
- the nominal voltage (U_o)
- the prospective short-circuit current at the origin of the installation (kA)
- the earth-fault loop impedance (Z_e) of that part of the system external to the installation
- the type and rating of the overcurrent device acting at the origin of the installation.

Remember

It is not possible for anyone to remember all the information relating to BS 7671. Therefore an essential part of the toolbox for someone involved in the inspection and test of installations is a copy of these documents

The following information should be provided as part of the design information, which should be checked by the person carrying out the inspection. You may have to sign the electrical installation certificate to confirm that the installation has been designed, installed and tested to comply with BS 7671 with reference to:

- the type and composition of circuits, including points of utilisation, number and size of conductors and type of cable used, including the installation method
- the method used to meet the requirements for fault protection
- the information to be able to identify each protective device, isolation and switching and its location
- any circuit or equipment that may be vulnerable to test.

Contract specification and drawings

You will need these in order to conclude that the job you have completed has been carried out in line with the requirements of the specifications. The position of outlets and level of voltage needed in wiring will be affected by these documents.

Manufacturer's instructions

These tell us the correct size of cable, appropriate control equipment, the size of protective device and any functional earthing requirements that may be required for safe operation of equipment.

Statutory legislation

You should be familiar with the details of many of these pieces of legislation. However, there are some areas for some of these regulations that relate specifically to inspection. We will look at these areas in more detail.

Electricity at Work Regulations 1989

We covered the provisions of the Electricity at Work Regulations 1989 in chapter 1 on pages 4–6 and at Level 2. The requirements of these Regulations are those that BS 7671 is written to mirror. Testing and inspection of installations is not just to confirm compliance with BS 7671 but to confirm compliance with EAWR. For inspection purposes, EAWR requires that:

- only competent persons should be engaged in the testing procedure (Regulation 16)
- suitable precautions for safe isolation need to be taken (Regulation 14)
- padlocks and isolock systems should be used so that control of isolation remains under the control of the inspector (Regulation 13).

Electricity Safety, Quality and Continuity Regulations 2002

These Regulations were covered fully on pages 3–4. As the requirements include reference to streetlights, traffic signals and bollards etc. they could therefore become part of the requirements for someone involved in the test and inspection of these installations.

Regulation 29 gives area supply boards the power to refuse to connect their supply to an installation that in their opinion is not constructed, installed and protected to an appropriately high standard. This Regulation would only be enforced if the installation failed to meet the requirements of BS 7671.

A survey should be carried out to identify all items of switchgear and control gear and their associated circuits.

During the survey a note should be made of any known changes in either the structure of the building, the environmental conditions or of any alterations or additions to the installation which may affect the suitability of the wiring or the method of installation.

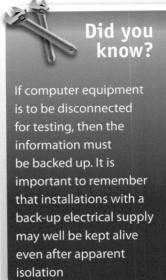

Did you know?

If computer equipment is to be disconnected for testing, then the information must be backed up. It is important to remember that installations with a back-up electrical supply may well be kept alive even after apparent isolation

Detailed inspection requirements

Anticipation of danger

As stated previously, when carrying out the inspection the opportunity should be taken to identify any equipment that may be damaged if subjected to high-test voltages. As well as computer equipment this may include safety systems such as fire or intruder alarms that could well have electronic components susceptible to test voltages.

Joints and connections

It may be impossible to inspect every joint and termination in an electrical installation. Therefore, where necessary, a sample inspection should be made. Provided the switchgear and distribution boards are accessible as required by the Regulations, then a full inspection of all conductor terminations should be carried out and any signs of overheating or loose connections should be investigated and included in the report. For lighting points and socket outlets a suitable sample should be inspected in the same way.

Conductors

The means of identification of every conductor, including protective conductors, should be checked and any damage or deterioration to the conductors, their insulation, protective sheathing or armour should be recorded. This inspection should include each conductor at every distribution board within the installation and a suitable sample of lighting points, switching points and socket outlets.

Flexible cables and cords

Where flexible cables or cords form part of the fixed installation the inspection should include:

- examination of the cable or cord for damage or deterioration
- examination of the terminations and anchor points for any defects
- checking the correctness of the installation with regard to additional mechanical protection or the application of heat resistant sleeving where necessary.

Switches

The *IEE Guidance Notes 3 (Inspection & Testing)*, recommends that a random sample of at least 10 per cent of all switching devices be given a thorough internal visual inspection to assess their electrical and mechanical condition. Should the inspection reveal excessive wear and tear or signs of damage due to arcing or overheating then, unless it is obvious that the problem is associated with that particular switch, the inspection should be extended to include all remaining switches associated with the installation.

Protection against thermal effects

Although sometimes difficult due to the structure of the building, the presence of fire barriers and seals should be checked wherever reasonably practicable.

Protection against direct and indirect contact

Separate Extra Low Voltage (SELV) is commonly used as a means of protection against both direct and indirect contact. When inspecting this type of system, the points to be checked include the use of a safety isolating transformer, the need to keep the primary and secondary circuits separate and the segregation of exposed conductive parts of the SELV system from any connection with the earthing of the primary circuit or from any other connection with earth.

Basic protection

Inspection of the installation should confirm that all the requirements of the Regulations have been met with regard to basic protection against direct contact with live conductors. This means checking to ensure there has been no damage or deterioration of any of the insulation within the installation, no removal of barriers or obstacles and no alterations to enclosures that may allow access to live conductors.

Fault protection

The method used for fault protection must be established and recorded on the Inspection Schedule. Where earthed equipotential bonding and automatic disconnection of the supply is used, a check on the condition of the main equipotential bonding conductor and the satisfactory connection of all other protective conductors with earth are essential.

Remember

An RCD must not be used as the sole means of basic protection

Protective devices

A check must be made that each circuit is adequately protected with the correct type, size and rating of fuse or circuit breaker. A check should also be made that each protective device is suitable for the type of circuit it is protecting and the earthing system employed, e.g. will the protective device operate within the disconnection time allowed by the Regulations and is the rating of the protective device suitable for the maximum prospective short circuit current likely to flow under fault conditions?

Enclosures and mechanical protection

The enclosures of all electrical equipment and accessories should be inspected to ensure that they provide protection not less than IP2X or IPXXB, and where horizontal top surfaces are readily accessible they should have a degree of protection of at least IP4X. IP2X represents the average finger of 12mm diameter and 80mm in length and can be tested by a metal finger of these dimensions. IP4X provides protection against entry by strips greater than 1.0mm thickness or solid objects exceeding 1.0mm in diameter.

Visual inspection

On completion of this section the candidate will be able to list the relevant items associated with visual inspection prior to commissioning.

Initial inspection

The following text provides a detailed description of the procedures required to carry out an initial inspection of an electrical installation. Substantial reference has been made to the IEE Wiring Regulations (BS 7671), the *IEE On-Site Guide* and *IEE Guidance Note 3*, and it is recommended that wherever possible these documents are referred to should clarification be required.

Inspection requirements

In order to meet the requirements for the inspection process we should also include the checking of the following relevant items:

1. **Requirements for both basic and fault protection**

   Separate extra low voltage (SELV) is the most common method of providing both basic and fault protection. Requirements for this type of system include:

   - an isolated source of supply, e.g. a safety-isolating transformer to BS 3535 (also numbered BS EN 60742 1996)

   - electrical separation, which means no electrical connection between the SELV circuit and higher voltage systems

Remember

Although the major part of any inspection will be visual, other human senses may be employed, e.g. a piece of equipment with moving parts may generate an unusual noise if it is not working correctly, or an electrical device which overheats will be hot to touch as well as giving off a distinctive smell. The senses of hearing, touch and smell will assist in detecting these

- no connection with earth or the exposed conductive parts or protective conductors of other systems.

2. **Specialised systems**

Non-conducting locations and earth-free situations are specialised systems which would only normally be used where specified and controlled by a suitably qualified electrical engineer. Although it is useful to be aware of the nature of these systems you are unlikely to be asked to carry out an inspection and test of these types of installation at this stage of your career.

3. **Prevention of mutual detrimental influence**

Account must be taken of the proximity of other electrical services of a different voltage band and of non-electrical services and influences, e.g. fire alarm and emergency lighting circuits must be separated from other cables and from each other, and Band 1 and Band 2 circuits must not be present in the same enclosure or wiring system unless they are either segregated or wired with cables suitable for the highest voltage present. Mixed categories of circuits may be contained in multicore cables, subject to certain requirements. This could also mean checking that water taps have not been fitted directly above a socket outlet.

Band 1 circuits are circuits that are nominally extra-low voltage, i.e. not exceeding 50 volts a.c. or 120 volts d.c., such as telecommunications or data and signalling. Band 2 circuits are circuits that are nominally low voltage, i.e. exceeding extra-low voltage but not exceeding 1000 volts a.c. between conductors or 600 volts a.c. between conductors and earth.

4. **Isolating and switching devices**

BS 7671 requires that effective means suitably positioned and ready to operate should be provided so that all voltage may be cut off from every installation, every circuit within the installation and from all equipment, as may be necessary to prevent or remove danger. This means that switches and/or isolating devices of the correct rating must be installed as appropriate to meet the above requirements. It may be advisable where practicable to carry out an isolation exercise to check that effective isolation can be achieved. This should include switching off, locking-off and testing to verify that the circuit is dead and no other source of supply is present.

5. **Under voltage protection**

Sometimes referred to in starters as no-volt protection, suitable precautions must be taken where a loss or lowering of voltage or a subsequent restoration of voltage could cause danger. The most common situation would be where a motor-driven machine stops due to a loss of voltage and unexpectedly restarts when the voltage is restored (unless precautions such as the installation of a motor starter containing a contactor are employed). Regulations require that where unexpected restarting of a motor may cause danger, the provision of a motor starter designed to prevent automatic restarting must be provided.

6. Selection of equipment appropriate to external influences

All equipment must be selected as suitable for the environment in which it is likely to operate. Items to be considered are ambient temperature, presence of external heat sources, presence of water, likelihood of corrosion, ingress of foreign bodies, impact, vibration, flora, fauna, radiation, building use and structure.

7. Access to switchgear and equipment

The Electricity at Work Regulations 1989 and BS 7671 state that every piece of equipment that requires operation or attention must be installed so that adequate and safe means of access and working space are provided.

8. Presence of diagrams, charts and other similar information

Checks should be made for layout drawings, distribution charts and information on circuits vulnerable to a particular test. All distribution boards should be provided with a distribution board schedule that provides information regarding types of circuits, number and size of conductors and type of wiring etc. These should be attached within or adjacent to each distribution board.

9. Erection methods

Correct methods of installation should be checked, in particular fixings of switchgear, cables and conduit, etc. which must be adequate and suitable for the environment.

Inspection checklists

To ensure that all the requirements of the Regulations have been met, inspection checklists should be drawn up and used as appropriate to the type of installation being inspected. Examples of suitable checklists are given in Table 8.03 which follows.

Switchgear (tick if satisfactory)	
All switchgear is suitable for the purpose intended	
Meets requirements of the appropriate BS EN standards	
Securely fixed and suitably labelled	
Suitable glands and gland plates used (526.1)	
Correctly earthed	
Conditions likely to be encountered taken account of, i.e. suitable for the environment	
Correct IP rating	
Suitable as means of isolation	
Complies with the requirements for locations containing a bath or shower	
Need for isolation, mechanical maintenance, emergency and functional switching met	
Fireman switch provided, where required	
Switchgear suitably coloured, where necessary	

(continued over)

All connections secure	
Cables correctly terminated and identified	
No sharp edges on cable entries, screw heads etc. which could cause damage to cables	
All covers and equipment in place	
Adequate access and working space	
Wiring accessories (general requirements) (tick if satisfactory)	
All accessories comply with the appropriate British Standard	
Boxes and other enclosures securely fastened	
Metal boxes and enclosures correctly earthed	
Flush boxes not projecting above surface of wall	
No sharp edges which could cause damage to cable insulation	
Non-sheathed cables not exposed outside box or enclosure	
Conductors correctly identified	
Bare protective conductors sleeved green and yellow	
All terminals tight and contain all strands of stranded conductor	
Cord grips correctly used to prevent strain on terminals	
All accessories of adequate current rating	
Accessories suitable for all conditions likely to be encountered	
Complies with the requirements for locations containing a bath or shower	
Cooker control unit sited to one side and low enough for accessibility and to prevent trailing flexes across the radiant plates	
Cable to cooker fixed to prevent strain on connections	
Socket outlet (tick if satisfactory)	
Complies with appropriate British Standard and is shuttered for household and similar installations	
Mounting height above floor or working surface is suitable	
All sockets have correct polarity	
Sockets not installed in bath or shower zones unless they are shaver-type socket or SELV	
Sockets not within 3m of zone 1	
Sockets controlled by a switch if the supply is direct current	
Sockets protected where floor mounted	
Circuit protective conductor connected directly to the earthing terminal of the socket outlet on a sheathed wiring installation	
Earthing tail provided from the earthed metal box to the earthing terminal of the socket outlet	
Socket outlets not used to supply a water heater with uninsulated elements	

(continued over)

Lighting controls (tick if satisfactory)

Light switches comply with appropriate British Standard

Switches suitably located

Single-pole switches connected in phase conductor only

Correct colour-coding of conductors

Correct earthing of metal switch plates

Switches out of reach of a person using bath or shower

Switches for inductive circuits (discharge lamps) de-rated as necessary

Switches labelled to indicate purpose where this is not obvious

All switches of adequate current rating

All controls suitable for their associated luminaire

Lighting points (tick if satisfactory)

All lighting points correctly terminated in suitable accessory or fitting

Ceiling roses comply with appropriate British Standard

No more than one flexible cord unless designed for multiple pendants

Devices provided for supporting flex used correctly

All switch wires identified

Holes in ceiling above ceiling rose made good to prevent spread of fire

Ceiling roses not connected to supply exceeding 250 V

Flexible cords suitable for the mass suspended

Lamp holders comply with appropriate British Standard

Luminaire couplers comply with appropriate British Standard

Conduits (general) (tick if satisfactory)

All inspection fittings accessible

Maximum number of cables not exceeded

Solid elbows used only as permitted

Conduit ends reamed and bushed

Adequate number of boxes

All unused entries blanked off

Lowest point provided with drainage holes where required

Correct radius of bends to prevent damage to cables

Joints and scratches in metal conduit protected by painting

Securely fixed covers in place adequate protection against mechanical damage

(continued over)

Rigid metal conduits (tick if satisfactory)	
Complies to the appropriate British standard	
Connected to the main earth terminal	
Line and neutral cables contained within the same conduit	
Conduits suitable for damp and corrosive situations	
Maximum span between buildings without intermediate support	

Rigid non-metallic conduits (tick if satisfactory)	
Complies with the appropriate British Standard	
Ambient and working temperature within permitted limits	
Provision for expansion and contraction	
Boxes and fixings suitable for mass of luminaire suspended at expected temperatures	

Flexible metal conduit (tick if satisfactory)	
Complies with the appropriate British Standard	
Separate protective conductor provided	
Adequately supported and terminated	

Trunking (tick if satisfactory)	
Complies to the appropriate British Standard	
Securely fixed and adequately protected against mechanical damage	
Selected, erected and rooted so that no damage is caused by ingress of water	
Proximity to non-electrical services	
Internal sealing provided where necessary	
Hole surrounding trunking made good	
Band 1 circuits partitioned from band 2 circuits, or insulated for the highest voltage present	
Circuits partitioned from band one circuits, or wired in mineral-insulated and sheathed cable	
Common outlets for band 1 and band 2 circuits provided with screens, barriers or partitions	
Cables supported for vertical runs	

(continued over)

Metal trunking (tick if satisfactory)	
Line and neutral cables contained in the same metal trunking	
Protected against damp corrosion	
Earthed	
Joints mechanically sound, and of adequate earth continuity with links fitted	
Circuit protective conductors (enter circuit details from specifications)	
1.	
2.	
3.	
4.	
5.	
6.	
7.	

Table 8.03 Inspection checklists

Did you know?

Where an item is not relevant to a specific installation, N/A should be placed in the tick box

Instruments

On completion of this topic area the candidate will be able to identify and select instruments suitable for testing and commissioning.

Types of instrument
Low-resistance ohmmeters

Where low resistance measurements are required when testing earth continuity, ring-circuit continuity and polarity, then a low-reading ohmmeter is required. This may be a specialised low-reading ohmmeter or the continuity scale of a combined insulation and continuity tester. Whichever type is used it is recommended that the test current should be derived from a source of supply not less than 4 V and no greater than 24 V with a short circuit current not less than 200 mA (instruments manufactured to BS EN 61557 will meet the above requirements). The ohmmeter should be able to measure from 0.01 ohm to 2 ohms. Digital instruments should have a resolution of 0.01 ohm.

Errors in the reading obtained can be introduced by contact resistance or by lead resistance. Although the effects of contact resistance cannot be eliminated entirely and may introduce errors of 0.01 ohm or greater, lead resistance can be eliminated either by clipping the leads together and zeroing the instrument before use (where this facility is provided) or measuring the resistance of the leads and subtracting this from the reading obtained.

Insulation-resistance ohmmeters

Insulation resistance should have a high value and therefore insulation resistance meters must have the ability to measure high resistance readings. The test voltage required for measuring insulation resistance is given in BS 7671 Table 61, as shown in Table 8.04.

Circuit nominal voltage (volts)	Test voltage d.c. (volts)	Minimum insulation resistance (megohms)
SELV and PELV	250	≥0.5
Up to and including 500 V with the exception of the above	500	≥1.0
Above 500 V	1000	≥1.0

Table 8.04 Test voltage required for measuring insulation resistance (from BS 7671)

The photograph top right on page 353 shows a typical modern insulation and continuity tester that will measure both low values of resistance, for use when carrying out continuity and polarity tests, and also high values of resistance when used for insulation resistance tests.

Did you know?

BS 7671 requires that the instruments used for measuring insulation resistance must be capable of providing the test voltages stated above while maintaining a test current of 1 mA. Instruments that are manufactured to BS EN 61557 will satisfy the above requirements

Instruments of this type are usually enclosed in a fully insulated case for safety reasons and have a range of switches to set the instrument correctly for the type of test being carried out, i.e. continuity or insulation. The instrument also has a means of selecting the voltage range required e.g. 250 V, 500 V, 1000 V.

Other features of this particular type of instrument are the ability to lock the instrument in the 'on' position for hands-free operation and an automatic nulling device for taking account of the resistance of the test leads.

Earth-loop impedance testers

Earth-loop impedance testers of the type shown in the photograph have the capability to measure both earth-loop impedance and also prospective short-circuit current, depending on which function is selected on the range selection switch. The instrument also has a series of LED warning lights to indicate whether the polarity of the circuit under test is correct.

The instrument gives a direct digital read-out of the value of the measurement being taken at an accuracy of plus or minus 2 per cent.

RCD testers

Instruments for testing residual current devices (RCD), such as the one shown in the photograph, have two selection switches. One switch should be set to the rated tripping current of the RCD (e.g. 30 mA, 100 mA etc.); the other should be set to the test current required, i.e. 50 per cent or 100 per cent of the rated tripping current or 150 mA for testing 30 mA RCDs when being used for supplementary protection.

Modern insulation and continuity tester

Earth-loop impedance tester

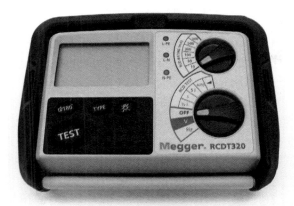

RCD tester

A modern innovation by manufacturers is the production of an all-in-one instrument that has the ability to carry out the most common tests required by the Regulations. These are:

- continuity tests (including polarity tests)
- insulation resistance tests
- earth-loop impedance tests
- RCD tests
- measurement of prospective short-circuit current.

The photograph below shows an example of this type of instrument, which by manipulation of the function and range switches will perform all of the above tests.

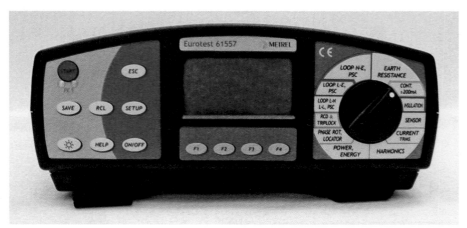

All-in-one RCD tester

Tong testers

Tong testers, or clamp meters, are used to measure the value of current flowing in an a.c. circuit without the need to interrupt the supply to connect it in series with the load. The meter allows the jaws of the instrument to open and be positioned over the conductor of the circuit – but just one of the conductors. This instrument works by magnetic induction. The magnetic field around the conductor induces an e.m.f. into the jaws. When the jaws are closed the circuit is complete and the instrument measures the current. If both the line and neutral conductors are held within the jaws, the magnetic field produced by the line conductor will be cancelled out by the magnetic field in the neutral, which will be equal in size and opposite in polarity. In this case the instrument would fail to register any current.

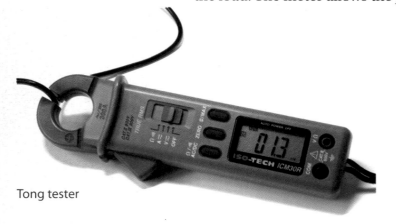

Tong tester

Instrument accuracy

On completion of this topic area the candidate will be able to explain the importance of regularly verifying the accuracy of the test instruments

Electrical test instruments

BS EN 61010 covers basic safety requirements for electrical test instruments, and all instruments should be checked for conformance with this standard before use. Older instruments may have been manufactured in accordance with BS 5458 but, provided these are in good condition and have been recently calibrated, there is no reason why they cannot be used.

Instruments may be analogue (i.e. fitted with a needle that gives a direct reading on a fixed scale) or digital, where the instrument provides a numeric digital visual display of the actual measurement being taken. Insulation and continuity testers can be obtained in either format while earth-fault loop impedance testers and RCD testers are digital only.

Calibration and instrument accuracy

To ensure that the reading being taken is reasonably accurate, all instruments should have a basic measurement accuracy of at least 5 per cent. In the case of analogue instruments a basic accuracy of 2 per cent of full-scale deflection should ensure the required accuracy of measured values over most of the scale.

All electrical test instruments should be calibrated on a regular basis. The time between calibrations will depend on the amount of usage that the instrument receives, although this should not exceed 12 months in any circumstances. Instruments have to be calibrated in laboratory conditions against standards that can be traced back to national standards; therefore, this usually means returning the instrument to a specialist test laboratory.

On being calibrated the instrument will have a calibration label attached to it stating the date the calibration took place and the date the next calibration is due (see Figure 8.03). It will also be issued with a calibration certificate detailing the tests that have been carried out and a reference to the equipment used. The user of the instrument should always check to ensure that the instrument is within calibration before being put to use.

A further adhesive label (see the example shown in Figure 8.04) is often placed over the joint in the instrument casing, stating that the calibration is void should the seal be broken. A broken seal will indicate whether anyone has deliberately opened the instrument and possibly tampered with the internal circuitry.

Did you know?

The basic standard covering the performance and accuracy of electrical test instruments is BS EN 61557, which incidentally also requires compliance with the safety requirements of BS EN 61010

Remember

Although older instruments often generated their own operating voltage by use of an in-built hand-cranked generator, almost all modern instruments have internal batteries for this purpose. It is essential, therefore, that the condition of the batteries is checked regularly; this includes checking for the absence of corrosion at the battery terminals

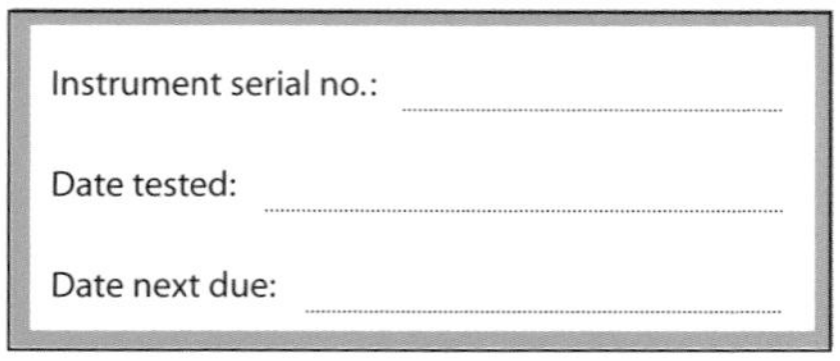

Instrument serial no.: ....

Date tested: ....

Date next due: ....

Figure 8.03 Adhesive calibration label

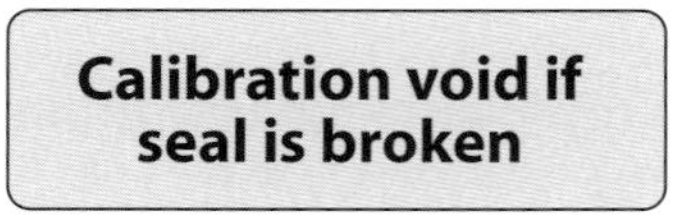

Calibration void if seal is broken

Figure 8.04 Calibration seal

Instruments that are subject to any electrical or mechanical misuse (e.g. if the instrument undergoes an electrical short circuit or is dropped) should be returned for recalibration before being used again.

Electrical test instruments are relatively delicate and expensive items of equipment and should be handled in a careful manner. When not in use they should be stored in clean, dry conditions at normal room temperature. Care should also be taken of instrument leads and probes to prevent damage to their insulation and to maintain them in good, safe working condition.

Testing resistors

Resistors must be removed from a circuit before testing, otherwise readings will be false. To measure the resistance, the leads of a suitable ohmmeter should be connected to each resistor connection lead and a reading obtained which should be close to the preferred value and within the tolerance stated.

Resistors as current limiters

A resistor is often provided in a circuit to limit, restrict or reduce the current flowing in the circuit to some level that better suits the ratings of some other component in the circuit. For example, consider the problem of operating a solenoid valve from a 36 V d.c. supply, given the information that the energising current of the coil fitted to the valve is 100 mA and its resistance is 240 Ω.

Note that the coil, being a wound component, is actually an inductor. However, we are concerned here with the steady d.c. current through the coil and not the variation in coil current at the instant the supply is connected, so we can ignore the effects of its inductance and consider only the effects of its resistance.

If the solenoid valve were connected directly across the 36 V supply, as shown in Figure 8.05, then from Ohm's law the steady current through its coil would be:

$$I = \frac{V}{R}$$

$$= \frac{36}{240}$$

$$= 0.15 \text{ A or } 150 \text{ mA}$$

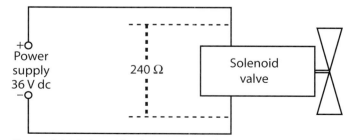

Figure 8.05 Solenoid valve connected across 36V supply

As the coil was designed to produce an adequate magnetic 'pull' when energised at 100 mA, any increase in the energising current is unnecessary and may in fact be highly undesirable due to the resulting increase in the power, which would be dissipated (as heat) within the coils.

Note: The power dissipated in the coil of the solenoid valve, when energised at the recommended current of 100 mA is:

$P = I^2 \times R$

$\quad = 0.1^2 \times 240 = 2.4$ watts

Now if the current were to be 0.15 A on the 36 V supply it would be:

$P = V \times I$

$\quad = 36 \times 0.15$

$\quad = 5.4$ watts

Thus connecting the coil directly across a 36 volt supply would result in the power dissipation in it being more than doubled. If the valve is required to be energised for more than very brief periods of time, the coil could be damaged by overheating.

Some extra resistance must therefore be introduced into the circuit so that the current through the coil is limited to 100 mA even though the supply is 36 V.

For a current of 100 mA to flow from a 36 V supply, the total resistance R_t connected across the 36 V must, from Ohm's law, be:

$R_t = \dfrac{36}{0.1}$

$\quad = 360\ \Omega$

Remember there is already 240 Ω in the coil.

A resistor of value 120 Ω must therefore be fitted in series with the coil to bring the value of R_t to 360 Ω. This limits the current through the coil to 100 mA when the series combination of coil and resistor is connected across the 36 V supply as shown in Figure 8.06.

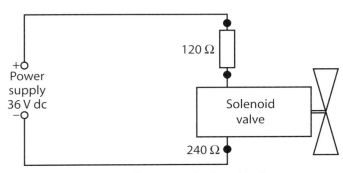

Figure 8.06 Series combination of coil and resistor connected across 36 V supply

Resistors for voltage control

Within a circuit it is often necessary to have different voltages at different stages and we can achieve this by using resistors.

For example, if we physically opened up a resistor and connected its ends across a supply, we would find that, if we then measured the voltage at different points along the resistor, the values would vary along its length.

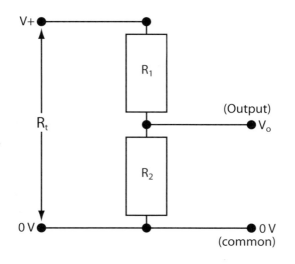

Figure 8.07 Series circuit for voltage control

However, reality will stop us from doing this, as resistors are sealed components. But we can create the same tapping effect by combining two resistors in series as shown in Figure 8.07, and then our tapping becomes a connection point made between the two resistors.

If we look at Figure 8.07, we can see that the series combination of resistors R_1 and R_2 is connected across a supply that is provided by two rails. One is shown as V_+ (the positive supply rail or, in other words, our input) and the other as $0\,V$ (or common rail of the circuit).

The total resistance of our network (R_t) will be:

$$R_t = R_1 + R_2$$

We know, using Ohm's law, that $V = I \times R$ and the same current flows through both resistors. Therefore, for this network, we can see that $V_+ = I \times R_t$, and the voltage dropped across resistor R_2, $V_o = I \times R_2$.

We now have two expressions, one for V_+ and one for V_o. We can find out what fraction V_o is of V_+ by putting V_o over V_+ on the left-hand side of an equation and then putting what we said each one is equal to in the corresponding positions on the right-hand side. This gives us the following formula:

$$\frac{V_o}{V_+} = \frac{I \times R_2}{I \times R_t}$$

As current is common on the right hand side of our formula, they cancel each other out. This leaves us with:

$$\frac{V_o}{V_+} = \frac{R_2}{R_t}$$

To establish what V_o (our output voltage) actually is, we can transpose again, which would give us:

$$V_o = \frac{R_2 \times V_+}{R_t}$$

Finally, we can replace R_t by what it is actually equal to, and this will give us the means of establishing the value of the individual resistors needed to give a desired output voltage (V_o). By transposition, our final formula now becomes:

$$V_o = \frac{R_2}{R_1 + R_2} \times V_+$$

This equation is normally referred to as the potential divider rule.

In reality R_1 and R_2 could each be a combination (series or parallel) of many resistors.

However, as long as each combination is replaced by its equivalent resistance so that the simplified circuit looks like Figure 8.07, then the potential divider rule can be applied.

The potential divider circuit is very useful where the full voltage available is not required at some point in a circuit and, as we have seen, by a suitable choice of resistors in the potential divider, the desired fraction of the input voltage can be produced.

In applications where the fraction produced needs to be varied from time to time, the two resistors are replaced by a variable resistor (also known as a potentiometer, which is often abbreviated to the word pot), which would be connected as shown in Figure 8.08.

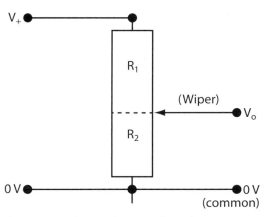

Figure 8.08 Circuit diagram for voltage applied across potentiometer

The potentiometer has a resistor manufactured in the form of a track, the ends of which effectively form our V_+ and $0\,V$ connections. Our output voltage (V_o) is achieved by means of a movable contact that can touch the track anywhere on its length and this is called the wiper. We have therefore effectively created a variable tapping point.

To compare this with our potential divider, we can say that the part of the track above the wiper can be regarded as R_1 and that part below the wiper as R_2. The fraction of the input voltage appearing at the output can therefore be calculated for any setting of the wiper position by using our potential divider equation:

$$V_o = \frac{R_2}{R_1 + R_2} \times V_+$$

Obviously, when the wiper is at the top of the track, R_1 becomes zero and the equation would give the result that V_o is equal to V_+. Equally, with the wiper right at the bottom of the track, R_2 now becomes zero and therefore V_o also becomes zero, which is not too surprising as the wiper is now more or less directly connected to the $0\,V$ rail.

This sort of circuit finds practical application in a wide variety of control functions, such as volume or tone controls on audio equipment, brightness and contrast controls on televisions and shift controls on oscilloscopes.

Power ratings

Resistors often have to carry comparatively large values of current, so they must be capable of doing this without overheating and causing damage. As the current has to be related to the voltage, it is the power rating of the resistor that needs to be identified.

The power rating of a resistor is thus really a convenient way of stating the maximum temperature at which the resistor is designed to operate without damage to itself. In general, the more power a resistor is designed to be capable of dissipating, the larger physically the resistor is. The resulting larger surface area aids heat dissipation.

Resistors with high power ratings may even be jacketed in a metal casing provided with cooling ribs and designed to be bolted flat to a metal surface – all to improve the radiation and conduction of heat away from the resistance element.

Power is calculated by:

$$P = V \times I$$

Instead of V we can substitute I × R for V and V/R for I. We can then use the following equations to calculate power:

$$P = I^2 \times R \qquad \text{or} \qquad P = \frac{V^2}{R}$$

What would the power rating of the 50 Ω resistor in Figure 8.09 be?

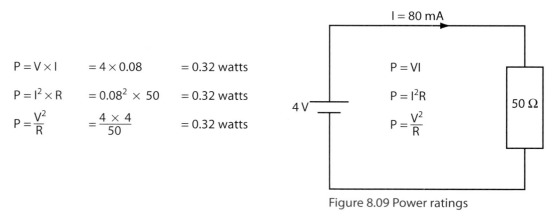

$$P = V \times I \qquad = 4 \times 0.08 \qquad = 0.32 \text{ watts}$$

$$P = I^2 \times R \qquad = 0.08^2 \times 50 \qquad = 0.32 \text{ watts}$$

$$P = \frac{V^2}{R} \qquad = \frac{4 \times 4}{50} \qquad = 0.32 \text{ watts}$$

Figure 8.09 Power ratings

Normally only one calculation is required. Typical power ratings for resistors are shown in Table 8.05.

Carbon resistors	0 to 0.5 watts
Ceramic resistors	0 to 6 watts
Wire wound resistors	0 to 25 watts

Table 8.05 Typical power ratings for resistors

Manufacturers also always quote a maximum voltage rating for their resistors on their data sheets. The maximum voltage rating is basically a statement about the electrical insulation properties of those parts of the resistor that are supposed to be insulators (e.g. the ceramic or glass rod which supports the resistance element or the surface coating over the resistance element).

If the maximum voltage rating is exceeded there is a danger that a flashover may occur from one end of the resistor to the other. This flashover usually has disastrous results. If it occurs down the outside of the resistor it can destroy not only the protective coating but, on film resistors, the resistor film as well.

If it occurs down the inside of the resistor the ceramic or glass rod is frequently cracked (if not shattered) and, of course, this mechanical damage to the support for the resistance element results in the element itself being damaged as well.

Testing

On completion of this topic area the candidate will be able to describe how to undertake the required tests for continuity, insulation resistance, polarity, earth fault loop impedance, earth electrode resistance, functional testing including operation of RCDs.

Continuity

Circuit protective conductors (cpcs) including main and supplementary equipotential bonding conductors

Regulations state that every protective conductor, including each bonding conductor, should be tested to verify that it is electronically sound and correctly connected. The test described below will check the continuity of the protective conductor and measure $R_1 + R_2$ which, when corrected for temperature, will enable the designer to verify the calculated earth fault loop impedance (Z_s). For this test you need a low-reading ohmmeter.

Test method 1. Before carrying out this test (shown in Figure 8.10) the leads should be 'nulled out'. If the test measurement does not have this facility, the resistance of the leads should be measured and deducted from the readings. The line conductor and the protective conductor are linked together at the consumer unit or distribution board. The ohmmeter is used to test between the line and earth terminals at each outlet in the circuit. The measurement at the circuit's extremity should be recorded and is the value of $R_1 + R_2$ for the circuit under test. On a lighting circuit the value of R_1 should include the switch wire at the luminaires. This method should be carried out before any supplementary bonds are made.

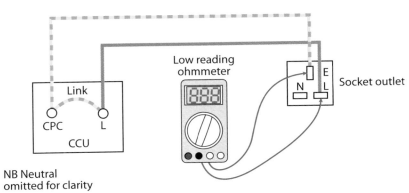

NB Neutral omitted for clarity

Figure 8.10 Test method 1

Test method 2. One lead of the continuity tester is connected to the consumer's main earth terminals (see Figure 8.11). The other lead is connected to a trailing lead, which is used to make contact with protective conductors at light fittings, switches, spur outlets etc. The resistance of the test leads will be included in the result; therefore the resistance of the test leads must be measured and subtracted from the reading obtained (since the instrument does not have a nulling facility). In this method the protective conductor only is tested and this reading (R_2) is recorded on the installation schedule.

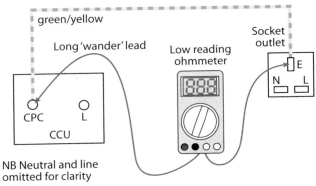

NB Neutral and line omitted for clarity

Figure 8.11 Test method 2

Test of the continuity of supplementary bonding conductors

Test method 2, as described on page 361, is used for this purpose. The ohmmeter leads are connected between the point being tested, between simultaneously accessible extraneous conductive parts, i.e. pipe work, sinks etc. or between simultaneously accessible extraneous conductive parts and exposed conductive parts (metal parts of the installation). The test will verify that the conductor is sound. To check this, move the probe to the metalwork to be protected, as shown in Figure 8.12. This method is also used to test the main equipotential bonding conductors.

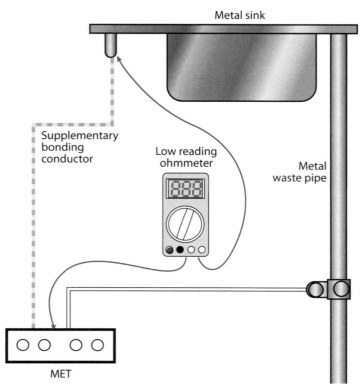

Figure 8.12 Test of the continuity of supplementary bonding conductors

Where ferrous enclosures have been used as the protective conductors, e.g. conduit, trunking, steel-wire armouring etc., the following special precautions should be followed.

- Perform the standard ohmmeter test using the appropriate test method described above. Use a low resistance ohmmeter for this test.

- Inspect the enclosure along its length to verify the integrity.

- If there is any doubt as to the soundness of this conductor, a further test using a line-earth loop impedance tester can be carried out after connection to the supply.

Continuity of ring final circuit conductors

A test is required to verify the continuity of each conductor including the circuit protective conductor (cpc) of every ring final circuit. The test results should establish that the ring is complete and has no interconnections. The test will also establish that the ring is not broken. Figure 8.13 shows a ring circuit illustrating these faults.

It may be possible, as an alternative and in order to establish that no interconnected multiple loops have been made in the ring circuit, for the inspector to check visually each conductor throughout its entire length. However, in most circumstances this will not be practicable and the following test method for checking ring circuit continuity is recommended.

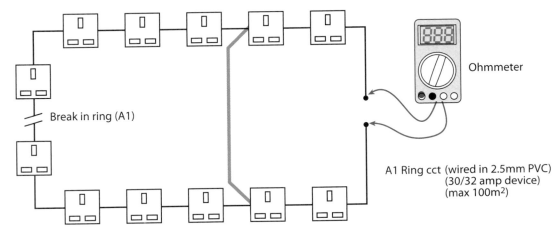

Figure 8.13 Test of continuity of ring final circuit conductors

The line, neutral and protective conductors are identified and their resistances are measured separately (see Figure 8.14). The end-to-end resistance of the protective conductor is divided by four to give R_2 and is recorded on the installation schedule form. A finite reading confirms that there is no open circuit on the ring conductors under test.

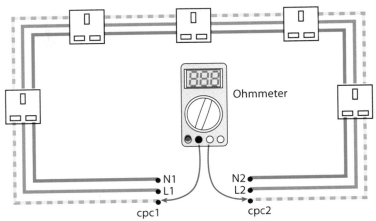

Figure 8.14 Measurement of phase, neutral and protective conductors

The resistance values obtained should be the same (within 0.05 ohms) if the conductors are the same size. If the protective conductor has a reduced csa, the resistance of the protective loop will be proportionally higher than that of the line or neutral loop. If these relationships are not achieved then either the conductors are incorrectly identified or there is a loose connection at one of the accessories.

The line and neutral conductors are then joined together so that the outgoing line conductor is linked to the returning neutral conductor and vice versa, as in Figure 8.15 (page 364). The resistance between line and neutral conductors is then measured at each socket outlet. The readings obtained from those sockets wired into the ring will be substantially the same and the value will be approximately half the resistance of the line or the neutral loop resistance. Any sockets wired as spurs will have a proportionally higher resistance value corresponding to the length of the spur cable.

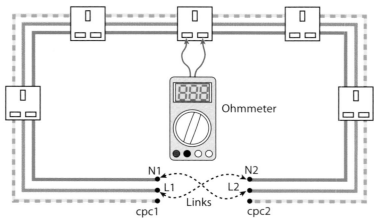

Figure 8.15 Line and neutral conductors connected together

The latter exercise is then repeated but with the line and cpc cross-connected as in Figure 8.16. The resistance between line and earth is then measured at each socket. The highest value recorded represents the maximum $R_1 + R_2$ of the circuit and is recorded on the test schedule. It can be used to determine the earth-loop impedance (Z_s) of the circuit, to verify compliance with the loop impedance requirements of the Regulations.

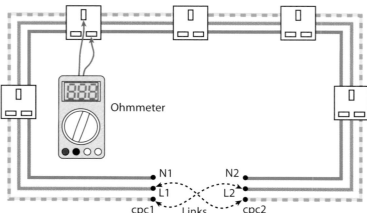

Figure 8.16 Line and cpc cross-connected

Insulation resistance

Insulation resistance tests are to verify, for compliance with BS 7671, that the insulation of conductors, electrical accessories and equipment is satisfactory and that electrical conductors and protective conductors are not short-circuited nor do they show a low insulation resistance (which would indicate defective insulation). In other words, we are testing to see whether the insulation of a conductor is so poor as to allow any conductor to 'leak' to earth or to another conductor. Before testing ensure that:

- pilot or indicator lamps and capacitors are disconnected from circuits to avoid an inaccurate test value being obtained

- voltage sensitive electronic equipment, such as dimmer switches, delay timers, power controllers, electronic starters for fluorescent lamps, emergency lighting, residual current devices etc. are disconnected so that they are not subjected to the test voltage

- there is no electrical connection between any line and neutral conductor (e.g. lamps left in).

To illustrate why we remove lamps, consider Figure 8.17. Should we leave the lamp in? The coil that is the lamp filament is effectively creating a short circuit between the line and neutral conductors.

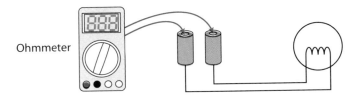

Ohmmeter

Figure 8.17 Insulation resistance test

The test equipment to be used would be an insulation resistance tester meeting the criteria as laid down in BS 7671, with insulation resistance tests carried out using the appropriate d.c. test voltage as specified in Table 61 of BS 7671 (shown on page 352 in Table 8.04). The installation will be deemed to conform with the Regulations if, with the main switchboard and each distribution circuit tested separately and with all its final circuits connected and current-using equipment disconnected, it has an insulation resistance not less than that specified in Table 61.

The tests should be carried out with the main switch off, all fuses in place, switches and circuit breakers closed, lamps removed, and fluorescent and discharge luminaires and other equipment disconnected. Where the removal of lamps and/or the disconnection of current-using equipment is impracticable, the local switches controlling such lamps and/or equipment should be open (see Regulation 612.3).

Simple installations that contain no distribution circuits should be tested as a whole; however, to perform the test in a complex installation it may need to be subdivided into its component parts. Although an insulation resistance value of not less than 1 megohm complies with the Regulations, where an insulation resistance value of less than 2 megohms is recorded the possibility of a latent defect exists. If this is the case, each circuit should be separately tested and its measured insulation resistance should be greater than 2 megohms.

Test 1: Insulation resistance between live conductors

- *Single-phase circuits* – test between the line and neutral conductors at the appropriate switchboard.
- *Three-phase circuits* – before this test is carried out you must ensure that the incoming neutral has been disconnected so there is no connection with earth.

Now make a series of tests between live conductors in turn at the appropriate switchboard as follows:

- between brown line and to the black line, grey line and neutral (blue) grouped together
- between black line and to the grey line and neutral (blue) grouped together
- between grey line and neutral (blue).

Where it is not possible to group conductors in this way, conductors may be tested singly. Figure 8.18 illustrates the testing of insulation resistance on a single-phase lighting circuit. But remember, on a two-way circuit operate the two-way switch and retest to make sure that you have tested all of the strappers!

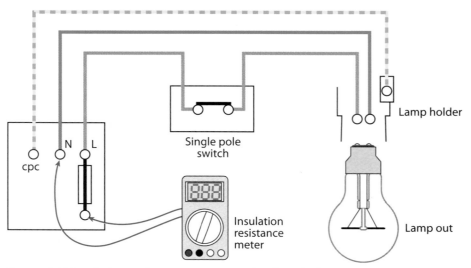

Figure 8.18 Insulation resistance test on single-phase lighting circuit

Test 2: Insulation resistance from earth to line and neutral connected together

- *Single-phase circuits* – test between the line and neutral conductors and earth at the appropriate distribution board. Where any circuits contain two-way switching, the two-way switches will need to be operated and another insulation resistance test carried out, including the two-way strapping wire which was not previously included in the test.

- *Three-phase circuits* – measure between all line conductors and neutral bunched together, and earth. Where a low reading is obtained (less than 2 megohms) it may be necessary to test each conductor to earth separately. Figure 8.19 shows the test of insulation resistance between earth to line and neutral connected together on a socket outlet circuit.

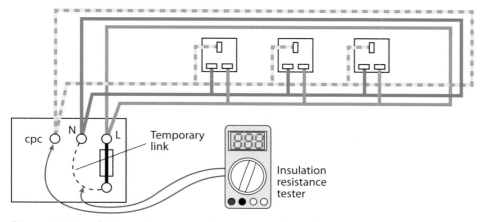

Figure 8.19 Insulation resistance test between earth to line and neutral connected together on socket outlets

Some electricians prefer, or find it quicker, to test between individual conductors rather than to group them together. As an example, Figure 8.20 shows the ten readings that would need to be taken for a three-phase circuit.

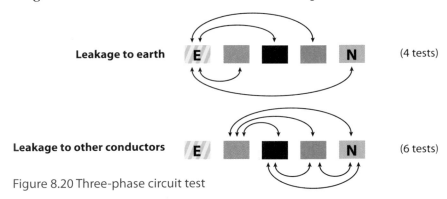

Figure 8.20 Three-phase circuit test

Polarity

A test needs to be performed to check the polarity of all circuits. This must be done before connection to the supply, with either an ohmmeter or the continuity range of an insulation and continuity tester.

The purpose of the test is to confirm that all protective devices and single-pole switches are connected to the line conductor, and that the line terminal in socket outlets and the centre contact of screw-type lamp holders are also connected to the line conductor. (In addition, a check should be made to ensure that the polarity of the incoming supply is correct, otherwise the whole installation would have the wrong polarity.)

In essence, having established the continuity of the cpc in an earlier test, we now use this as a long test lead, temporarily linking it with the cpc at the distribution board and then making our test across the line and earth terminals at each item in the circuit under test. Remember to close lighting switches before carrying out the test.

Figure 8.21 shows the test for polarity of a lighting circuit using a continuity tester.

For ring circuits, if the tests required by Regulation 612.2.2 ring circuit continuity have been carried out, the correct connections of line, neutral and cpc conductors will have been verified and no further testing is required. For radial circuits the $R_1 + R_2$ measurements should also be made at each point, using this method.

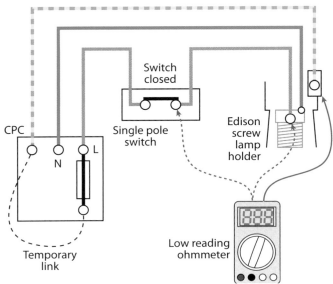

Figure 8.21 Polarity test of lighting circuit using continuity tester

Earth electrode resistance (Regulation 612.7)

Where the earthing system incorporates an earth electrode as part of the system, the electrode resistance to earth needs to be measured. In previous years, the metal pipes of water mains were used; however the change to more plastic pipe work means this practice can no longer be relied upon.

Some of the types of accepted earth electrode are:

- earth rods or pipes
- earth tapes or wires
- earth plates
- underground structural metalwork embedded in foundations
- lead sheaths or other metallic coverings of cables
- metal pipes.

The resistance to earth depends upon the size and type of electrode used; remember that we want as good a connection to earth as possible. The connection to the electrode must be made above ground level.

Measurement by standard method

When measuring earth electrode resistances to earth where low values are required, as in the earthing of the neutral point of a transformer or generator, test method 1 below may be used, using an earth electrode resistance tester.

Test method 1

Before this test is undertaken, the earthing conductor to the earth electrode must be disconnected either at the electrode or at the main earthing terminal. This will ensure that all the test current passes through the earth electrode alone. However, as this will leave the installation unprotected against earth faults, switch off the supply before disconnecting the earth.

The test should be carried out when the ground conditions are at their least favourable, i.e. during a period of dry weather, as this will produce the highest resistance value.

The test requires the use of two temporary test electrodes (spikes) and is carried out in the following manner.

- Connect the earth electrode to terminals C_1 and P_1 of a four-terminal earth tester. To exclude the resistance of these test leads from the resistance reading, individual leads should be taken from these terminals and connected separately to the electrode. However, if the test lead resistance is insignificant, the two terminals may be short-circuited at the tester and connection made with a single test lead, the same being true if you are using a three-terminal tester.

- Connection to the temporary 'spikes' is now made, as shown in Figure 8.22. The distance between the test spikes is important. If they are too close together their resistance areas will overlap. In general, you can expect a reliable result if the distance between the electrode under test and the current spike is at least ten times the maximum length of the electrode under test, e.g. 30 metres away from a 3m long rod electrode. (Resistance area is important where livestock are concerned, as the front legs of an animal may be outside and the back legs of the same animal inside the resistance area, thus creating a potential difference. As little as 25 V can be lethal to an animal, so it is important to ensure that all of the electrode is well below ground level and RCD protection is used).

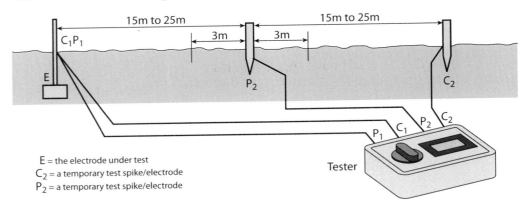

E = the electrode under test
C_2 = a temporary test spike/electrode
P_2 = a temporary test spike/electrode

Figure 8.22 Earth electrode test

- We then take three readings:

  1. with the potential spike initially midway between the electrode and current spike

  2. at a position 10 per cent of the electrode-to-current spike distance back towards the electrode

  3. at a position 10 per cent of the distance towards the current spike.

- By comparing the three readings a percentage deviation can be determined. We do this calculation by taking the average of the three readings, finding the maximum deviation of the readings from this average in ohms, and expressing this as a percentage of the average. The accuracy of the measurement using this technique is on average about 1.2 times the percentage deviation of the readings. It is difficult to achieve an accuracy better than 2 per cent, and you should not accept readings that differ by more than 5 per cent. To improve the accuracy of the measurement to acceptable levels, the test must be repeated with larger separation between the electrode and the current spike.

Once the test is completed, make sure that the earthing conductor is re-connected.

Test method 2

Guidance Note 3 to BS 7671 lists a test method 2. However, this is an alternative method for use on RCD protective TT installations only.

Earth-fault loop impedance

When designing an installation, it is the designer's responsibility to ensure that, should a line-to-earth fault develop, the protection device will operate safely and within the time specified by BS 7671. Although the designer can calculate this in theory, it is not until the installation is complete that the calculations can be checked.

It is necessary, therefore, to determine the earth-fault loop impedance (Z_s) at the furthest point in each circuit and to compare the readings obtained with either the designer's calculated values or the values tabulated in BS 7671 or the *IEE On-Site Guide*.

The earth-fault loop is made up of the following elements:

- the line conductor from the source of the supply to the point of the fault
- the circuit protective conductor
- the main earthing terminal and earthing conductor
- the earth return path (dependent on the nature of the supply, TN-S, TN-C-S etc.)
- the path through the earthed neutral of the supply transformer
- the secondary winding of the supply transformer.

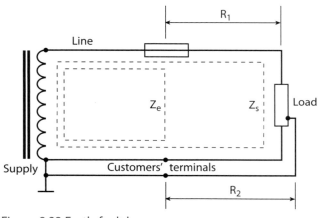

Figure 8.23 Earth-fault loop

The value of earth-fault loop impedance may be determined by:

- direct measurement of Z_s
- direct measurement of Z_e at the origin of the circuit and adding to this the value of $R_1 + R_2$ measured during continuity tests: $Z_s = Z_e + (R_1 + R_2)$
- obtaining the value of Z_e from the electricity supplier and adding to this the value of $R_1 + R_2$ as above. However, where the value of Z_e is obtained from the electricity supplier and is not actually measured, a test must be carried out to ensure that the main earthing terminal is in fact connected to earth, using an earth-loop impedance tester or an approved test lamp.

Direct measurement of Z_s

Direct measurement of earth-fault loop impedance is achieved by use of an earth-fault loop impedance tester, which is an instrument designed specifically for this purpose. The instrument operates from the mains supply and, therefore, can only be used on a live installation.

The instrument is usually fitted with a standard 13 A plug for connecting to the installation directly through a normal socket outlet, although test leads and probes are also provided for taking measurements at other points on the installation.

Earth-fault loop impedance testers are connected directly to the circuit being tested and care must be taken to prevent danger. Should a break have occurred anywhere in the protective conductor under test then the whole of the earthing system could become live. It is essential, therefore, that protective conductor continuity tests be carried out prior to the testing of earth-fault loop impedance. Communication with other users of the building and the use of warning notices and barriers is essential.

Measurement of Z_e

The value of Z_e can be measured using an earth fault loop impedance tester at the origin of the installation. However, as this requires the removal of covers and the exposure of live parts, extreme care must be taken and the operation must be supervised at all times. The instrument is connected using approved leads and probes between the line terminal of the supply and the means of earthing, with the main switch open or with all sub-circuits isolated. To remove the possibility of parallel paths, the means of earthing must be disconnected from the equipotential bonding conductors for the duration of the test. With the instrument correctly connected and the test button pressed, the instrument will give a direct reading of the value of Z_e. It must be remembered to reconnect all earthing connections on completion of the test.

Verification of test results

The values of earth loop impedance obtained (Z_s) should be compared with one of the following:

- for circuits wired in standard PVC-insulated cables, the values given in the *IEE On-Site Guide* after correction for ambient temperature where necessary
- the earth-fault loop impedance values determined by the designer
- the values given in Tables 41.2, 41.3 or 41.4 of BS 7671 *IEE Wiring Regulations* after being corrected for the difference in temperature between normal conductor operating temperature and the ambient temperature when the test takes place
- rule of thumb.

Standard PVC circuits

The measured loop impedances given in the *IEE On-Site Guide* (repeated in Tables 8.06 to 8.10 – pages 372–374) have been determined to ensure compliance with the required disconnection times where conventional final circuits are installed. The values assume an ambient temperature, when carrying out the tests, of between 10–20°C. At ambient temperatures other than this, correction factors shown in Table 8.06 should be applied.

Design values

Where the designer of the installation provides calculated values of earth-loop impedance, the measured values should be compared with these.

Tables 41.2, 41.3 and 41.4

You may need to refer to BS 7671 2001 for measured values of earth-loop impedance given in Tables 41.2, 41.3 and 41.4 of BS 7671, which should not be exceeded when the conductors are at their normal operating temperature. If the conductors are at a different temperature when tested, the reading should be adjusted accordingly.

Rule of thumb

As an approximation or 'rule of thumb', the measured value of earth-fault loop impedance (Z_s) at the most remote outlet should not exceed three-quarters of the relevant value given in Tables 41.2, 41.3 or 41.4.

Warning: If the circuit being tested is protected by an RCD, the test procedure may cause the RCD to operate causing unwanted isolation of the circuit. Certain types of test instrument may be used that are specifically designed to overcome this problem; otherwise it will be necessary to measure the value of R_1 and R_2 with the circuit isolated and add this to the value of Z_e measured at the incoming terminals.

Did you know?

A 0.4 second disconnection time is used for circuits up to 32A. Circuits over 32A may use a 5 second disconnection time.

(a) 0.4-second disconnection time NP = not permitted

Protective conductor (sq mm)	Fuse rating (amps)				
	5	15	20	30	45
1.0	8.00	2.14	1.48	NP	NP
1.5	8.00	2.14	1.48	0.91	NP
2.5 to 16.0	8.00	2.14	1.48	0.91	0.50

(b) 5-second disconnection time NP = not permitted

Protective conductor (sq mm)	Fuse rating (amps)				
	5	15	20	30	45
1.0	14.8	4.46	2.79	NP	NP
1.5	14.8	4.46	3.20	2.08	NP
2.5	14.8	4.46	3.20	2.21	1.20
4.0 to 16.0	14.8	4.46	3.20	2.21	1.33

Table 8.06 *IEE On-Site Guide* Table 2A. Maximum measured value of earth-loop impedance (in ohms) when the overcurrent protective device is a fuse to BS 3036

(a) 0.4-second disconnection time NP = not permitted

Protective conductor (sq mm)	Fuse rating (amps)							
	6	10	16	20	25	32	40	50
1.0	7.11	4.26	2.26	1.48	1.20	0.69	NP	NP
1.5	7.11	4.26	2.26	1.48	1.20	0.87	0.67	NP
2.5 to 16.0	7.11	4.26	2.26	1.48	1.20	0.87	0.69	0.51

(b) 5-second disconnection time NP = not permitted

Protective conductor (sq mm)	Fuse rating (amps)							
	6	10	16	20	25	32	40	50
1.0	11.28	6.19	3.20	1.75	1.24	0.69	NP	NP
1.5	11.28	6.19	3.49	2.43	1.60	1.12	0.67	NP
2.5	11.28	6.19	3.49	2.43	1.92	1.52	1.13	0.56
4.0	11.28	6.19	3.49	2.43	1.92	1.52	1.13	0.81
6.0 to 16.0	11.28	6.19	3.49	2.43	1.92	1.52	1.13	0.87

Table 8.07 *IEE On-Site Guide* Table 2B. Maximum measured earth-fault loop impedance (in ohms) when the overcurrent device is a fuse to BS 88

(a) 0.4-second disconnection time NP = not permitted

Protective conductor (sq mm)	Fuse rating (amps)				
	5	15	20	30	45
1.0	8.72	2.74	1.42	0.80	NP
1.5	8.72	2.74	1.42	0.96	0.34
2.5 to 16.0	8.72	2.74	1.42	0.96	0.48

(b) 5-second disconnection time NP = not permitted

Protective conductor (sq mm)	Fuse rating (amps)				
	5	15	20	30	45
1.0	13.68	4.18	1.75	0.80	NP
1.5	13.68	4.18	2.34	1.20	0.34
2.5	13.68	4.18	2.34	1.54	0.53
4.0	13.68	4.18	2.34	1.54	0.70
6.0 to 16.0	13.68	4.18	2.34	1.54	0.80

Table 8.08 *IEE On-Site Guide* Table 2C. Maximum measured earth-fault loop impedance (in ohms) when the overcurrent protective device is a fuse to BS 1361

Ambient temperature (degrees Celsius)	Correction factors
0	0.96
5	0.98
10	1.00
25	1.06
30	1.08

Table 8.09 Ambient temperature correction factors

MCB type	MCB rating (amps)												
	5	6	10	15	16	20	25	30	32	40	45	50	63
1	9.60	8.00	4.80	3.20	3.00	2.40	1.92	1.60	1.50	1.20	1.06	0.96	0.76
2	5.49	4.57	2.74	1.83	1.71	1.37	1.10	0.91	0.86	0.69	0.61	0.55	0.43
B	–	6.40	3.84	–	2.40	1.92	1.54	–	1.20	0.96	0.86	0.77	0.61
3 or C	3.84	3.20	1.92	1.28	1.20	0.96	0.77	0.64	0.60	0.48	0.42	0.38	0.30
D	1.92	1.60	0.96	0.64	0.60	0.48	0.38	0.32	0.30	0.24	0.22	0.19	0.15

Table 8.10 *IEE On-Site Guide* Table 2D. Maximum measured earth-fault loop impedance (in ohms) when overcurrent protective device is an MCB to BS 3871 or BS EN 60898

Operation of residual current devices (RCDs)

Although the majority of residual current devices (RCDs) in use are designed to operate at earth leakage currents not exceeding 30 mA, other values of operating current are available for use in special circumstances. All RCDs are electromechanical devices that must be checked regularly to confirm that they are in working order. This can be done at regular intervals by simply pressing the test button on the front of the device.

Where an installation incorporates an RCD, Regulation 514.12.2 requires a notice to be fixed in a prominent position at or near the origin of the installation.

The integral test button incorporated in all RCDs only verifies the correct operation of the mechanical parts of the RCD and does not provide a means of checking the continuity of the earthing conductor, the earth electrode or the sensitivity of the device. This can only be done effectively by use of an RCD tester specifically designed for testing RCDs as described below.

Test method

The test must be made on the load side of the RCD between the line conductor of the protected circuit and the associated circuit protective conductor. The load being supplied by the RCD should be disconnected for the duration of the test. The test instrument is usually fitted with a standard 13 A plug top. The easiest way of making these connections, wherever possible, is by plugging the instrument into a suitable socket outlet protected by the RCD under test.

The test instrument operates by passing a simulated fault current of known value through the RCD and then measures the time taken for the device to trip. Although different types of RCD have different requirements (time delays etc.), for general purpose RCDs the test criteria are as follows.

- With an earth leakage current equivalent to 50 per cent of the rated tripping current of the RCD the device should not open (this is to ensure that nuisance tripping, due to leakage currents less than the rated tripping current, will not occur).

- For general purpose RCDs to BS 4293, an earth leakage current equivalent to 100 per cent of the rated tripping current of the RCD should cause the device to open in less than 200 milliseconds.

- Individual RCD-protected socket outlets should also meet the above criteria.

- For general purpose RCDs to BS EN 61008 or RCBOs to BS EN 61009, an earth leakage current equivalent to 100 per cent of the rated tripping current of the RCD should cause the device to open in less than 300 milliseconds, unless it is of 'Type S' which incorporates an intentional time delay, in which case it should trip within the time range of 130 to 500 milliseconds.

- Where an RCD with an operating current not exceeding 30 mA is used to provide supplementary protection against direct contact, a test current of five times the rated trip current (e.g. 5 × 30 = 150 mA) should cause the RCD to open in less than 40 milliseconds.

A further mechanical function check should be carried out to ensure that the operating switch of the residual current device is functioning correctly. Results of all the above tests should be recorded on the appropriate test results schedule.

Safety

Under certain circumstances these tests can result in potentially dangerous voltages appearing on exposed and extraneous conductive parts within the installation, therefore suitable precautions must be taken to prevent contact by persons or livestock with any such part. Other building users should be made aware of the tests being carried out and warning notices should be posted as necessary.

Safety tip

Prior to RCD tests being carried out, it is essential for safety reasons that the earth loop impedance of the installation has been tested to check that the earth-return path is sound and that all the necessary requirements have been met. BS 7671 stipulates the order in which tests should be carried out

Compliance with BS 7671

On completion of this topic area the candidate will be able to explain the need to compare test values with those in BS 7671 and state the actions to be taken in the event of unsatisfactory results.

Continuity

When testing the continuity of circuit protective conductors or bonding conductors we should always expect a very low reading, which is why we must always use a low-reading ohmmeter. Main and supplementary bonding conductors should have a reading of not more than 0.05 ohms while the maximum resistance of circuit protective conductors can be estimated from the value of ($R_1 + R_2$) given in Table 9A of the *IEE On-Site Guide*. These values will depend upon the cross-sectional area of the conductor, the conductor material and its length.

A very high (end-of-scale) reading would indicate a break in the conductor itself or a disconnected termination that must be investigated. A mid-range reading may be caused by the poor termination of an earthing clamp to the service pipe; e.g. a service pipe which is not cleaned correctly before fitting the clamp or corrosion of the metal service pipe due to its age and damp conditions.

Poor values of earth continuity mean that, in the event of an earth fault, insufficient current will flow causing the protective device to operate within the times required by BS 7671. The smallest sizes of steel conduit and trunking will have resistance values of 0.005 ohms/metre and values of cpc will depend on their size. However, a 1.00mm^2 conductor will have resistance of 0.018 Ω/metre while 2.5mm^2 is approximately 0.007 Ω/metre. If higher values than expected are found:

- confirm the correct size of cpc
- install a larger one if necessary
- check the effectiveness of the connections
- look for corrosion on the termination
- make sure any paint has been removed from any clamps that have been used.

Polarity

Incorrect polarity is one of the most dangerous situations that can occur in an installation. The aim of tests for correct polarity is to confirm that all single pole switches and protective devices are fitted in the line conductor.

Switches fitted in the neutral conductor would result in the supply to the electrical equipment still able to be turned on and off. However, when the switch is in the off position the supply would still be live up to the terminals of the equipment. If this

Did you know?

A reading only slightly higher than the required reading may be possible to correct by replacing the conductor with one of a larger cross-sectional area were a lighting circuit, then anyone turning the light off to change the bulb would be at risk of an electric shock.

If the result of reverse polarity was to effectively put the protective devices in the neutral conductor then, in the event of an earth fault, the fault current would not pass through the protective device. Therefore automatic disconnection of the supply would not occur and this could give rise to:

- potential fatal electric shock

- fire

- death by smoke inhalation as a result of fire.

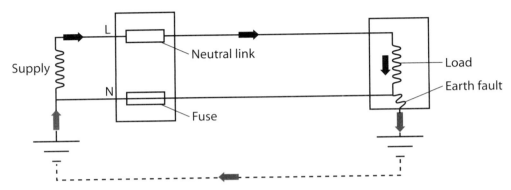

Figure 8.24 Current action in Earth fault

In Figure 8.24 the red arrows indicate the path the current will take in the event of an earth fault. If, as a result of reverse polarity, the fuse/protective device is connected in the neutral conductor as shown above, then the fault current will not pass through the protective device and therefore will continue to flow, leading to a hazard. If tests confirm a reversed polarity:

- remove the supply immediately, if connected

- alter the terminations to comply with polarity so that single pole switches and protective devices are connected in the line conductor, and the centre contact of ES lamp holders are also connected to the line conductor.

Insulation resistance

The level of insulation resistance of an installation will depend upon its size, complexity and the number of circuits connected to it. When testing a small domestic installation you may expect an insulation resistance reading in excess of 200 megohms. For a large industrial or commercial installation with many sub-circuits, each providing a parallel path, you will get a much smaller reading. It is recommended that, where the insulation resistance reading is less than 2 megohms, individual distribution boards or even individual sub-circuits be tested separately to identify any possible cause of poor insulation values.

A zero value of insulation resistance would indicate either:

- a possible short circuit between live conductors
- an earth fault, if a bare conductor is in contact with earth at some point in the installation.

Both of these must be investigated immediately.

A low value reading of below 1 megohm suggests a latent weakness in the insulation. This could possibly be due to the ingress of dampness or dirt in such items as distribution boards, joint boxes or lighting fittings etc. Other causes of low insulation resistance can be the infestation of equipment by rats, mice or insects.

Earth-fault loop impedance

The earth-fault loop path is made up of those parts of the supply system external to the premises being tested (Z_e) and the line conductor and circuit protective conductor within the installation ($R_1 + R_2$), the total earth-fault loop impedance being $Z_s = Z_e + (R_1 + R_2)$.

If the value of impedance measured is higher than that required by the tables in Chapter 41 of BS 7671, then this indicates that, in the event of an earth fault, disconnection of the protective device may not happen. If it does take place, it certainly will not be within the time constraints on protective device activation required by BS 7671. Therefore, people living in or working on the installation would be at great risk.

Options available would be to:

- check the effectiveness of the terminations
- increase the size of the cpc
- change the type of protective device used
- add a RCD to protect the circuit, providing it is electrically sound.

Functional testing of residual current devices (RCDs)

Where a residual current device (RCD) fails to trip when pressing the integral test button, this would indicate a mechanical fault within the device itself, which should therefore be replaced.

When a residual current device fails to trip when being tested by an RCD tester, this would suggest that there is a fault with the RCD and that it should be replaced. It may be that there is an issue with the cpc; however a test of the earth-loop impedance would prove whether this is satisfactory or not.

If the RCD does trip out, but not within the time specified, then a check should be made that the test instrument is set correctly for the nominal tripping current of the device under test. If the correct tripping current was selected then this indicates that the RCD would fail to give the protection required and, therefore, would need replacing.

A RCD is fitted to the circuit for safety and protection. If the device is not working, then the installation is not protected and people and livestock are at risk of electric shock.

Certification

On completion of this topic area the candidate should be able to describe the certification process for completed installation.

Electrical Installation Certificate

BS 7671 requires that, following the inspection and testing of all new installations, alterations and additions to existing installations or periodic inspections, an Electrical Installation Certificate, together with a schedule of test results, should be given to the person ordering the work.

Examples of such forms are provided in appendix 6 of the Regulations and are also referred to in the *IEE On-Site Guide* and *IEE Guidance Note 3*. Other professional organisations, such as the Electrical Contractors Association (ECA) and the National Inspection Council for Electrical Installation Contractors (NICEIC), produce their own versions of these forms for their own members' use. The different types of forms available are as follows:

- Periodic Inspection Report
- Electrical Installation Certificate (Type 1)
- Electrical Installation Certificate (Type 2)
- Inspection Schedule
- Schedule of Test Results
- Electrical Installation – Minor Works Certificate.

Periodic Inspection Report

The Periodic Inspection Report is for use when carrying out a routine periodic inspection and test of an existing installation; it is not for use when alterations and additions are made. An inspection schedule and a schedule of test results should accompany the Periodic Inspection Report. The general requirements for the content and completion of all the above documents are stated in the introduction to Appendix 6 of BS 7671 and are summarised below.

- Electrical Installation Certificates and Electrical Installation – Minor Works Certificates must be completed and signed by a competent person or persons in respect of the design, the construction, and the inspection and test of the installation.
- Periodic Inspection Reports must be completed and signed by a competent person in respect of just the inspection and test of the installation.
- Competent persons must have a sound knowledge and relevant experience of the type of work being undertaken and of the technical requirements of BS 7671. They must also have a sound knowledge of the inspection and testing procedures contained in the Regulations and must use suitable testing equipment.

- Electrical Installation Certificates must identify who is responsible for the design, construction, inspection and testing, whether this is new work or an alteration or addition to an existing installation. As explained below this may be one person or it may be a number of different organisations.

- Minor Works Certificates must also identify who is responsible for the design, construction, inspection and testing of the work carried out.

- Periodic Inspection Reports must indicate who is responsible for carrying out the periodic inspection and testing of the existing installation.

- A schedule of test results must be issued with all Electrical Installation Certificates and periodic inspection reports.

- When signing the above forms on behalf of a company or other business organisation, individuals should state for whom they are acting.

- For larger or more complex installations, additional documentation, such as inspection checklists, may be required for clarification purposes.

The design, construction, inspection and testing of an installation may be the responsibility of just one person, in which case the shortened version of the Electrical Installation Certificate may be used.

On larger or more complex installations the design, construction, inspection and testing may be the responsibility of a number of different organisations, in which case all the parties concerned should be identified and must sign the certificate in the appropriate place. This may be the case where a design consultant is employed to carry out the design of the installation and an electrical contractor is employed to carry out the construction.

Many larger companies now employ dedicated inspectors whose job it is to inspect and test the work of other electricians within the company. In this case, the certificate may be signed by the person carrying out the installation work as the constructor, and the person carrying out the inspection and test as the inspector.

Provided that each party is clearly identified in the appropriate place, this is quite acceptable. In other cases, the inspection and testing of the installation may be sub-contracted to a second contractor, in which case they should sign the certificate for that section of the work.

The certificate also provides accommodation for the identification and signatures of two different design organisations. This is for use where different sections of the installation are designed by different parties, e.g. the electrical installation may be designed by one consultant and the fire-alarm system or other specialist service may be designed by another. If this should be the case then the details and signatures of both parties are required. BS 7671 provides advice notes for the completion of each type of form and also provides guidance notes for the person receiving the certificate. These are particularly helpful for non-technical persons.

Electrical Installation Certificate (Type 1)

These are used when inspecting and testing a new installation, a major alteration or addition to an existing installation where the design, construction, and inspection and testing of the installation are the responsibility of different organisations. See Figure 8.25.

For design

I/We being the person(s) responsible for the design of the Electrical Installation (as indicated by my/our signatures) particulars of which are described above, having exercised reasonable skill and care when carrying out the design hereby CERTIFY that the design work for which I/we have been responsible is to the best of my/our knowledge and belief in accordance with BS7671 (2008) amended to .... except for the departures (if any), detailed below.

Details of departures from BS 7671 (Regulations 120.3 and 120.4):

....

....

....

The extent of liability of the signatory is limited to the work described above.
For the DESIGN of the installation:

Signature .... Date: .... Name: .... Designer No. 1

Signature .... Date: .... Name: .... Designer No. 1

For construction

I/We being the person(s) responsible for the construction of the Electrical Installation (as indicated by my/our signatures) particulars of which are described above, having exercised reasonable skill and care when carrying out the construction hereby CERTIFY that the construction work for which I/we have been responsible is to the best of my/our knowledge and belief in accordance with BS7671 (2008) amended to .... except for the departures (if any), detailed below.

Details of departures from BS 7671 (Regulations 120.3 and 120.4):

....

....

....

The extent of liability of the signatory is limited to the work described above.
For the CONSTRUCTION of the installation:

Signature .... Date: .... Name: .... Constructor

For instruction and testing

I/We being the person(s) responsible for the inspection and testing of the Electrical Installation (as indicated by my/our signatures) particulars of which are described above, having exercised reasonable skill and care when carrying out the inspection and testing hereby CERTIFY that the work for which I/we have been responsible is to the best of my/our knowledge and belief in accordance with BS7671 (2008) amended to .... except for the departures (if any), detailed below.

Details of departures from BS 7671 (Regulations 120.3 and 120.4):

....

....

....

The extent of liability of the signatory is limited to the work described above.
For the INSPECTION and TEST of the installation:

Signature .... Date: .... Name: .... Inspector

Figure 8.25 Electrical Installation Certificate (Type 1)

As stated previously, the full certificate requires not only the signatures of those persons responsible for the design, construction, inspection and testing of the installation but also their full details including the company they represent. See Figure 8.26.

Particulars of the Signatories to the Electrical Installation Certificate

Designer No.1

Name : .... Company : ....
Address : ....
....
Postcode : .... Tel No.: ....

Designer No. 2
(if applicable)

Name : .... Company : ....
Address : ....
....
Postcode : .... Tel No.: ....

Constructor

Name : .... Company : ....
Address : ....
....
Postcode : .... Tel No.: ....

Inspector

Name : .... Company : ....
Address : ....
....
Postcode : .... Tel No : ....

Figure 8.26 Signatories to a full certificate

Electrical Installation Certificate (Type 2)

Client details : ....

Address of Installation : ....

....

.... **Postcode :** ....

Description and Extent of Installation :	New Installation
Description :	
....	**Addition**
Extent :	
....	**Alteration**

For design, Construction and Inspection and Testing

I being the person responsible for the Design, Construction and Inspection and Testing of the Electrical Installation (as indicated by my signature below) and particulars of which are described above, having exercised reasonable skill when carrying out the Design, Construction and Inspection & Test, hereby CERTIFY that the said work for which I have been responsible is to the best of my knowledge and belief in accordance with BS 7671, amended to .... (date) except for the departures, if any, detailed below.

Departures from BS 7671 (if any)

....

....

The extent of liability of the signatory is limited to the work described above as the subject of this certificate.

Name : .... Position : ....
Signature : .... Date : ....
For and on behalf of : ....
Address : ....
.... Tel no. : ....

Next Inspection

I recommend that this installation is further inspected and tested after an interval of not more than : ....

Supply characteristics and earthing arrangements

Earthing arrangements	Number and Type of live conductors	Nature of supply	Protective device characteristics
TN-S	1 phase – 2 wire	Nominal voltage :	Type : ....
TN-C-S	1 phase – 3 wire	Frequency :	Current rating : ....
TT	2 phase – 3 wire	Prospective fault current :	
	3 phase – 4 wire	External loop impedance :	

Particulars of the installation

Means of earthing
Suppliers facility

Maximum demand: .... kVA/Amperes

Earth electrode

Details of earth electrode (where applicable)
Type: .... Location: .... Resistance to earth: ....

Main protective conductors

Earthing conductor: Material: .... csa: .... Connection verified: Yes/No

Main equipotential bonding conductors Material: .... csa: .... Connection verified: Yes/No

To incoming water and/or gas services: ....

To other elements ....

Main switch or circuit breaker

BS. Type: .... No of poles: Current rating: A Voltage rating: V

Location: .... Fuse rating or setting: A

RCD operating current: .... mA RCD operating time .... mS

Comments on existing installation: (in cases of alterations or additions) ....

....

Schedules

The attached Inspection Checklists and Schedule of Test Results form part of this certificate.

Attached: .... Inspection Checklists and .... Schedule of Test results

Figure 8.27 Electrical Installation Certificate (Type 2 short form)

This is a shortened version of the Type 1 form. It is used when inspecting and testing a new installation, major alteration or addition to an existing installation where one person is responsible for the design, construction, inspection and testing of the installation.

This certificate is intended for use with new installations or alterations or additions to an existing circuit where new circuits have been installed. The original certificate is to be given to the person ordering the work and a copy retained by the contractor. This certificate is not valid unless accompanied by a Schedule of Test Results.

The signatures on the certificate should be those of the persons authorised by the companies carrying out the work of design, construction, inspection and testing.

The recommended time to the next inspection should be stated, taking into account the frequency and quality of maintenance that the installation is likely to receive.

The prospective fault current should be the greater of either that measured between live conductors or that between phase conductor(s) and earth.

Guidance notes

Guidance notes for those receiving the certificate are as follows.

- This certificate has been issued to confirm that the electrical installation work to which it relates has been designed, constructed, inspected and tested in accordance with BS 7671 (*IEE Wiring Regulations*).

- As the person ordering the work you should have received the original certificate and the contractor should have retained a copy. However, if you are not the user of the installation (the occupier), then you should pass the certificate or a full copy (including schedules) to the user immediately.

- The original certificate should be retained in a safe place to be shown to anyone carrying out further work on the installation at a later date (this includes periodic inspection). If you later vacate the property, this certificate will demonstrate that the installation did comply with BS 7671 at the time the certificate was issued. The Construction (Design and Management) (CDM) Regulations require that copies of the Installation Certificate together with any attached schedules are retained in the project Health and Safety File.

- To ensure the safety of the installation it will need to be inspected at regular intervals throughout its life. The recommended period between inspections is stated on the certificate.

- This certificate is only intended to be used for new installations or for new work such as an alteration or addition to an existing installation. For the routine inspection and testing of an existing installation a periodic inspection report should be issued.

- This certificate is only valid if it is accompanied by the appropriate schedule of test results.

Inspection Schedule

The Inspection Schedule provides confirmation that a visual inspection has been carried out as required by Part 6 of BS 7671 and lists all the inspection requirements of Regulation 611.3. The completed Inspection Schedule is attached to, and forms part of, the Electrical Installation Certificate.

Each item on the Inspection Schedule (Figure 8.28) should be checked and either ticked as satisfactory or ruled out if not applicable. In almost all cases, basic protection against direct contact will be by the presence of insulation and the enclosure of live parts. Fault protection will be by protection equipotential bonding and automatic disconnection of supply (together with the presence of earthing and bonding conductors). Therefore, other means of protection referred to on page 384 can be ruled out as not applicable.

1. Connection of conductors
2. Identification of conductors
3. Routing of cables in safe zones, or protection against mechanical damage, in compliance with Section 522
4. Selection of conductors for current-carrying capacity and voltage drop, in accordance with the design
5. Connection of single-pole devices for protection or switching in line conductors only
6. Correction connection of accessories and equipment
7. Presence of fire barriers, suitable seals and protection against thermal effects
8. Methods of protection against electric shock
   (a) both basic protection and fault protection, i.e.:
      – SELV
      – PELV
      – Double insulation
      – Reinforced insulation
   (b) basic protection (including measurement of distances, where appropriate), i.e.:
      – protection by insulation of live parts
      – protection by a barrier or an enclosure
      – protection by obstacles
      – protection by placing out of reach
   (c) fault protection:
      (i) automatic disconnection of supply
         presence of earthing conductor
         presence of circuit protective conductors
         presence of supplementary bonding conductors
         presence of earthing arrangements for combined protective and functional purposes
         presence of adequate arrangements for alternative source(s), where applicable
         FELV
         choice and setting of protective and monitoring devices (for fault and/or overcurrent protection)
      (ii) non-conducting location (including measurement of distances, where appropriate)
         absence of protective conductors
      (iii) earth-free local equipotential bonding
         presence of earth-free protective bonding conductors
      (iv) electrical separation
   (d) additional protection
9. Prevention of mutual detrimental influence
10. Presence of appropriate devices for isolation and switching correctly located
11. Presence of undervoltage protective devices
12. Labelling of protective devices, switches and terminals
13. Selection of equipment and protective measures appropriate to external influences
14. Adequacy of access to switchgear and equipment
15. Presence of danger notices and other warning signs
16. Presence of diagrams, instructions and similar information
17. Erection methods.

Figure 8.28 Inspection Schedule

Some example points are given below.

- SELV is an extra-low-voltage system which is electrically separate from earth and from other systems. Where this method is used the particular requirements of the *IEE Wiring Regulations* must be checked (see Regulation 414).
- Basic protection against direct contact by the use of obstacles or placing out of reach is usually only employed in special circumstances. Where they are used the requirements of the Regulations must be met (see Regulation 417).
- The use of class II equipment as a means of protection is infrequently adopted and only when the installation is to be supervised and under the control of a qualified electrical engineer (Regulation 412 applies).
- Non-conducting locations also require special precautions to be taken and are not applicable to domestic locations (see Regulation 418.1).
- Earth-free equipotential bonding as a means of protection is also only used in special circumstances; again it is not applicable in domestic installations.
- For the requirements of systems using electrical separation as a means of protection see Regulation 418.3.

Contractor: Address/Location of distribution board

Test Date: Type of Supply: TN-S/TN-C-S/TT

Ze at origin: .... ohms

Signature PFC: ....kA

Method of fault protection: .... Confirmation of supply polarity ☐

Instruments
loop impedance: ....
continuity:
insulation:
RCD tester:

Equipment vulnerable to testing: ....

Description of Work: ....

Circuit description	Overcurrent Device Short-circuit capacity: ....kA		Wiring Conductors		Test Results									
					Continuity			Insulation Resistance		Polarity	Earth Loop Impendance	Functional Testing		Remarks
	type	Rating I_n A	live mm^2	cpc mm^2	$(R_1 + R_2)$* Ω	R2* Ω	Ring	Live/ Live MΩ	Live/ Earth MΩ		Zs Ω	RCD time ms	Other	
1	2	3	4	5	6	7	8	9	10	11	12	13	14	15

Deviations from BS 7671: IEE Wiring Regulations and special notes:

* Complete column 6 or 7

Figure 8.29 Schedule of Test Results

On completion of the Inspection Schedule it should be signed and dated by the person responsible for carrying out the inspection.

Schedule of Test Results

The Schedule of Test Results is a written record of the results obtained when carrying out the electrical tests required by Part 6 of BS 7671. The following notes give guidance on the compilation of the schedule, which should be attached to the Electrical Installation Certificate.

1. The type of supply should be determined from either the supply company or by inspection of the installation.
2. Z_e should be measured with the main bonding disconnected. If Z_e is determined from information provided by the supply company, rather than measured, then the effectiveness of the earth must be confirmed by testing.
3. The value of prospective fault current recorded on the results schedule should be the greater of that measured between live conductors (short-circuit current) or line conductor and earth (earth-fault current).
4. In the column headed 'Overcurrent device', details of the main protective device including its short circuit should be inserted.

5. Continuity tests are to include a check for correct polarity of the following:

    - every fuse and single-pole device to be connected in the line conductor only

    - all centre-contact lamp holders to have the outer contact connected to the neutral conductor

    - wiring of socket outlets and other accessories correctly connected.

6. Continuity of protective conductors – every protective conductor, including bonding conductors, must be tested to verify that it is sound and correctly connected.

7. Continuity of ring circuit conductors – a test must be carried out to verify the continuity of each conductor including the protective conductor of every ring circuit. A satisfactory reading should be indicated by a tick in the appropriate columns.

8. The sum of the resistances of the line conductor and the protective conductor ($R_1 + R_2$) should be recorded and, after appropriate correction for temperature rise, may be used to determine Z_s ($Z_s = Z_e + (R_1 + R_2)$). Where Test method 2 is used, the maximum value of R_2 should be recorded.

9. Insulation resistance – where necessary all electronic devices should be disconnected from the installation so they are not damaged by the tests themselves. Details of such equipment should be recorded on the schedule of test results so that the person carrying out routine periodic testing in the future is aware of the presence of such equipment.

10. The insulation resistance between conductors, and between conductors and earth, shall be measured and the minimum values recorded in the appropriate columns on the schedule of test results. The minimum insulation resistance for the main switchboard and each distribution (board tested separately with all final circuits connected but with all current-using equipment disconnected) should be not less than that given in the table 10.1 of *IEE On-Site Guide* (Table 10.1).

11. Following the energising of the installation, polarity should be re-checked before further testing.

12. Earth-loop impedance (Z_s) – this may be determined either by direct measurement at the furthest point of a live circuit or by measuring Z_e at the origin of the circuit and adding to it the measured value of $R_1 + R_2$. Alternatively, the value of Z_e may be obtained by inquiry from the electricity supply company. Values of Z_s should be less than the values (corrected where necessary) given in appendix 2 of the *IEE On-Site Guide*.

13. Functional testing – the correct operation of all RCDs and RCBOs must be tested by using an instrument that simulates a fault condition. The time taken to trip the device is automatically recorded by the instrument and should be written in the appropriate space on the schedule of test results. At rated tripping current, RCDs should operate within 200 milliseconds.

MINOR ELECTRICAL INSTALLATION WORKS CERTIFICATE
(REQUIREMENTS FOR ELECTRICAL INSTALLATIONS – BS 7671 [IEEE WIRING REGULATIONS])
To be used only for minor electrical work which does not include the provision of a new circuit

PART 1: Description of minor works

1. Description of the minor works

2. Location/Address

3. Date minor works completed

4. Details of departures, if any, from BS 7671: 2008

PART 2: Installation details

1. System earthing arrangement   TN-C-S ☐   TN-S ☐   TT ☐

2. Method of fault protection

3. Protective device for the modified circuit   Type ....   Rating ....A

Comments on existing installation, including adequacy of earthing and bonding arrangements (see Regulation 131.8):

PART 3: Essential Tests

Earth continuity satisfactory   ☐

Insulation resistance: Line/neutral   ....MΩ
                     Line/earth   ....MΩ
                     Neutral/earth   ....MΩ

Earth fault loop impedance.....................................Ω

Polarity satisfactory   ☐

RCD operation (if applicable). Rated residual operating current IDN ....mA and operating time of ....ms (at IDN)

PART4: Declaration

I/We CERTIFY that the said works do not impair the safety of the existing installation, that the said works have been designed, constructed, inspected and tested in accordance with BS 7671:2008 (IEE Wiring Regulations), amended to .... (date) and that the said works, to the best of my/our knowledge and belief, at the time of my/our inspection, complied with BS 7671 except as detailed in Part 1 above.

Name: ....

For and on behalf of: ....

Address: ....

....

....

Signature: ....

Position: ....

Date: ....

Figure 8.30 Minor Works Certificate

Electrical Installation – Minor Works Certificate

Minor works are defined as an addition to an electrical installation that does not extend to the installation of a new circuit, e.g. either the addition of a new socket outlet or of a lighting point to an existing circuit. This certificate includes space for the recording of essential test results and does not require the addition of a schedule of test results.

Scope

The Minor Works Certificate shown in Figure 8.30 is only to be used for additions to an electrical installation that do not include the introduction of a new circuit (see BS 7671 Regulation 633.1).

Part 1 – Description

- Description of the minor works – this requires a full and accurate description in order that the work carried out can be readily identified.

- Location/address – state the full address of the property and also where the work carried out can be located.

- Date of completion.

- Departures from BS 7671 (if any) – departures from the Regulations would only be expected in the most unusual circumstances.

Part 2 – Installation details

- Earthing arrangements – the type of supply should be indicated by a tick, i.e. TN-C-S, TN-S, TT etc.

- Method of fault protection – the method of fault protection should be clearly identified, e.g. earthed equipotential bonding and automatic disconnection of supply using fuse/circuit breaker or RCD.

- Protective device – state the type and rating of the protective device used.

- Comments on existing installation – if the existing installation lacks either an effective means of earthing or adequate main equipotential bonding conductors, this must be clearly stated. Any departures from BS 7671 may constitute non-compliance with either the Electricity Supply Regulations or the Electricity at Work Regulations and it is important that the client is informed in writing as soon as possible.

Part 3 – Essential tests

- Earth continuity – it is essential that the earth terminal of any new accessory is satisfactorily connected to the main earth terminal.

- Insulation resistance – the insulation resistance of the circuit that has been added to must comply with the requirements of BS 7671 (Table 61).

- Earth-fault loop impedance – the earth-fault loop impedance must be measured to establish that the maximum permitted disconnection time is not exceeded.

- Polarity – a check must be made to ensure that the polarity of the additional accessory is correct.

- RCD operation – if the additional work is protected by an RCD, then it must be checked to ensure that it is working correctly.

Part 4 – Declaration

- The certificate must be completed and signed by a competent person in respect of the design, construction, inspection and testing of the work carried out (in the case of minor works this will invariably be the responsibility of one person).

- The competent person must have a sound knowledge and experience relevant to the work undertaken and to the technical standards laid down in BS 7671. This person should also be fully aware of the inspection and testing procedures contained in the Regulations and must use suitable test instruments.

- When signing the form on behalf of a company or other organisation, individuals should state clearly whom they represent.

Guidance notes

Guidance notes for those receiving the certificate are as follows.

- This certificate has been issued to confirm that the work to which it relates has been designed, constructed, inspected and tested in accordance with BS 7671.

- You should have received the original certificate and the contractor should have retained a duplicate. If you are not the owner of the installation then you should pass the certificate or a copy of it to the owner.

- The Minor Works Certificate is only to be used for additions, alterations or replacements to an installation that do not include the provision of a new circuit. This may include the installation of an additional socket outlet or lighting point to an existing circuit, or the replacement or relocation of an accessory such as a light switch. A separate certificate should be issued for each existing circuit on which work of this nature has been carried out. The Minor Works Certificate is not valid where more extensive work is carried out, when an Electrical Installation Certificate should be issued.

- The original certificate should be retained in a safe place and be shown to any person inspecting the installation or carrying out further work in the future. If you later vacate the property, this certificate will demonstrate to the new owner that the minor works covered by the certificate met the requirements of the Regulations at the time the certificate was issued.

Requirements for testing

On completion of this topic area the candidate will be able to describe the requirements for testing with reference to types of test equipment, sequence and procedures for tests and application of BS 7671.

The sequence of tests

Regulation 612 lists the sequence in which tests should be carried out. If any test indicates a failure to comply, that test and any preceding test, the results of which may have been influenced by the fault indicated, must be repeated after the fault has being rectified.

Initial tests should be carried out in the following sequence where applicable, before the supply is connected or with the supply disconnected as appropriate.

1. Continuity of circuit protective conductors (cpcs) including main and supplementary equipotential bonding circuits

2. Continuity of ring final circuit conductors

3. Insulation resistance

4. Protection by separation of circuits

5. Protection by barriers or enclosures provided during erection

6. Insulation of non-conducting floors and walls

7. Polarity

8. Earth electrode resistance

With the electrical supply connected, re-test polarity before further testing.

9. Earth-fault loop impedance

10. Additional protection

11. Prospective fault current

12. Check of phase sequence

13. Functional testing

14. Verification of voltage drop

The test results should be recorded on an installation schedule similar to the diagram in Figure 8.31.

The list above indicates the order in which the tests should be conducted. This is to ensure that the results obtained are reliable. Tests 1–8 are dead tests, to ensure that it is safe to put the supply on. It is therefore important to prove the continuity of protective conductors before beginning insulation resistance tests.

Forms of completion or periodic inspection, inspection, test and an installation schedule (including test results) should be provided to the person ordering the work.

Contractor		Address/location of dist board		Type of supply		Instruments	
Test date				Ze at origin		RCD tester	
		Equipment vulnerable to testing				Continuity	
Signature				PSSC		Insulation	
						Others	

Circuit	No. of points	Overcurrent device				Circuit conductors		Test results								Earth loop impe- dance	Functional testing			polarity	remarks
		Type	Rating	Short Circuit Cap		Live	CPC	Continuity			Insulation resistance						RCD	Time ms	Other		
								$R_1 + R_2$	R_2		Line/line	Line/ neutral	Line/ earth	Neutral/ earth							
1	2	3	4	5		6	7	8	9		10	11	12	13		14	15	16	17	18	19
1																					
2																					
3																					
4																					
5																					
6																					
7																					
8																					
9																					
10																					
11																					
12																					

Main bonding check	Size		Gas
Gas			
Water			Earth electrode resistance
Other			

Figure 8.31 Installation schedule

These tests are fully described in *Guidance Note 3 to BS 7671*.

Types of test equipment

BS 7671 gives guidance on the standard of instruments that are required to undertake the tests. It also specifies that instruments used should comply with BS EN 61010, BS EN 61557 and BS EN 61010, with test leads conforming to HSE guidance note GS38. Here we will look at the requirements these instruments have for testing.

Applied voltage testers

These should be able to provide an output current that does not exceed 5 mA and a required test voltage of 3750 V a.c. The instruments should be capable of maintaining a test voltage continuously for at least one minute, and should have a means of indicating when a failure of the insulation has occurred.

Earth-fault loop impedance testers

These should have a resolution of 0.01 ohm for circuits rated up to 50 amps. Earth-fault loop impedance testers may also have a facility for measuring prospective short-circuit current. The basic measuring principle is the same as for the earth-fault loop impedance tester. The current is calculated by dividing the supply voltage by the measured earth-loop impedance value.

Earth electrode resistance testers

These either have four or three terminals, depending upon configuration, and are best used with long leads and temporary test spikes that are placed in the ground.

RCD testers

These test instruments have to be capable of delivering the full range of test currents required for a maximum period of two seconds and give a read out calibrated in seconds.

Insulation resistance test

Figure 8.32 shows an insulation resistance test being conducted on a twin and earth cable between the line and cpc at the distribution board end of the cable. The reading obtained should be greater than 100 MΩ, indicating that the insulation resistance is satisfactory and that the supply is safe to put on. But what would the instrument indicate in the situation shown in Figure 8.33?

In the situation below an insulation resistance test is again being conducted between line and cpc. However, this time a nail has penetrated the sheath of the cable, breaking the cpc and touching the line conductor (the neutral conductor has been removed for clarity). When the test is done the instrument may read greater than 100 MΩ, indicating that the insulation resistance is acceptable and that it is safe to connect the supply. However, we can clearly see that it is not.

Beyond the break in the cpc, the line and cpc are connected. If the supply was now connected to the cable we would have a potentially lethal situation, as all the metal work connected to the cpc will become live. Automatic disconnection will not take place as the break in the cpc means there is no longer a return path. The metalwork will remain live until someone touches it – which could result in a fatal electric shock.

In this case, if we had conducted a continuity test of the cpc before the insulation resistance test, we would have identified that the cpc was broken. Action could then have been taken to remedy the situation.

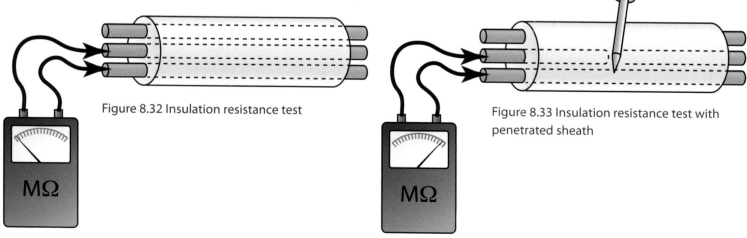

Figure 8.32 Insulation resistance test

Figure 8.33 Insulation resistance test with penetrated sheath

MΩ

MΩ

Procedures for dealing with the customer and reports

On completion of this topic area the candidate will be able to describe the procedures for dealing with the customer and the reports.

The customer

Following initial verification, BS 7671 requires that an Electrical Installation Certificate, together with a schedule of test results and an inspection schedule, should be given to the person ordering the work. Until this has been done, the Regulations have not been met.

Sometimes the person ordering the work is not the end-user, e.g. the builder of a new housing estate sells the individual houses to various occupiers. In these cases it is recommended that copies of the inspection and test certificates, together with a test results schedule, are passed on to the new owners.

Handover to customer

Handover of the installation to the customer is the final task. This should include a tour of the installation, an explanation of any specific controls or settings and, where necessary, a demonstration of any particularly complicated control systems. The operation and maintenance manuals produced for the project should be formally handed to the customer at this stage, including copies of the Electrical Installation Certificate, the Schedule of Test Results and the Inspection Schedule.

Activity

1. Obtain copies of the Electrical Installation Certificates, Inspection Schedules and Schedules of Test Results for your home or the site that you are currently working on. If the building is old enough these will also be accompanied by one or more Periodic Inspection Reports. If you can, compare these with the current installation to see if any alterations or additions have been made since they were prepared.

2. Have a look at any test equipment you have access to and see when and how it was last calibrated and when it is next due for calibration.

3. If you have not done any 'real' inspection and testing before, you might like to 'work shadow' someone while they are carrying out an inspection and testing of an installation.

FAQ

Q Do I have to inspect and test the electrical work that I do?

A Yes. BS7671 states that: 'Any addition or alteration to an electrical installation must be inspected and tested to verify, as far as is reasonably practicable, that the requirements of the Regulations have been met.'

Q Do I need a particular qualification to be an inspector and tester?

A BS 7671 simply states that you need to be a 'competent person' in order to carry out inspection and testing. However, some bodies (such as the NICEIC or Part P Registration) may impose requirements of their own (such as a C&G 2391) before they will accept your signature on any certificates.

Q Do my instruments need to be calibrated annually?

A You should always have confidence that your instruments are in good condition and give the correct results. You can do this yourself by regularly testing your instruments against a known value and keeping a record of the results. Some clients or registration bodies will insist that your instruments are sent to a laboratory and a calibration certificate issued (usually annually).

Q Should I test an existing installation before I start working on it?

A According to BS 7671 you are not allowed to make extensions or alterations to an installation unless it is in a satisfactory condition. Therefore, some pre-installation inspection and testing is necessary to determine if the installation is safe to be altered. This testing should be carried out even if you are only replacing an item of equipment. It is particularly important to check the cable size, protective device, polarity, and earthing and bonding arrangements.

Q I need to fit a spur to an existing ring circuit. If the existing socket has two cables connected to it, then it's OK to spur off it, isn't it?

A Not necessarily. An existing socket outlet may have two cables connected to it because it is on the ring. Another possibility is that it is already a spur that is feeding another spur, hence the two cables. In this case, it already contravenes the current edition of BS 7671, so you cannot add another one. You would need to carry out a ring circuit continuity test first to establish whether the ring circuit was, in fact, a compliant ring circuit in the first place.

Knowledge check

1. How often is it recommended that a hospital has an electrical inspection?

2. There are four reasons why we carry out an initial inspection. What are they?

3. Name five things that should be covered by an initial inspection.

4. Name all the initial tests that should be carried out on an electrical installation, placing them in the correct sequence and indicate which are 'live' tests.

5. Describe what is meant by the term 'indirect contact'.

6. When carrying out an insulation resistance test, why is it important to remove all GLS lamps before applying the test?

7. Why is it unacceptable only to test an RCD by pressing the test button located on the front of the RCD?

8. A 30 mA RCD is providing supplementary protection against direct contact. A test current of what should cause the RCD to trip within how many milliseconds?

9. What other items of documentation must accompany an Electrical Installation Certificate for it to be valid?

10. What is a Minor Works Certificate and when should it be used?

chapter 9

Fault diagnosis and repair

Unit 3 outcome 1

After completion of testing it becomes the duty of the electrician to resolve any faults that have been found in the system. To do so, we need to know how to diagnose any faults. In other words, we need to be able to use the evidence we have gathered to identify exactly what the problem is in order to fix it. A full knowledge of electro-technical systems, combined with your own experiences, will be essential when diagnosing faults.

Once a problem has been successfully identified it then becomes our task to rectify it. When carrying out repairs it is vital to remember the importance of safe working practices, while also working within any limits that may be placed on repairs – such as costs and downtime.

This chapter builds on topics covered earlier in this book and at Level 2.

On the completion of this outcome the candidate will be able to:

- identify electro-technical systems and equipment utilising single- and three-phase supplies

- state the safe working procedures to be used before undertaking fault diagnosis

- state the basic principles of fault finding

- list and describe the stages of fault finding and rectification

- list circumstances where faults may occur

- describe typical systems

- describe the factors which influence repair or replacement

- state factors which effect rectification

- recognise situations where special precautions should be applied.

Identify electro-technical systems and equipment

On completion of this topic area the candidate will be able to identify electro-technical systems and equipment utilising single- and three-phase supplies within installations.

Lighting circuits

We covered lighting circuits fully at Level 2, and you should be familiar with the operation of these systems. As such we will recap only the essential points here. You will remember there are numerous switching arrangements for lighting circuits. Table 9.01 shows a brief summary.

Cabling	Switching arrangements
• Wiring in conduit and/or trunking • Wiring using multicore/composite cables	• One-way switching • Two-way switching • Intermediate switching

Table 9.01 Switching arrangements

Cabling arrangements

Wiring in conduit and trunking is carried out using PVC single-core insulated cables (Ref. 6491X). The line conductor is taken directly to the first switch and looped from switch to switch for all the remaining lights in the circuit. The neutral conductor is taken directly to the lighting outlet (luminaries) and looped between all the remaining luminaries on the circuit. The switch wire is run between the switch and the luminaire it controls.

Wiring using multicore or composite cables is normally a sheathed multicore twin and earth or a three-core and earth (Ref. 6242Y and 6243Y respectively). It may use a joint box. Here, only one cable is run to each wiring outlet. All conductors need to be contained within a non-combustible enclosure at wiring outlets. A cpc must be installed, insulated and terminated suitably.

Switching arrangements

One-way switching is the most basic circuit possible. Here a switch contact links the supply with the luminaire and can be held in place mechanically until opened. It is possible to set up a second light, feeding from the same switch.

Two-way switching is used when the light needs to be controlled from more than one location, for example opposite ends of a long corridor. In this system the switch feed is feeding one two-way switch, with the switch wire going from the other two-way switch to the luminaries. Two wires, known as 'strappers', link the two switches together. By connecting the two two-way switches together, we have the ability at each switch to energise or de-energise the switch wire going to the light.

Intermediate switching is used when more than two switch locations are needed, for example a long corridor with other corridors coming off it. These intermediate switches are wired in the 'strappers' between the two-way switches. This cross-connects the 'strapping' wires, allowing us to route the supply to any terminal.

To use the example of a long hotel main corridor with other minor corridors coming off it, if we have an intermediate switch at each junction with the main corridor, anyone joining or leaving the main corridor now has the ability to switch luminaires on or off.

Multicore cables

A multicore cable (commonly referred to as twin and earth) is basically a three-core cable consisting of a line, neutral and earth conductor. These cables are used on one-way lighting and power circuits. The two cores are copper conductors with colour insulation, either brown or blue (for line and neutral) or two browns (for use in lighting circuit with switch wires). There is a bare non-insulated cpc (earth) in between. A green and yellow sleeve is applied to this at terminations.

When two-way switching with multicore cables, a three-core and earth cable is required. This has three coloured (brown, black and grey) insulated conductors and a bare circuit protective conductor. This type of cable is normally only stocked in 1.5mm and is used in a lighting circuit for converting an existing one-way switching arrangement into a two-way switch.

Further information, including diagrams, on of all of the above can be found at Level 2.

Power circuits

We covered the principles of rings, radials and spurs at Level 2. As you will be familiar with the concepts behind these circuits, we will only recap the principles here.

The standard power circuit arrangements found in domestic premises are:

- ring circuits
- permanently connected equipment
- non-fused spurs
- fused spurs
- radial circuits.

In ring circuits the line, neutral and circuit protective conductors are connected to their respective terminals at the consumer unit, looped into each socket outlet in turn and then returned to their respective terminals in the consumer unit, thus forming a ring. Each socket outlet has two connections back to the mains supply. Consideration must always be given to the loading of the ring main and ensuring the assessed load is balanced.

Permanently connected equipment should be locally protected by a fuse which does not exceed 13 A and be controlled by a switch complying with BS 7671 Chapter 46 or by a circuit breaker not exceeding 16 A rating.

Remember

Any number of intermediate switches may be used between two-way switches and they are all wired into the 'strappers'. In other words, an intermediate switch has two positions.

Non-fused spurs can be installed on a ring circuit at the terminals of socket outlets, junction boxes or the distribution board. This supplies only one single or one twin-socket outlet. The total number of spurs should not exceed the total number of socket outlets. The conductor must be the same size as used on the ring circuit.

A fused spur is connected through a fused connection unit. The fuse should be related to the current-carrying capacity of the cable used for the spur. When sockets are wired from this, the minimum size of the conductor is 1.5mm² for rubber- or PVC-insulated cables with copper conductors, and 1mm² for mineral-insulated cables with copper conductors. The total number of fused spurs is unlimited, subject to loading.

In a radial circuit the conductors do not form a loop but finish at the last outlet. As with the ring circuit, the number of outlets in any circuit is unlimited in a floor area up to the maximum allowed. In each case this will be determined by the estimated load and shock protection constraints.

Control systems

Timers and programmers come in a wide variety of shapes and sizes and can control simple and/or complex switching operations in all aspects of modern living. Simple timers include:

Did you know?

Manufacturer Danfoss Randall is one of many that supply programmers

- plug-in clock timers that switch lights or heaters on and off
- timers that delay the operation of contactors in motor control circuits
- timers that switch Christmas lights on and off rapidly or in a pattern.

More complex timers can be programmed to carry out multiple functions when external inputs are received or a certain time is reached. One of the most common programmers, which most of you will see each day, is the one used to switch the heating and hot water on at home. It is mains operated and, in the event of a power failure, a rechargeable battery takes over and maintains the settings for up to two days.

These programmers can be set to switch on/off either heating or hot water or both at several different times each day, seven days per week. The component parts of a typical central heating/hot-water system consist of a multi-way connection box, circulating pump, boiler, room thermostat, hot-water tank thermostat, changeover/diverter valve and, of course, the programmer itself.

The control circuits for these are many and varied depending on the type of heating system installed. For typical control circuits the use of Danfoss Randall Control Packs with WB12 wiring centres is recommended. The WB12 includes terminal-to-terminal wiring details for Heatshare (HSP) and Heatplan (HPP) packs.

The supply to the programmer, and hence the rest of the circuit, is usually via a fused spur with a 3 A fuse fitted. The various cables are installed to the respective component parts via the central connecting block and the programmer itself.

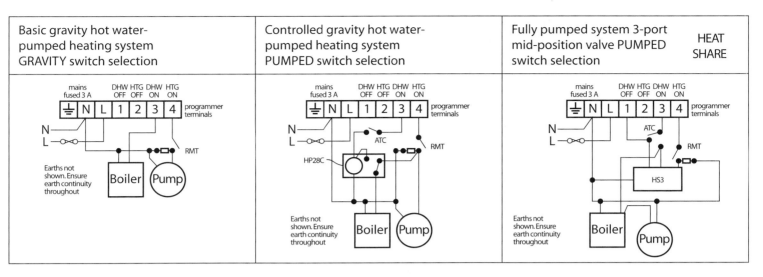

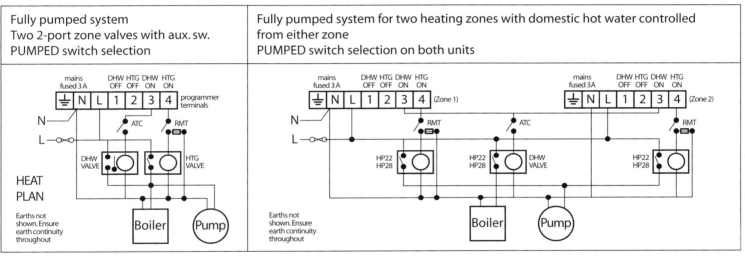

Figure 9.01 Wiring diagram for timer and programming domestic system

When the clock part of the programmer switches on to, say, central heating, then the supply is fed via the room thermostat to the boiler and the pump until the thermostat switches itself – and thus the heating – off. The same principle applies when hot water is selected, only this time the thermostat on the hot-water cylinder cuts out.

On the job: Customer technical query

You have been asked to install an additional light fitting in the kitchen of a house. The central heating in the house is controlled through a central programmer located in the kitchen and an electronic digital thermostat mounted on a wall in the entrance hall. The owners ask how the electronic thermostat works with the programmer.

What could you tell them?

Components

We covered the electro-technical systems for installation components comprehensively at Level 2. As a reminder, the main types of installation are:

- heating
- environmental control
- emergency management systems.

- security and alarm systems
- CCTV and communication systems.

Motor starters

We covered the properties of motor starters comprehensively in chapter 6 on pages 250–258; please refer back to these pages for more information.

Safe working practices

On completion of this topic area the candidate will be able to state the safe working procedures to be applied before undertaking fault diagnosis.

Fault diagnosis

We covered the principles of isolation and test instruments in chapter 8. Please refer back to this chapter for more information.

Many items of electrical equipment when installed are provided with a local means of switching or isolation. This is not only convenient for the user (as control is readily accessible) but also is a safety feature and indeed a requirement of BS 7671 *IEE Wiring Regulations* (Chapter 53), which states that 'equipment should be readily controlled, protected and isolated'. A good example of local isolation is when a motor is provided with a control unit as this can include the starter, switch and lock-off facility.

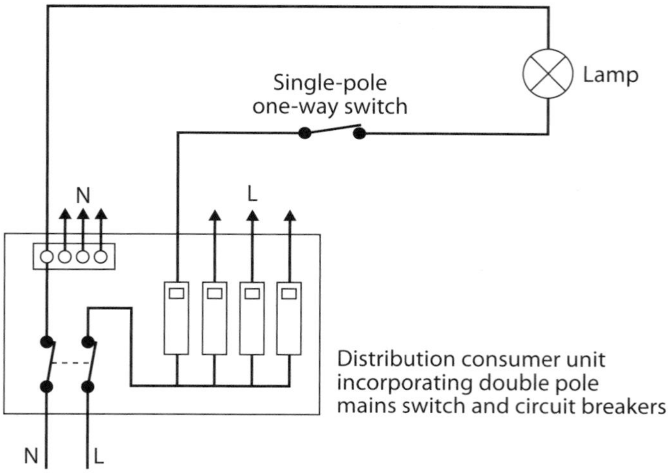

Single-pole one-way switch

Lamp

N   L

Distribution consumer unit incorporating double pole mains switch and circuit breakers

N   L

Figure 9.02 One-way lighting circuit with mains switch and breakers

The types of equipment used in circuits to provide switching and isolation of the circuits, and indeed complete installations, can be categorised as having one or more of the following functions:

- control
- isolation
- protection.

A simple example of the control, isolation and protection functions can be represented in any basic simple circuit. As an example, Figure 9.02 shows a one-way lighting circuit supplied from a distribution board having a mains switch and circuit breakers.

From Figure 9.02 we can see the following.

- The distribution consumer unit combines all three functions of control, protection and isolation.

- The main double-pole switch can provide the means for switching off the supply and, when locked off, can provide complete isolation of the installation.

- The circuit breakers in the unit provide protection against faults and overcurrents in the final circuits and, when switched off and locked off, the circuit breakers can provide isolation of each individual circuit.

- The one-way switch has only one function, which is to control the circuit enabling the luminaire to be switched on or off.

On- and off-load devices

Not all devices are designed to switch circuits on or off. It is important to know that, when a current is flowing in a circuit, the operation of a switch (or disconnector) to break the circuit will result in a discharge of energy across the switch terminals.

You may have experienced this when entering a dark room and switching on the light; for an instant you may see a blue flash from behind the switch plate. This is actually the arcing of the current as it dissipates and makes contact across the switch terminals. A similar arcing takes place when circuits are switched off or when protective devices operate, thus breaking fault current levels.

Fundamentally, an isolator is designed as an **off-load device** and is usually only operated after the supply has been made dead and there is no load current to break. An **on-load device** can be operated when current is normally flowing and is designed to make or break load current.

An example of an on-load device could be a circuit breaker, which is not only designed to make and break load current but has been designed to withstand high levels of fault current. Remembering the three previous functions, it is important to install a device that meets the needs at a particular part of a circuit or installation. Some devices can meet the needs of all three functions; however, some devices may only be designed to meet a single function.

All portable appliances should be fitted with the simplest form of isolator: a fused plug. When unplugged from the socket outlet, this provides complete isolation of the appliance from the supply.

Restoration of the supply

After the fault has been rectified, which may have resulted in either parts being replaced or simple reconnection of conductors, and before the supply is restored, it is important that the circuit is tested for functionality. These tests may be simple manual rotation of a machine or the sequence of tests as prescribed in the *IEE Wiring Regulations*. For example, a simple continuity test will check resistance values, open and closed switches and their operation.

Basic principles of fault diagnosis

On completion of this topic area the candidate will be able to state the basic principles of fault finding and understand the need for a logical approach.

Understanding of electrical systems, installation and equipment

Experience and understanding of electrical installations takes years to learn and a vast amount of knowledge is gathered during your apprenticeship. However, putting that knowledge into practice only really occurs when you take up responsibilities as an electrician. When analysing faults on electrical installations, even the trained electrician will need to ask questions to help rectify faults.

Hopefully by now you should have come across many different types of wiring systems, equipment, enclosures and protective devices. Equally, you should have now used the procedures for safe isolation and locking-off procedures many times. However, when a fault occurs, the art of finding information and asking questions can be a daunting task.

Understanding of electrical systems can generally be categorised as follows:

- voltage – 230 volt single-phase or 400 volt three-phase
- installation type – domestic, industrial or commercial etc.
- system type – lighting, power, fire alarm system, emergency lighting, heating etc.

If a fault occurs on a system which is in the process of being installed then the information and data for the system should be at hand. However, the big problem starts when you are asked to rectify a fault on a system that you have had no experience or knowledge of. In this situation you will need to access drawings and data to familiarise yourself with layout and distribution boards etc. The type of information that you will require when called out to a fault can be listed as:

- type of supply – single- or three-phase
- nominal voltage – 230 volt or 400 volt
- type of earthing supply system – TT, TN-S, TN-C-S
- type of earthing arrangement – EEBAD
- types of protective device – HRC, MCB, RCD etc.
- ratings of devices
- location of incoming supply services
- location of electrical services
- distribution board schedules
- location drawings
- design and manufacturer's data.

Optimum use of personal and other people's experience of systems and equipment

Your own knowledge of an installation can be an asset, not only in the fault finding process but also to your company for future business. If your company can problem-solve quickly and be relied upon by a client when a problem occurs, they will be the first to be called.

The fault may not exist on the installation wiring but on auxiliary equipment such as refrigeration and air conditioning. Unless you have specialist-trained knowledge of such equipment, it is not safe to attempt to rectify the faults and your company may not be insured to work on such systems. However, if the fault is on the circuit supplying these systems, the fault may be investigated and rectified. You may also have to work alongside other specialists, assisting each other in commissioning and testing such systems. On no account should you attempt to investigate or rectify faults on their systems without knowledge or experience and training.

Equally, you will often be asked to repair faults on wiring and equipment that have been caused by inexperienced personnel attempting to install or repair circuits or equipment.

When you first arrive at any installation to investigate and rectify a fault, you should always use the personnel present to help you to obtain any background to the fault and relevant information. If possible, seek the specialist knowledge of the person responsible for the electrical installation or general maintenance.

The different people that you will be dealing with to obtain special and essential knowledge and information could be:

- the electrician who may have previously worked on the system
- design engineer
- works engineer
- shift engineer
- maintenance electrician.
- machine operator
- home owner
- site foreman
- shop manager
- school caretaker.

These persons may have access to:

- operating manuals
- wiring and connection diagrams
- manufacturer's product data/ information
- maintenance records
- inspection and test results
- installation specifications
- drawings
- design data
- site diary.

Their experience of the installation and day-to-day knowledge is essential. It will help you to solve, replace or repair the fault more efficiently.

Logical approach

When looking to identify a fault, a logical approach will result in the shortest possible time spent on the task. This will save not only time but money and will reduce the level of frustration that can be experienced by the client.

Always check the obvious things first. If a client complains that a luminaire does not work, why spend time checking the wiring and switches when it may be as simple as the lamp needing replacing? If the lamp holder is an Edison screw type, does it just need screwing in properly? Checking these basics first can save you a lot of time and the client a lot of expense.

If the RCD has tripped, then there is either an earth fault or a faulty RCD. In this case turn off any MCBs or remove any fuses from circuits protected by the RCD, then turn the RCD on. If the device resets, turn on the MCBs, or replace the fuses one by one. If the RCD trips this is indication of a fault on that particular circuit.

In the above scenario ensure that all the loads are disconnected or unplugged from the circuit and retry resetting the RCD. By plugging in or turning on the loads in turn the faulty load will trip the RCD and the equipment can be identified and action taken to remedy the situation. If the RCD resets, the fault is in the equipment that has been unplugged. If the RCD fails to reset then the fault is in the circuit wiring and an insulation resistance test on the cables will prove which.

Logical fault diagnosis and rectification

On completion of this topic area the candidate will be able to list and describe the stages of logical fault diagnosis and rectification.

Diagnosis and rectification principles

To be able to competently investigate, diagnose and find faults on electrical installations and equipment is one of the most difficult aspects of being an electrician. Therefore, if the electrician is to be successful in fault finding and diagnosis, a thorough working knowledge of the installation or equipment is essential. Consequently, the person carrying out the procedure should take a reasoned approach and apply logical steps or stages to the investigation and subsequent remedy. The electrician should also realise their knowledge limitations and seek expert advice and support where necessary.

In an ideal world an electrician should not embark on any testing or fault finding without some forward planning. However, an emergency or dangerous fault, because of its very nature, may allow very little time for planning the remedy or repair. That said, faults that are visible and straightforward can be easily repaired, and some careful planning and liaison with the client will limit disruption.

Inspection and testing prior to energising a new installation can be important and can save embarrassment by rectifying problems before circuits are energised.

Remember

It is good business practice to carry out tasks as efficiently as possible. The effect on your business reputation and trading should be impressive

Remember

The *IEE Wiring Regulations* on testing can be a useful guide, but the *Electricity at Work Regulations* and safe use of test equipment should be followed during the process

For example, pre-energising tests can be extremely important when checking the function of switching circuits, particularly two-way and intermediate switching.

The electrician should have knowledge of inspection and testing in order to fault find competently – in particular, the correct use of test instruments, the choice of instrument for testing and knowledge of each instrument's range and limitations.

Logical stages of diagnosis and rectification

Some faults can be rectified very easily, especially when the electrician has a working knowledge of the installation. But there are many occasions when you may be called out to a repair and there is no information for you. It is on these occasions that your years of training and knowledge of wiring systems have to be used. However, this knowledge is not always sufficient so a sequential and logical approach to rectifying a fault and the gathering of information is needed. Such an approach could be as follows.

- **Identify the symptom** – this can be done by establishing the events that led up to the problem or fault on the installation or equipment.

- **Gather information** – this is achieved by talking to people and obtaining and looking at any available information. Such information could include manufacturer's data, circuit diagrams, drawings, design data, distribution board schedules and previous test results and certification.

- **Analyse the evidence** – carry out a visual inspection of the location of the fault and cross-reference with the available information. Interpret the collected information and decide what action or tests need to be carried out. Then determine the remedy.

- **Check the supply** – confirm supply status at the origin and locally. Confirm the circuit or equipment when the fault is isolated from the supply.

- **Check protective devices** – check the status of protective devices. If they have operated, this would determine location of the fault on the circuit or equipment.

- **Isolation and test** – confirm isolation prior to carrying out the sequence of tests.

- **Interpret information and test results** – by interpreting the test results, the status of protective devices and other information, the fault may be identified and remedied/rectified.

- **Rectify the fault** – this may be done quite simply, or replacement of parts may be needed.

- **Carry out functional tests** – before restoring the supply, the circuit or equipment will need to be tested not only electrically but also for functionality.

- **Restore the supply** – care must be taken that the device has been reset or repaired to the correct current rating. Make sure the circuit or equipment is switched off locally before restoring the supply.

Remember

Information is a necessity and no fault finding should commence without the relevant information and background to the fault being made available to the person carrying out the work. Some faults can be recognised by simply analysing test results and on further investigation can be remedied simply

In summary, the logical sequence of events would be as follows.

1. Identify the symptoms
2. Gather information
3. Analyse the evidence
4. Check the supply
5. Check protective devices
6. Isolation and test
7. Interpret information and results
8. Rectify the fault
9. Carry out functional tests
10. Restore the supply

Information required for fault location

We have already said that some information regarding the fault will be gathered by asking people questions relating to the events leading up to the fault. If the fault occurred on the installation when the electrician was working on it, then the information will be to hand. Most often we will be called out to a breakdown and asked to carry out repairs. It is when we are called out to a fault or repair that we will need information such as:

- operating manuals
- wiring and connection diagrams
- manufacturer's product data/information
- maintenance records
- inspection and test results
- installation specifications
- drawings
- design data
- site diary.

Interpretation of test results

Some faults can be recognised at the installation stage when the testing and inspection process is carried out. We should be able to recognise typical test results for each non-live test and interpret the type of fault that may exist.

Non-live tests

The *IEE Wiring Regulations* BS 7671 Part 6, Chapter 61, lists non-live and live tests. The non-live tests are carried out on the installation wiring circuits before they are ever energised. These tests will confirm the integrity of the circuit and they can be listed as:

- continuity of protective conductors
- continuity of ring final circuit conductors
- insulation resistance
- polarity.

Continuity of protective conductors

Protective conductors can be found in final circuit wiring and from the main earthing terminals to all exposed and extraneous metalwork that is to be found in the installation. It is important to verify the continuity of such conductors because we rely upon them for safety in the event of an earth fault and the operation of protective devices and fuses. A low-resistance ohmmeter would be used on conductors to verify a low resistance value. The reading obtained will depend on the length and area of the conductor, as resistance is proportional to length and inversely proportional to cross-sectional area.

If you are testing for the continuity of equipotential bonding the test current should be a minimum of 200 mA (Regulation 612.2.1). A typical reading for earth continuity would be 0.01–0.05 ohms. If a reading greater than 0.05 ohms was obtained, although on the upper side of acceptable, this may need investigating. Although it may be acceptable in relation to the conductor type and size, terminations may require checking for effectiveness.

Continuity of ring final circuit conductors

This test, carried out on the circuit wiring, is designed to highlight open circuits and interconnections within the wiring. A low-resistance ohmmeter is used to carry out this test. Typical readings will be 0.01 to 0.1 ohms dependent on the conductor size and length of circuit wiring. If the circuit is wired correctly then the readings on the instrument will be the same at each point of test. If variable readings are found at each point of the test, this may indicate an open circuit or interconnections. If these types of fault exist, the consequence can be an overload on part of the circuit wiring.

Insulation resistance

This test is designed to confirm the integrity of the insulation resistance of all live conductors between each other and earth. The type of instrument is known as an insulation resistance tester. Typical test results would be in excess of 100 megohms, i.e. 100 million ohms.

Table 61 of the *IEE Wiring Regulations* indicates minimum values between one-quarter of a million and one million ohms. These are minimum values and, in practice, if these values existed the circuit may need further investigation. Testing should be carried out with a test appliance capable of supplying the test voltage indicated in Table 61 when loaded with 1 mA. Such minimum values exist because the test voltage is divided by the delivered value of current of the instrument, which will indicate the actual insulation resistance of the test.

$$\frac{500}{1\ mA} = 0.5\ megohms$$

Depending on where the test takes place, variable values of resistance may be recorded. In the case of a test on the supply side of a distribution board (Figure 9.03), a group value reading will indicate circuits connected in parallel (Figure 9.04).

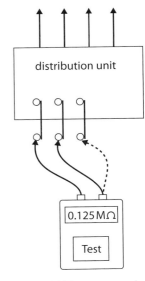

Figure 9.03 Test on supply side of distribution board

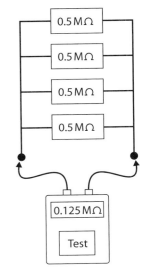

Figure 9.04 Test on circuits connected in parallel

From Figures 9.03 and 9.04 it can be seen that when testing grouped circuit conductors, a parallel reading would indicate a poor reading, but if each circuit were tested individually then the actual reading would be acceptable.

Polarity

The polarity test can be carried out using a low-resistance ohmmeter, as used for typical continuity testing. This polarity can be confirmed during the continuity test on the circuit and protective conductors. Typical test results will depend on the resistance of the conductors under test but will generally be low, around 0.01 to 1.0 ohms. Values in excess of this may indicate an open circuit or incorrect polarity.

Functional testing

This is mentioned in Regulations 612.10 and 612.13 of the *IEE Wiring Regulations*, which state that where a residual current device provides fault protection and additional protection, the effectiveness of the device should be confirmed by a test simulating typical fault-current conditions.

Such a test would use an appropriate RCD test instrument and the test should confirm the operation of the device independently of the device's integral test button. Residual current circuit breakers or devices are rated in milliamperes and the test should show that the device operates within the milliampere ratings of the device and within the time constraints. Typically, if the rated current of the device was 30 mA then the instrument should prove operation at this value within 200 milliseconds.

Regulation 612.13.12 states that assemblies such as switchgear and control gear, drives, controls and interlocks should be subjected to a functional test, to show that they are properly mounted, adjusted and installed in accordance with relevant requirements of the *IEE Wiring Regulations*. Typical functional tests should be applied to:

- lighting controllers and switches etc.
- motors, fixings, drives and pulleys etc.
- motor controllers
- controls and interlocks
- main switches
- isolators.

Earth-fault loop impedance testing

Overcurrent protective devices must, under earth-fault conditions, disconnect fast enough to reduce the risk of electric shock. This can be achieved if the actual value of the earth-loop impedance does not exceed the tabulated values given in BS 7671. The purpose of the test, therefore, is to determine the actual value of the loop impedance (Z_s) for comparison with those values.

The test procedure requires that all main equipotential bonding should be disconnected or MET to avoid parallel paths to earthed material. The test instrument is then connected by the use of 'flying' leads to the line, neutral and earth terminals at the remote end of the circuit under test. Press to test and record the results.

Once Z_s and the voltage have been established at the remote point in the circuit, you can divide the voltage by Z_s to give you the fault current to earth. Apply this calculation to the time current characteristic graphs shown in BS 7671 and you will be able to determine the actual disconnection time of the circuit.

The limitation and range of instruments

The correct choice of instrument for each particular test should now be known, but for those who are still unsure refer to Table 9.02.

Test	Instrument	Range
Continuity of protective conductors	Low-resistance ohmmeter	0.005 to 2 ohms or less
Continuity of ring final circuit conductors	Low-resistance ohmmeter	0.05 to 0.8 ohms
Insulation resistance	High-reading ohmmeter	1 megohm to greater than 200 megohms
Polarity	Ohmmeter Bell/Buzzer	Low resistance None

Table 9.02   Limitation and range of test instruments

Some instruments can provide more than one facility and many manufacturers provide such instruments. Inspectors and testers should know the range of operation of their instruments, be conversant with their operation and understand the instructions on the instrument.

When carrying out testing procedures, and before any instrument is used, some simple checks should be carried out to ensure that the instrument is operating within its range and calibration. A suitable checklist for use prior to testing is given below.

☑ Check the instrument and leads for damage or defects.

☑ Zero the instrument.

☑ Check the battery level.

☑ Select the correct scale for testing. If in doubt, ask or select the highest range available.

☑ Check the calibration date and record the serial number.

☑ Check leads for open and closed circuit prior to test.

☑ Record test results.

☑ After test, leave selector switches in off position. (Some analogue instruments turn off automatically to save battery life.)

☑ Always store instruments in their cases in secure, dry locations when not in use.

Specific types of fault

On completion of this topic area the candidate will be able to list the circumstances where faults may occur in an electro-technical system.

The type of equipment we install as electricians is normally chosen by a designer in line with the needs of the client's specification. We therefore often take it for granted that the equipment and wiring we fit is right for the job and that the designer has not only met the needs of the client but also the environment.

The manufacturer and the types of cables and accessories that you install must comply with BS and BS EN Standards and must be installed in accordance with BS 7671 *IEE Wiring Regulations*. Chapter 52 of the Regulations covers selection and erection of equipment, in particular:

- selection of type of wiring system
- external influences
- current-carrying capacity of conductors
- cross-sectional area of conductors
- voltage drop
- electrical connections
- minimising spread of fire
- proximity to other services
- selection and erection in relation to maintainability.

The items listed in this area of the Regulations should have been accounted for in the initial design but it is the responsibility of the installer to enforce them. This is achieved by adopting good practice when installing; and understanding the consequences (faults) that would occur if such good practice (and Chapter 52) were ignored.

Cable interconnections

Cable interconnections are used on many occasions in electrical installations. Their use is generally seen to be poor planning, and good design would limit their use. However, they are used in one or more of the following ways:

- lighting circuit joint boxes for line, neutral, switch wire and strappers
- power circuits, ring final socket outlet wiring and spurs etc.
- street lighting and underground cables where long runs are needed
- general alterations and extensions to circuits when remedial work is being done
- rectification of faults or damage to wiring.

Where they must be used, they should be mechanically and electrically suitable. They must also be accessible for inspection as laid down by Regulation 526.3, which states that: 'Every connection and joint shall be accessible for inspection, testing and maintenance'. There is an exception to this Regulation when any of the following is used:

- a joint buried in the ground
- a compound or encapsulated joint (commonly underground)
- a connection between a cold tail and heating element (under floor)
- a joint made by welding, soldering, brazing or compression tool
- a joint forming part of equipment.

The joints in non-flexible cables should be made by soldering, brazing, welding, mechanical clamps or of a compression type. The devices used should relate to the size of the cable and be insulated to the voltage of the system used. Connectors used must be of the appropriate British Standard and the temperature of the environment must be considered when choosing the connector.

Where cables having insulation of dissimilar characteristics are to be jointed – for example, XLPE thermosetting with general purpose PVC – the insulation of the joint must be to the highest temperature of the two insulators. The most common types of terminating devices used are:

- plastic connectors
- porcelain connectors
- soldered joints
- Screwits
- uninsulated connectors
- compression joints
- junction boxes.

Screwit

RB4 junction box

Cable interconnections are usually seen as the first point to investigate in the event of a fault. This is because they are recognised as the weak link in the wiring system and because they are usually readily accessible.

The mechanical and electrical connection when joining two conductors together relies on good practice by the installer to ensure that the connection is sound. The terminating device used should:

- be the correct size for the cross-section of the cable conductor
- be at least the same current rating as the circuit conductor
- have the same temperature rating as the circuit conductor
- be suitable for the environment.

Safety tip

Follow the safe isolation procedures before removing any cover or lid from any termination point

The most common fault at this point would probably be due to a poor/loose connection, which would produce a resistive joint and excessive heat that could lead to insulation breakdown or eventually fire. When a fault occurs at a cable interconnection it may not necessarily be due to the production of heat but to the lack of support or strain being placed on the conductors, which may lead to the same outcome as above.

Cable terminations, seals and glands

Terminations

The same rules as for cable interconnections apply to the termination of cables, i.e. good mechanical and electrical connections. The same consequences will occur if they are not adhered to. Hence the following.

- Care must be taken not to damage the wires.
- BS 7671 requires that a cable termination of any kind should securely anchor all the wires of the conductor that may impose any appreciable mechanical stress on the terminal or socket.
- A termination under mechanical stress is liable to disconnect, overheat or spark.
- When current is flowing in a conductor a certain amount of heat is developed and the consequent expansion and contraction may be sufficient to allow a conductor under stress, particularly tension, to be pulled out of the terminal or socket.
- If the PF improvement capacitor became disconnected in a luminaire, the circuit current would increase.
- One or more strands or wires left out of the terminal or socket will reduce the effective cross-sectional area of the conductor at that point. This may result in increased resistance and probably overheating.

Poorly terminated conductors in circuits that continue to operate correctly are commonly known as latent defects. A fault caused at a poorly connected terminal would be known as a high-resistance fault.

Types of terminals

There are a wide variety of conductor terminations. Typical methods of securing conductors in accessories include pillar terminals, screw heads, nuts and washers, and connectors as shown in Figures 9.05, 9.06 and 9.07.

Figure 9.05 Pillar terminal

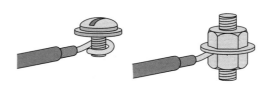

Figure 9.06 Screwhead, nut and washer terminals

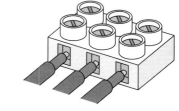

Figure 9.07 Strip connectors

Seals and entries

Where a cable or conductor enters an accessory or piece of equipment, the integrity of the conductor's insulation and sheath, earth protection and the enclosure or accessory should be maintained. Some cables and wiring systems have integrated mechanical protection:

- PVC/PVC twin and cpc
- PVC/steel-wire armoured
- PVC covered mineral-insulated copper cable (MICV)
- FP 200.

FP 200 cable

Twin and cpc PVC-insulated and sheathed cable

MCIV cable with pot and seal

When these and similar cables are installed, their design capabilities should not be degraded and special glands and seals, which are produced by the makers, should be used. Where PVC/PVC cables enter accessories, the accessory itself should have no loss of integrity and there should be no damage to any part of the cable.

When carrying out a visual inspection of an installation, either at the completion of works or at a periodic inspection, the checking of terminations of cables and conductors is an integral part of the inspection. Some items listed for checking in Regulation 611.3 include:

- connection of conductors
- identification of conductors
- routing of cables
- selection of conductors in accordance with the design
- correct connection of accessories and equipment.

These checks are made to comply with Regulation 611.2(iii), which states: 'The inspection shall be made to verify that the installed electrical equipment is not visibly damaged or defective so as to impair safety.'

We should also routinely check:

- for correct entry of cable into the accessory
- that the correct type of gland and seal has been used
- that accessory or enclosure seals and building structure/cable routes are sealed.

Accessories including switches, control equipment, contactors, electronic and solid-state devices

Faults can appear on most items of equipment that we install. The most common fault is due to wear of the terminal contacts, which constantly make and break during normal operation. All items of equipment should be to BS or BSEN and must be type tested by the manufacturer.

Some accessories break down due to excessive use and, through experience, the electrician can usually recognise them. During regular inspections and maintenance of an installation the inspector will check 10 per cent of accessories visually, which will usually be items that are constantly being used, such as:

- entrance hall switches
- socket outlets in kitchens for kettles
- cleaner's socket
- any item which is regularly being switched on or off.

Control equipment

This is also termed switchgear and is found at the origin of the supply after the metering equipment. It can take many forms; a few examples are:

- domestic installation – double pole switch in consumer unit
- industrial installation – single-phase double pole switch fuse or three-phase triple pole and neutral switch fuse
- large industrial installation moulded case circuit breakers – polyphase
- switch rooms industrial/commercial – as above; also oil and air blast types.

The switchgear is categorised not only in normal current ratings for load in amperes but also for fault current ratings in 1000 amperes or kA.

If a fault occurred on the outgoing conductors from these devices then the level of fault current would be dependent on the location. The basic rule for fault current level is that the nearer to the supply the fault occurs, the greater the level of fault current. This is due to there being less impedance (resistance) to restrict the current flow at that point, whereas the further away from the origin of the supply the greater the impedance (resistance) to limit the current flow.

The operation of this type of control gear, either manually or automatically due to a fault, would leave all circuits fed from such equipment dead. Manual operation of such devices is usually for maintenance reasons; it should be noted that permission of the client by prior arrangement would need to be obtained before commencing such work.

Protective devices are usually an integral part of switchgear, which will be one of the following:

- high-breaking capacity fuses (HBC or HRC)
- moulded-case circuit breakers (MCCB).

Specialist information and manufacturer's instructions are needed for replacing fuses that have operated and when re-setting tripped circuit breakers.

In older installations you may come across switchgear that is prone to fault due to lack of maintenance. For example, oil-type circuit breakers have their contact breaker points submerged in a special mineral-based oil. The oil, which quenches arcing, assists in breaking high levels of load and fault currents. If the oil viscosity and level is not regularly checked, this could lead to high levels of fault current circulating, leading to equipment damage and insulation breakdown or even fire.

Moulded-case circuit breaker

Contactors

Contactors are commonly used in electrical installations. They can be found in:

- motor control circuits
- control panels
- electronic controllers
- remote switching.

The basic principle of operation of a contactor is the use of the solenoid effect. A magnetic coil is used in the energised or de-energised mode and spring-loaded contacts for auxiliary circuits are either made or broken when the coil has been energised or de-energised.

Operating coils used in motor control circuits are energised when the start button is operated. The most common fault associated with their use is coils burning out. This can be due to age or the wrong voltage type being used, i.e. a 230 volt coil being used across a 400 volt supply.

Remote switching can utilise the contactor with great effect and convenience because the contacts of large, load-switching contactors can be used to carry large loads, i.e. distribution boards and large lighting loads such as used in car parks and floodlights etc. The benefit of the system is that a local 5 A switch in an office or reception area can be operated when entering or leaving, thus energising the coil of the contactor and allowing heavy loads to be carried through the contactor's contacts.

Electronic and solid-state devices

Solid state and electronic equipment works within sensitive voltage and current ranges in millivolts and milliamperes, and is consequently susceptible to mains voltage and heat.

Did you know?

As contacts are regularly being made and broken, a build-up of carbon appears on the contact points. On larger contactors these can simply be cleaned periodically using fine emery paper to remove the residue and make the electrical contact point more efficient

Remember

On smaller contactors used in electronic circuits it is not possible to clean contact points. When breakdown occurs it is normal practice to replace rather than repair

In the case of resistors on circuit boards, excessive heat produces more resistance and eventually an open circuit when the resistor breaks down. We can use a low-resistance ohmmeter to check resistor values.

In the case of capacitors, voltages in excess of their working voltage will cause them to break down, which will result in a short circuit. Such equipment obviously requires specialist knowledge but some basic actions by the electrician when carrying out tests and inspections can prevent damage to this type of equipment. When testing, such equipment should be disconnected prior to applying tests to the circuits. The reason for this is that some test voltages can damage these sensitive components and their inclusion in the circuit would also give an inaccurate reading.

Transient voltages are discussed later in this chapter. They are a major cause of faults on components and equipment of this type. These voltages can arise from supply company variations or lightning strikes, and therefore most large companies protect their equipment from such voltages and employ specialist companies to install lightning protection and filtering equipment.

Instrumentation

Switch room panels have integrated metering equipment. In this part of the installation please note that **large currents are being monitored**. This monitoring allows the consumer to view current and voltage values at particular times of the day. Some panels will have equipment which monitors and records these values, allowing the consumer to plot peak energy times and budget for energy costs.

As these instruments measure large currents, transformers are used to reduce current and voltage values, and this reduces the size of instrument needed. These transformers are known as CTs and VTs. However, great care must be taken when servicing and repairing such equipment as large voltages can exist across the transformer terminals. The common remedial work on such equipment is the replacement of faulty instruments and burnt-out transformers. It is essential that safe isolation procedures are followed.

Protective devices

These devices operate in the event of a fault occurring on the circuit or equipment that the device is protecting. Typical faults include:

- short circuits between live conductors – line to neutral single-phase, line to line three-phase
- earth faults – between any live conductor and earth
- overcurrents.

The most common reason for a device not operating is that the wrong type or rating of device has been used. Therefore, when replacing or re-setting a device after a fault has occurred, it is important to replace the device or repair it with the same type and rating. This should only be done after the fault has been investigated and corrected.

Did you know?

When a protective device has operated correctly and this device was the nearest device to the fault, **discrimination** is said to have occurred

The consequence of switching on a device with an un-repaired fault is for the device to operate again. This type of activity will cause damage to the circuit wiring, the equipment and possibly the device.

Luminaires

The most common fault with luminaires is expiry of lamp life, which obviously only requires replacement. Discharge lighting systems employ control gear which on failure will need replacement as they cannot be repaired. Discharge-type lighting may have problems with the control circuit.

Common items of equipment in the luminaire control circuit are:

- the capacitor used for power factor correction; if this had broken down it would not stop the lamp operating but would prevent the luminaire operating efficiently
- the choke or ballast used to create high p.d. to assist in the lamp starting; a common item which could break down and needs replacement
- the starter, which is used to assist the discharge across the lamp when switched on at the start; this is the usual part to replace when the lamp fails to light.

Flexible cables and cords

This type of conductor is used to connect many items within an electrical installation, for example:

- pendant ceiling rose lamp holders
- flex outlet spur units
- fused plugs to portable appliances
- immersion heaters
- flexible connection to fans and motors.

Regulation 511.1 states that all flexible cables and cords shall comply with the appropriate British and Harmonised Standard. The most common faults that occur with this type of conductor are likely to arise from poor choice and suitability for the equipment and the environment. Common faults relating to flexible cables and cords are:

- poor terminations into an accessory – conductors showing etc.
- wrong type installed, i.e. PVC instead of increased temperature type
- incorrect size of conductor – usually too small for the load
- incorrectly installed when load bearing in luminaires.

Such problems should be identified during the visual inspection stage.

Did you know?

Many fluorescent luminaires have starter-less electronic control gear, which is not only quick-start with increased efficiency but also requires less maintenance. They have a longer lamp life but when the quick-start unit fails it will need replacement

Portable appliances and equipment

Flexible cables and cords are used to connect many items of fixed and portable equipment to the installation wiring, and most faults on cords and flexible cables usually relate to poor choice or installation. Portable appliances come pre-wired with a fused plug to BS 1363.

When supplying equipment with flexible cables and cords it is important that the requirements of the *IEE Wiring Regulations* are followed, but specifically:

- correct size conductor is used for the load
- correct type is used for the environment/temperature/moisture/corrosion
- correct termination method is used to ensure good connection and no stress
- correct type and rating of protective device is used to protect the cable and appliance.

Symptoms of faults

On completion of this topic area the candidate will be able to describe the typical symptoms of faults.

General categories of a fault

The consequences of a fault often depend on its location within the installation or in a specific circuit. Care when installing electrical systems and continual maintenance can help avoid such faults. The essential points to be covered are:

- position of faults (complete loss of supply at the origin of the installation and localised loss of supply)
- operation of overload and fault current devices
- transient voltages
- insulation failure
- plant, equipment or component failure
- faults caused by abuse, misuse and negligence
- prevention of faults by regular maintenance.

Position of faults

The knowledge of fault finding and the diagnosis of faults can never be complete because no two situations are exactly the same. Also, as technology advances and the systems we install are being constantly improved, then the faults that develop become more complicated to solve. Therefore an understanding of the electrical installation and the equipment we install becomes paramount.

To diagnose faults an electrician will adopt basic techniques and these can usually solve the most common faults that occur. However, there are occasions when it may be impractical to rectify a fault. This could be due to costs involved in down time or the cost of repair being more than the cost of replacement, in which case it could mean the renewal of wiring and/or equipment or components. Such situations and outcomes must be monitored, and the client should be kept informed and be made party to the decision to repair or replace.

Most faults should never exist. A fault can be compared to an accident: careful planning and thought can prevent accidents. It is therefore part of a designer's job to build into the design of the electrical installation fault protection and damage limitation, such as:

- installing more than one circuit; when a fault occurs on one of the two circuits the fault can be limited to that circuit

- installing fuses and protective devices to disable the fault and limit its effect

- ensuring the ability to access, maintain and repair; good maintenance will limit faults, and access to the installation allows for maintenance and repair.

It is easier to find faults on installations where there are plenty of circuits. Indeed it is a requirement of the *IEE Wiring Regulations* BS 7671 Part 3 under **Installation circuit arrangements**, where Regulation 314.1 states: 'Every installation shall be divided into circuits as necessary to:

- prevent the indirect energising of a circuit intended to be isolated (vi)

- facilitate safe operation, inspection, testing and maintenance. (ii)'

We can see that compliance with this Regulation will help us to locate faults more easily; usually by a process of elimination (operating each fuse individually) or by simply looking at the device to see which one has operated and, thus, identify a defective circuit or device.

Protective devices and simple fuses are designed to operate when they detect large currents due to excess temperature. In the case of a short-circuit fault, high levels of fault current can develop, causing high temperatures and breakdown of insulation. Such faults, if allowed to develop, can cause fires.

Meltdown of insulation due to overcurrent

Meltdown of plastic connector due to overcurrent

> **Remember**
>
> Always keep the client informed

When considering the consequences of a fault it is important to know that the location of the fault can limit its severity with regard to disruption and inconvenience.

Figure 9.08 could be a typical layout of equipment for a small industrial or commercial installation where a three-phase supply is required and the load is divided between the three final consumer units. It will assist you in understanding simple installation layouts and the consequence of a located fault.

Key to Figure 9.08

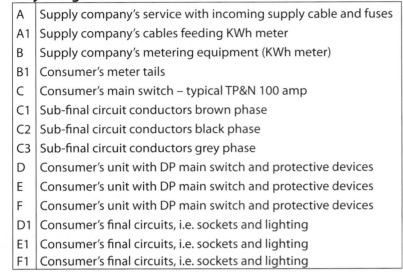

A	Supply company's service with incoming supply cable and fuses
A1	Supply company's cables feeding KWh meter
B	Supply company's metering equipment (KWh meter)
B1	Consumer's meter tails
C	Consumer's main switch – typical TP&N 100 amp
C1	Sub-final circuit conductors brown phase
C2	Sub-final circuit conductors black phase
C3	Sub-final circuit conductors grey phase
D	Consumer's unit with DP main switch and protective devices
E	Consumer's unit with DP main switch and protective devices
F	Consumer's unit with DP main switch and protective devices
D1	Consumer's final circuits, i.e. sockets and lighting
E1	Consumer's final circuits, i.e. sockets and lighting
F1	Consumer's final circuits, i.e. sockets and lighting

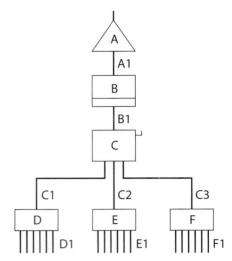

Figure 9.08 Typical three-phase supply for small installation

Complete loss of supply at the origin of the installation

With reference to Figure 9.08 it can be seen that a fault at points A to B1 could result in the complete loss of supply to the rest of the installation. The equipment at this point is mainly the property of the supply company; faults that appear on this equipment are usually the result of problems with the supply, such as an underground cable being severed by workmen.

However, faults on the installation wiring could result in the supply company's protective devices operating – although this is unusual and is only a consequence of poor design, misuse of the equipment or overloading of the supply company's equipment. As an example of this, if the supply company's protective device or fuses were rated at 100 amperes per phase and the designer loaded each phase up at 200 amperes, this would result in a 100 per cent overload. The consequence of this would be the operation of the mains protective devices and the whole installation being without power.

Localised loss of supply

Looking again at Figure 9.08, we can recognise what happens when parts of the installation lose their supply. If a fault appeared on cable C1, then the consumer unit (D) and all the final circuits from it would be dead. If a fault occurred on any one final circuit, then only that circuit's protective device should have operated and only that circuit would be dead.

If a fault occurred on any one final circuit it should not affect any other final circuit or any other cable leading up to that consumer unit.

Because the system in Figure 9.08 is three phase, this should mean that a fault on any one phase should only result in that phase being affected, again limiting the consequence of a fault. This further highlights the reason for Regulation 314.1 that we mentioned earlier, which stated: 'Every installation shall be divided into circuits as necessary to:

- avoid hazards and minimise inconvenience in the event of a fault (i)
- take account of damage that may arise from the failure of a single circuit (iii).'

Operation of overload and fault current devices

We have covered the principles and operation of these devices comprehensively at Level 2.

Transient voltages

A transient voltage can be defined as a variance or disturbance to the normal voltage state, the normal voltage state being the voltage band that the equipment is designed to operate.

It is the installation designer's responsibility to stipulate adequate sizes for the circuit conductors. But this sizing procedure will only prevent voltage drop during normal circuit conditions. Designers cannot prevent transient voltages as they are outside their control, but they can compensate by including equipment that will protect against such voltage variations and disturbances.

It is now becoming common practice to install filter systems to IT circuits, which provide stabilised voltage levels. These devices can also reduce the effect of voltage spikes by suppressing them. If transient voltages are not recognised they can cause damage to equipment and lead to a loss of data that could prove costly, especially if the user was an organisation such as a bank.

Some common causes of transient voltages are:

- supply company faults
- electronic equipment
- heavy current switching (causing voltage drops)
- earth-fault voltages
- lightning strikes.

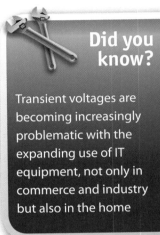

Did you know?

Transient voltages are becoming increasingly problematic with the expanding use of IT equipment, not only in commerce and industry but also in the home

Did you know?

If transient voltages are thought to exist on installations, monitoring equipment can be installed to record events over periods of time. The results of the monitoring can usually trace the cause

Insulation failure

Insulation is designed primarily to separate conductors and to ensure their integrity throughout their life. Insulation is also used to protect the consumer against electric shock, and is often used as a secondary protection against light mechanical damage, such as in the case of PVC/PVC twin and cpc cables. However, as previously mentioned, the breakdown of electrical insulation can lead to short-circuit or earth faults.

When insulation of conductors and cables fails, it is usually due to one or more of the following:

- poor installation methods
- poor maintenance
- excessive ambient temperatures.
- high fault-current levels
- damage by a third party

Poorly connected 13 A plug top resulting in dangerous working conditions

Burnt out starter

Plant, equipment and component failure

It is said that nothing lasts forever and this is certainly true of electrical equipment. Some faults that you attend will be the result of a breakdown simply caused by wear and tear, although planned maintenance systems and regular testing and inspections can extend the life of equipment. Some common failures on installations and plant are:

- switches not operating – due to age
- motors not running – new brushes required
- lighting not working – lamp's life expired
- fluorescent luminaire not working – new lamp or starter needed
- outside PIR not switching – ingress of water causing failure
- corridor socket outlet not working due to poor contacts created by excessive use/age.

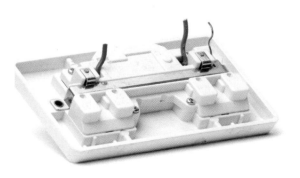

Poorly wired socket outlet

Faults caused by misuse, abuse and negligence

A common reason for faults on any electrical system or equipment is misuse – the system or equipment is simply not being used in the way it was designed to be used. Every item of electrical equipment comes with user instructions that usually cover procedures and precautions, and such instructions should be read carefully as misuse can lead to invalidation of the guarantee.

When an electrical installation is completed, a manual is normally handed over to the client that includes all product data and installation schedules and test results. This data will help the client when additions to the installation are made, inspections and tests are carried out or to assist in maintaining the installation.

Some faults are caused by carelessness during the installation process. Simple faults caused by poor termination and stripping of conductors can result in serious short circuits or overheating. Therefore good mechanical and electrical processes should always be carried out and every installation should be tested and inspected prior to being energised. Some examples of faults arising from misuse are given below.

Remember

If you are called out to repair/rectify a fault and you find that misuse of the system has led to the fault, then the guarantee for the installation may be invalidated by the fault. This may lead to the client having to pay for the repair costs

User misuse

Examples of user misuse include:

- using an MCB as a switch where constant use could lead to breakdown
- unplugging on-load portable appliances, thus damaging socket terminals
- damp accessories, for instance due to hosing walls in dry areas
- MCBs' nuisance-tripping, caused by connecting extra load to circuits.

Installer misuse

This can be caused by:

- poor termination of conductors – overheating due to poor electrical contact
- loose bushes and couplings – no earth continuity, therefore electric shock risk
- wrong size conductors used – excessive voltage drop – excess current which could lead to an inefficient circuit and overheating of conductors
- not protecting cables when drawing them into enclosures, causing damage to insulation
- overloading conduit and trunking capacities, causing overheating and insulation breakdown.

MCB

Poor termination of conductors causing electric shock risk

Prevention of faults by regular maintenance

A well designed electrical installation should serve the consumer efficiently and safely for many years, whereas an installation that is not regularly inspected and tested could lead to potential faults and areas of weakness not being discovered. Even a visual inspection can uncover simple faults which, if left, could lead to major problems in the future.

Large organisations in the commercial sector and industry have realised the need for regular inspections and planned maintenance schedules. Some major supermarkets choose to have regular maintenance take place when the stores are closed to the public, as store closure or part closure to carry out corrective electrical work could cost sales. A typical example of maintenance in shops is lamp cleaning, changing and repairs, and servicing to refrigeration and air conditioning. Large manufacturing companies usually carry out maintenance during planned holiday periods when the workforce is on annual leave; at these times machinery, equipment and the electrical installation wiring can be accessed without loss of production.

Carrying out repairs and servicing in this way gives electricians and engineers the opportunity to test installations and carry out remedial work and general servicing on equipment and machinery.

Repair or replacement

On completion of this topic area the candidate will be able to describe the factors which influence repair or replacement decisions.

Factors affecting the fault repair process

Having achieved the aim of locating the fault it may be a simple decision by the electrician to eliminate the cause and re-energise the circuit for the equipment or circuit to function again. However, in many cases there are issues that will not allow the power to be switched on to the circuit or equipment under scrutiny. This judgement in many cases cannot be made by the electrician and may need to be discussed by third parties, i.e. the client, the manufacturer and the person ordering the work. In this section we will look at the factors which influence this decision-making process as to whether the equipment or circuit is replaced or repaired and whether the fault can be remedied instantly or at a pre-determined future time.

Cost of replacement

When successfully finding a fault it would not make for good customer relations to go ahead and put the fault right without first considering the consequences to your company or client. There are points that need consideration such as: 'What will the cost of replacement be?'

This question can only be answered by liaison and consultation with your engineer and the customer, as an agreement must be reached on how to proceed. It may well be that some electrical contractors have an agreement or service contract in place where remedial works are carried out on a day work and contracted basis. This is where a prior contract agreement is signed and the contractor carries out maintenance and repair work at pre-determined rates. However, where this is not the case the question may arise: 'Will the replacement cost be exorbitant?'

It could be that the price quoted for replacement is high, which may result in the customer using your company's quotation to compare with other companies' prices. It may lead to negotiations where only part of a faulty piece of equipment is replaced or repaired, which may reduce the outlay costs to the customer. It is often the case that the customer is given a choice of options to help them decide what to do; they will compare these options and make the decision.

The options may be:

- full replacement
- part replacement
- full repair.

In offering such options it can be seen generally that the time taken to carry out a repair or part replacement may in many cases outweigh the time and cost of a full replacement.

Availability of replacement

In many cases where a complete replacement of equipment or machinery is needed it is because the fault has arisen due to the life span of the equipment being exceeded. In these cases the equipment or machinery may have been in operation for so long that a repair may not be possible and the item may even be obsolete. When this happens a suitable replacement must be found.

The most obvious decision to be made is that of finding a replacement that closely matches the original specification of the faulty item. But this is not always possible and the factor that, in reality, will affect the selection process becomes the availability of the equipment. This is important and will influence the decision to order. It is often the manufacturer with the replacement item in stock and a fast delivery that will be given the order to supply, as this will enable the electrician to remedy the fault quickly.

Downtime under fault conditions

It is often the case that the consequence of a fault on a circuit or item of equipment will be costing the client money. This can be due to:

- lost production where manpower and machinery are not working
- lost business where products or services are not being provided to the customer
- data loss because of power failure or disruption during repair.

Sometimes the fault may not be on the electrical installation but on the supply to the consumer's premises. In such cases the rectification of the fault is out of your company's responsibility and control. However, in such circumstances and where the loss of supply is prolonged, it may make good business sense, where possible, to provide a temporary supply to the customer.

Remember

It is often the case that the original manufacturer is no longer in business and therefore the problem becomes finding a suitable alternative manufacturer. Most manufacturers will have their own design team who will help in the process of finding a suitable alternative piece of equipment

Remember

Whatever the cause of the fault, the consequence is often loss of revenue either directly because of output or because of poor service to the customer, which in turn can affect the client's reputation and ultimately result in a future loss of business

Availability of resources and staff

We have already mentioned the availability of equipment due to choice, manufacturer and delivery time, but the availability of staff is often the cause of delay in rectifying faults and carrying out repairs.

Some faults and repairs may be minor and need little resource in terms of materials and labour. The response time for small faults could be as low as minutes and hours, but in the case of major industrial and commercial repairs the down time and repair time could be weeks or even months. The consequences of some faults may be major contract works such as rewiring or large equipment replacement. In such cases careful planning and decisions must be taken by the contractor's management and design team when deciding to commit the company to carry out such works.

The following are basic considerations that any company should make before commencing such contracts.

- **Availability of staff.** Does the company employ enough people to carry out and complete the work?

- **Competency.** Does the company have staff with relevant experience?

- **Cost implications.** Does the company havethe monetary capability for the contract?

- **Special plant and machinery.** Does the company have this type of resource available?

Legal responsibility

The legal responsibilities surrounding the replacement or repair of a fault must also be considered when making the decision to repair or replace. It is usual for a contract to be entered into by both parties as this would help to avoid, or at least assist, in settling future disputes. Such a contract would itemise:

- costs

- time

- guarantee period/warranty

- omissions from contract.

The costs should be agreed and adhered to. Most customers will not pay more than the initial agreed costs of repair unless an allowance was made in the contract for unforeseen work. Unforeseen work could result from something found when dismantling or taking out existing items, such as finding faulty or dangerous wiring underneath an item of heavy plant that is to be repaired.

If the initial costing accounted for repair and replacement of parts then it is essential that you do what you said you were going to do. For instance, if an inspection is subsequently carried out and it is evident that no work has taken place then you may be liable for legal proceedings to be taken against you for a breach of contract terms.

An accurate cost analysis should be considered when giving an estimate or fixed price for repair work, as it is usual practice to carry out repair work on a daily rate where the customer pays only for the time taken and materials used in rectifying the fault.

The customer should also be given a guarantee for the work done, both in respect of quality and also for the materials used.

When deciding whether to repair or replace, it may be deemed necessary to avoid writing a guarantee in the contract when the client insists on re-using existing items in order to reduce costs. It is often the case that old industrial or antiquated items would be prone to failure so to avoid legal action such items should not be included in the contract and should be written out of your guarantee.

Additional factors affecting rectification

Remember

Most companies will carry insurance to cover against any liabilities in respect of work done

On completion of this topic area the candidate will be able state factors which may affect the rectification process with regard to access, emergency supplies and demands for continuous supply.

Other factors

Careful planning can mean that most work is carried out efficiently with little or no disruption to the client. However, faults on installation equipment and wiring systems are certainly not planned and trained personnel can only minimise the disruption caused by the effects of faults.

When an electrician is called out to a fault they may not have the day-to-day knowledge or experience of the maintenance staff. However, through training and knowledge the electrician will use contingency plans to institute the procedures to rectify the fault with the minimum of disruption. Such contingency plans provide a common format when dealing with fault finding at any customer's premises. Your company will have its own procedures and these may include:

- signing in
- wearing identification badges
- locating supervisory personnel
- locating data drawings etc.
- liaising with the client and office before commencing any work
- following safe isolation procedures.

Special requirements

We have previously looked at the safety procedures and the actions needed for determining causes of faults. If we assume that safety requirements and the location of faults have been met and that the prime objective of the electrical contractor (the electrician) is to locate the fault, rectify and re-commission, then this will be done in liaison with the customer and with minimal disruption. However, the following topics are worthy of note.

Did you know?

In a factory environment, maintenance electricians are on hand to limit the effect of a fault and can usually solve problems and reduce the down time of a machine or circuit to a minimum. This is because of their day-to-day working knowledge of the electrical services and the equipment connected to it

Access to the system during normal working hours

There is nothing more costly or embarrassing than arriving at a customer's premises to carry out work of any kind, only to find the premises locked or unmanned because no prior arrangements were made for access. Access to the premises and the electrical installation or the specific item of equipment may prove a problem, especially if you are unfamiliar with the customer's premises.

When attending a fault or breakdown we must remember the procedures for ascertaining the information and familiarising ourselves with the installation. Remember the logical sequence is as follows.

1. Identify the symptoms.
2. Gather information.
3. Analyse the evidence.
4. Check supply.
5. Check protective devices.
6. Isolation and test.
7. Interpret information and test results.
8. Rectify the fault.
9. Carry out functional tests.
10. Restore the supply.

Remember

All electrical installation systems should have been designed and installed in accordance with the *IEE Wiring Regulations*, thus ensuring that the facility for isolating circuits and sections is inherent in the system

If you need to familiarise yourself with the installation, then access must be readily available not only at the fault repair location but also at the supply intake and isolation point. This has to be agreed prior to commencing any contracted work, whether a repair or a new installation. Such prior agreement must take into account those personnel who may have their normal day-to-day routine and activities affected. It is not always possible to carry out repair work during normal working hours. To avoid disruption it is often the case that such work is done outside of normal working hours. It is therefore important that arrangements for out-of-hours work are pre-arranged to account for security and especially access to the premises.

Whether the system can be isolated section by section

After access to the installation has been agreed, the repairs should be carried out using a logical approach. However, if the premises are still occupied and parts of the system are still being used, then it is important that the faulty circuit or plant to be worked on is isolated.

When only a small area or a single circuit or piece of equipment is affected, it is not good practice to switch off the mains supply and isolate all the installation. By analysing the installation drawings and data the area or items should be isolated individually to limit disruption.

Provision of emergency or stand-by supply

Often the disruption to the customer is not having an electrical supply to an area of the installation. If the work cannot be done outside of normal working hours, a temporary supply could be arranged. Some installations may have a stand-by service in the form of emergency lighting or a stand-by generator, but this is only usual in hospitals or where computers are being used and data is stored.

Client demand for a continuous supply

Where the client does not want any interruption in supply and it is not possible to work on the affected area during normal working hours, arrangements should be made for a planned maintenance or shut down for the repair to be carried out safely.

Special precautions

On completion of this topic area the candidate will be able to recognise situations where special precautions should be applied.

Special situations

Special precautions and a risk analysis should always be carried out before any work is commenced. It is a requirement of the Health and Safety at Work Act 1974 that precautions should be accounted for and that any risk is removed or persons carrying out any work activity should be protected from harm. Risk assessments are now commonplace and are part of the daily work routine.

Fibre optic cabling

We covered this type of cabling in detail at Level 2.

Fibre optics is becoming increasingly popular in the data and telecommunications industry, although it is unlikely that you will terminate or fault find on such systems. However, your company may be asked to install such systems and you may be asked to assist in their installation.

As its name suggests the cable is made up of high-quality minute glass fibres, which can carry vast amounts of transmitted signals and data using emitted signals of light. The levels of light transmitted can reach dangerously high levels. Therefore YOU SHOULD NEVER LOOK INTO THE END OF FIBRE OPTIC CABLE!

Remember

Any temporary supply which is provided should be wired in accordance with the *IEE Wiring Regulations*

Did you know?

Some formal risk assessments are recorded and noted in method statements. These can be seen in your company's health and safety statements

Safety tip

The levels of light carried by fibre optic cabling can be dangerously high, so you should never look into the end of a fibre optic cable

Antistatic precautions

Antistatic precautions are usually prevalent in areas that are hazardous due to the high risk of ignition or explosion. This is usually where flammable, ignitable liquids, gases or powders are stored or dispensed, for instance:

- petrol filling stations
- chemical works
- offshore installations
- paint stores.
- flour mills.

Petrol filling station

In such situations we must be aware that NO NAKED FLAMES or equipment which can cause a spark or discharge of energy can be used. There would obviously be a no smoking policy. It is important to adhere to the safety signs and regulations in such areas. When asked to work in such environments you should be given special induction and training to make you aware of the dangers and the risks.

Electrostatic discharge

Static is an electric charge caused by an excess or deficiency of electrons collected on a conducting or non-conducting surface, thus creating a potential difference measured in volts. Static charges are caused in several ways, as described below.

Friction

When two surfaces are rubbed together they behave as though electrons were rubbed off one surface and on to the other, leaving one surface with a deficiency of electrons, therefore being positively charged. The other surface acquires and retains electrons, therefore being negatively charged.

Separation

When self-adhesive plastic or cellulose tape is rapidly pulled off a reel it generates a static charge by tearing electrons from one surface to another.

Induction

Static is generated by induction when a charged item – for example, a folder of paper – is placed on a desk and on top of that is placed a conductor, say a printed circuit board. If the paper was positively charged, the charge from the paper is immediately induced into the printed circuit board, making it negatively charged.

All materials can be involved in the production of static. In the case of non-conductors (insulators such as paper, plastic and textiles) the charge cannot redistribute itself and therefore remains on the surface.

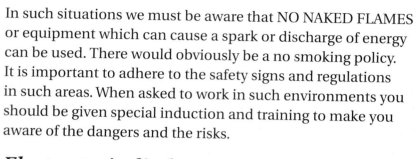

Did you know?

Every step, every move, every shuffle on a seat, generates static charges. Picking up, putting down, wrapping and unwrapping, even just handling an object creates static. So you can see that with all the everyday movements we make, while wearing a lot of clothes made from man-made fibres, you will generate large amounts of static

Protection from electrostatic discharge

The main effort must be to protect electrostatic sensitive devices by providing handling areas where they can be handled, stored, assembled, tested, packed and dispatched. In the special handling area there should be total removal of all untreated wrappings, paper, polystyrene, plastics, non-conductive bins, racks and trays, and soldering irons including tips should be earthed. Specifications, drawings and work instructions must be in protective bags, folders or bins. Wrist straps should be worn at all times in the static free area. These straps 'earth' the handler and provide a safe path for the removal of static electricity.

Static electricity

Damage to electronic devices due to over voltage

When testing or inspecting faults we must take account of electronic equipment and their rated voltage. It is usual to isolate or disconnect such equipment to avoid damage due to test voltages exceeding the equipment's rated voltage. Such voltage levels will cause components, control equipment or data and telecommunication equipment to be faulty. This could be costly not only in repairs but also to the reputation of your company, resulting in the loss of future contracts. The voltage of an insulation resistance tester can reach levels between 250 to 1000 volts.

Avoidance of shut down of IT equipment

On no account should any circuit be isolated where computer equipment is connected. This type of equipment is to be found in many commercial, industrial and even domestic installations. The loss of data is the most common occurrence when systems fail due to faults or power supply problems. Larger organisations will have data protection in the form of uninterruptible power supplies (UPS systems). Whether UPS systems are in place or not, it is still important to plan the isolation of circuits where IT systems are connected. This will give the client time to arrange data storage and download, which will allow systems to be switched off and repairs to be carried out.

Risk of high frequency on high capacitive circuits

Capacitive circuits could be circuits with capacitors connected to them or some long runs of circuit wiring which may have a capacitive effect. This is usual on long runs of mineral-insulated cables. When working on such circuits no work should commence until the capacitive effect has been discharged. In some cases it would be practical to discharge capacitors manually by shorting out the capacitor.

Danger from storage batteries

Batteries can be found on many installations and provide an energy source in the event of an emergency when there is a power failure. This can be seen where battery panels provide a back-up for the emergency lighting in a building when the power fails.

Batteries can also be found in alarm panels, IT UPS systems and emergency stand-by generators for starting.

Safety tip

On some systems where high frequencies exist there is a danger of exceptionally high voltages occurring across the terminals. On no account should you as a trainee ever attempt to tamper with capacitive circuits and, if in doubt, seek the assistance of your supervisor

Wherever you have cause to work with or near batteries, special care should be observed, particularly with regard to the following.

- Lead acid cells contain dilute sulphuric acid. This acid is harmful to the skin and eyes, can rot clothing and is highly corrosive. If there is any contact with the skin it should be immediately diluted and washed with water.

- Lead acid cells also emit hydrogen gas when charging and discharging, which when mixed with air is highly explosive.

- High voltages applied to cell terminals will damage the battery.

- When connecting cells, shorts across the terminals will produce arcing, which could cause an explosion.

- Cells should never be disconnected as this can also produce arcing, which could cause an explosion.

You can take precautions by:

- wearing gloves and suitable eye protection

- wearing a plastic apron and suitable boots when handling battery chemicals such as sulphuric acid or potassium hydroxide

- removing metal objects from pockets that could fall on to the battery, or bridge across its terminals

- keeping sources of ignition, such as flames, sparks, electrical equipment, hot objects and mobile phones, away from batteries that are being charged, have recently been charged or are being moved

- using suitable single-ended tools with insulated handles

- fitting temporary plastic covers over the battery terminals

- charging batteries in a dedicated, well-ventilated area

- sharing the load with a workmate when lifting batteries – they can be very heavy

- using insulated lifting equipment and checking there are no tools, cables or other clutter you could trip on

- washing your hands thoroughly after working with batteries, especially before eating, smoking or going to the toilet.

Always avoid:

- working with batteries unless you have been trained

- smoking

- wearing a watch, ring, chain, bracelet or any other metal item

- overcharging the battery – stop charging as soon as it is fully charged.

FAQ

Q I've always struggled with fault finding. Is there a technique I can use?

A Yes, over time you will develop a technique based on your ever-growing knowledge and experience as an electrician. Many manufacturers publish fault-finding charts which can be very useful. Schematic, block or 'functional flow' diagrams will help you to understand the sequence of control and therefore aid fault finding. A good general principle is to work backwards from the fault towards the supply. This will tell you where you have 'lost' the supply, and therefore probably where the fault is.

Q Which is the best test instrument to use for fault finding?

A This depends on the type of circuit. For basic mains circuits a simple two-lead voltage indicator is usually sufficient, as it is often just a question of 'is there a voltage at this point, yes or no'? Other circuits may require you to actually measure voltage and resistance, so you will need a suitable multimeter (with GS38 probes) for this. Measuring current for fault-finding purposes is quite rare but, if you have to, then a suitably ranged clamp-type ammeter is preferable to a contact type.

Q Will I be safe if I use GS38 test probes?

A Testing almost inevitably means working on or near live conductors and you should always use the correct equipment, insulated tools, GS38 probes, rubber mats etc. However, you must never be complacent. Many accidents have happened where cables have become detached from their terminals or enclosure lids have fallen onto live conductors causing severe arcing, flash-overs and even explosions! So, your GS38 probes might be alright, always be on the look-out for other objects which may cause unsafe incidents.

Activity

1. Have a look through the manufacturer's literature for items of equipment you have installed. See if they have published any fault-finding charts, block diagrams or other aids to fault-finding.

2. If you have access to suitable facilities, you could get someone to put faults on a circuit, such as a central heating control system for you to find. This is best done at a college or training centre under controlled conditions and should not result in equipment being damaged or unsafe situations being created.

Knowledge check

1. Using manufacturers' data, find out the 'manufacturers codes' for 1mm^2, 2.5mm^2 and 4mm^2 single-core PVC insulated cables and 1mm^2 PVC insulated and sheathed twin and earth and three-core and earth cables, and 2.5mm^2 PVC insulated and sheathed twin and earth cables.

2. List the three main functions of a domestic consumer unit.

3. Describe the danger of using an off-load device, such as a fuse carrier, to interrupt the current flowing in a circuit.

4. List the people with whom you may wish to consult before commencing fault-finding on an installation or some equipment.

5. List the ten logical steps to fault-finding.

6. A distribution board has a line-neutral insulation resistance of 1 MΩ. What would the insulation resistance of the board be if another circuit with the same insulation resistance were added to it?

7. In relation to switchgear, what would be the potential problem if the earth loop impedance were too low a value?

8. What additional safety precautions would you introduce when working in flammable areas such as petrol stations?

9. Why must special precautions be taken when working on sensitive electronic circuits?

10. List the hazards associated with storage batteries, such as lead-acid cells.

chapter 10

Restoring systems to working order

Unit 3 outcome 2

Diagnosing and rectifying problems also requires us to know how to repair the problem and, more importantly, what equipment we need to use to restore systems to fully working order. This application of rectification techniques will be a vital part of your skills as an electrician.

When carrying out repair work of this kind, it is important to remember the lessons we have already learned. Your working practices must be efficient and, above all, safe. Repaired materials will have to be recommissioned and then restored to working order. Efficiency and safety are as crucial to repair work as they are to installation.

This chapter builds on topics covered earlier in this book and at Level 2.

On the completion of this chapter the candidate will be able to:

- list relevant sources of information to facilitate recommissioning
- describe the requirements of testing and commissioning
- describe procedures to deal with clients and documentation
- select suitable instruments
- explain the importance of verifying test instruments, calibration and documentation
- describe how to carry out functional checks and tests to verify rectification and restoration of a system
- explain the need to comply with test values and state actions to take
- list procedures for dealing with waste
- describe the completion process.

Recommissioning

On completion of this topic area the candidate will be able to list the relevant information and documents required to facilitate the recommissioning of installations and equipment after completing fault diagnosis and repair.

The purpose of recommissioning

The purpose of recommissioning is to ensure that the rectification process used is:

- is electrically safe
- is mechanically sound
- complies with the specification
- functions
- complies with current legislation.

In order to do this, reference will need to be made to the following sources of information:

- BS 7671
- contract specifications
- distribution/wiring diagrams
- manufacturers' instructions
- relevant statutory legislation.

BS 7671

The *IEE Wiring Regulations*, the *IEE On-Site Guide* and Guidance Note 3 on *Inspection and Testing* are required to confirm the safe repair of the fault and the maximum values of the results to be gained from tests carried out on the repairs. In all these cases you must follow the conditions defined in BS 7671 in order to fully complete the tasks.

Contract specifications

These are required to ensure that whatever action has been taken to repair or replace the fault has not resulted in the installation of a lesser specification. For example, if the specification required three light switches to control lights along a long corridor (two-way and intermediate switching), then a repair will not be acceptable to the specification if it results in losing one of these switches.

Distribution or wiring diagrams

Distribution diagrams record the sequence of control within a system. This makes it essential to ensure that, after rectification work has been completed, the document is checked to establish that the sequence has not changed as a result. It also means that the document will have to be changed in the result of any alterations made as part of the work.

Did you know?

The distribution diagram will identify switching and control equipment for the functional testing of the installation

Manufacturers' instructions

The manufacturers' instructions ensure that the user has knowledge of the equipment, its supply voltage and the current setting of protective devices and cable sizes. Information is also given on the position and settings of limit switches and other components.

Relevant statutory regulations

We have covered these comprehensively, both in this book (chapter 1, pages 2–12) and at Level 2. The key pieces of legislation that will affect rectification are:

- Health and Safety at Work Act 1974
- Controlled Waste Regulations 1989
- Electricity at Work Regulations 1989
- Provision and Use of Work Equipment Regulations 1989
- Management of Health and Safety at Work Regulations 1999
- Electricity Safety, Quality and Continuity Regulations 2002

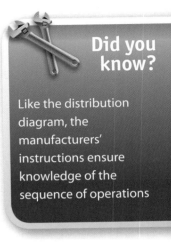

Did you know?

Like the distribution diagram, the manufacturers' instructions ensure knowledge of the sequence of operations

Requirements of testing and commissioning

On completion of this topic area the candidate will be able to describe the requirements of testing and commissioning with reference to types of test equipment, sequence and procedures for testing and application of current wiring regulations.

Types of test equipment

You can find a wide range of test equipment on the market, produced by many different manufacturers. In order to make the correct decision as to the type of test equipment to use, it is vital to know which is suitable for the variety of tests we plan to undertake. The Institution of Engineering Technology (IET), formerly the Institute of Electrical Engineers, publishes guidance notes to accompany BS 7671. The most important of these is *Guidance Note 3 on Inspection and Testing*. This details the performance standards that test equipment must meet in order to be sure that results gained are reliable.

The requirements needed for test equipment are discussed in chapter 8.

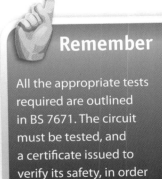

Remember

All the appropriate tests required are outlined in BS 7671. The circuit must be tested, and a certificate issued to verify its safety, in order to comply with BS 7671

Sequence of test

After rectification has been completed, the electrical integrity and safety of the circuit must be confirmed before it is re-energised.

Application of the current wiring regulations

The type of certificate issued after testing is dependent on the nature of the fault and the degree of work that has been carried out as part of the repair.

For example, what if the fault was in an installation constructed in 1950? At that point the 12th edition of the regulations was published and in force. While you carry out a repair you may notice deviations from the current edition of the regulations. In this case, what action should you take?

First, complete the fault rectification then test and commission the circuit worked on. Let us assume that this has involved rewiring the circuit, renewing the accessories and replacing the rewireable fuse with a MCB. In this case the new circuit would need to be tested and commissioned to confirm compliance.

The test certificate allows space for passing comments on the existing installation. Here you can make recommendations for bringing the installation up to the current standards. If the situation could be dangerous then the seriousness must be made known to the client.

Clients and reports

On completion of this topic area the candidate will be able to describe the procedures for dealing with the customer and the reports.

Reports

After completing the fault rectification process, the wiring regulations require that testing and commissioning take place.

Handover to client

The final task of any work is handover to the client. This should include a tour of the installation, with an explanation of any specific controls or settings. Where necessary a demonstration of the circuit in action, and any particularly complicated control systems, should be given.

In addition, an explanation should be given that fully outlines why the fault occurred and recommendations given that will help avoid a reoccurrence of the problem.

A copy of the installation certificate, along with recommendations for a re-inspection, a date for the installation and contact details of your organisation, should be left with the client.

Instrumentation

On completion of this topic area the candidate will be able to identify and select instruments suitable for fault diagnosis and rectification.

The instruments an electrician would use for fault diagnosis and rectification are identical to those used for initial testing of an installation. This topic was covered in full in chapter 8 pages 325–326.

Instrument accuracy

On completion of this topic area the candidate will be able to explain the importance of regularly verifying the accuracy of test instruments.

Electrical test instruments

This topic was explored in depth in chapter 8, pages 355–360. Please refer back to this section for more information. There are some crucial points to remember when we rely on the test results of instruments. We must:

- verify that the instrument works before we use it
- check the batteries
- check the instrument is not damaged
- check the leads for compliance to health and safety guidance note GS38
- confirm that the instrument complies with BS EN 61010 or older instruments with BS 5458
- have the instrument calibrated; this is usually every 12 months, but can be less if the instrument is used more frequently (this would generally be the case for a specialist calibration engineer)
- have a calibration sticker fitted on the instrument with the date that it was calibrated and a date for the next calibration
- keep calibration (certificates) records for each instrument
- store instruments in clean dry conditions.

Testing

On completion of this topic area the candidate will be able to describe how to undertake the functional checks and tests required to verify the rectification and restoration of a system and equipment.

Functional tests and checks

Testing and the methods employed were discussed fully in chapter 8, pages 361–375. Please refer back to this section for more information on this area.

Remember that on completion of the fault rectification process, the circuit or equipment worked on has to be electrically tested to ensure that the repair, and any new installation associated with the repair:

- is effective
- is safe
- complies with BS 7671.

In addition to this, the circuit or equipment worked on must function as the specification intended. After all dead electrical testing has been completed and the supply reconnected, functional testing of the circuit has to take place, with checks being made to:

- lighting, switches, PIRs, time clocks and anything associated with the control of the lighting circuit to confirm that the circuit works
- motors, specifically their control gear, including isolators, emergency stop buttons, electronic/manual drives, controls and any interlocks on guards or limit switches
- main switches and sub-distribution circuits for control and isolation
- RCDs and overcurrent protective devices, including earth-loop impedance tests at every socket outlet on the circuit worked on.

Compliance with BS 7671

On completion of this topic area the candidate will be able to explain the need to compare test values with those in BS 7671 and state the actions to be taken in the event of unsatisfactory results.

The topic was covered in detail in chapter 8, page 376–378. This outlined the need for the test results to comply with BS 7671 and the actions to be taken in the event of unsatisfactory results. We will briefly recap the essential points here.

Unsatisfactory test results

For continuity, very low values are required. These are: main equipotential bonds maximum 0.05 Ω; resistance of cpc for steel conduit/trunking, 0.005 Ω/m; resistance of 1.5mm^2 copper conductor, 0.012 Ω/m. High values of resistance will result in the protective device failing to operate within the times required by BS 7671.

Polarity

Incorrect polarity is one of the most dangerous situations that can occur in an installation. Tests for correct polarity confirm that all single-pole switches and protective devices are fitted in the line conductor.

If tests confirm reverse polarity you must immediately:

- remove the supply and inform the client
- inform the local area board if the problem is related to the supply
- if the supply is not connected, alter the terminations to comply with polarity so that single-pole switches and protective devices are connected in the phase conductor. Ensure the centre contact of ES lamp holders are also connected to the phase conductor.

Insulation resistance

The value of insulation resistance on 230 volt circuits should be above 1 MΩ when tested between live conductors and live conductors and earth.

Zero resistance indicates a dead short. Values up to 0.49 MΩ indicate a low insulation resistance and could be the result of the ingress of damp etc. Tests on the whole installation should produce values greater than 2 MΩ; any less requires further investigation.

Earth-fault loop impedance

Maximum values of earth-loop impedance are found in Chapter 41 of BS 7671 (which will need correcting for temperature) or the *IEE On-Site Guide*, which contains temperature corrected values. Value should not be exceeded if disconnection of the protective device is to meet the time requirements of the Regulations. These are 0.4 seconds for circuits ≤32A and 5.0 seconds for circuits >32A. If a high value is recorded then:

- check the effectiveness of the terminations
- increase the size of the cpc
- change the type of the protective device used
- add a RCD to protect the circuit, providing it is electrically sound.

To take no action would mean that people living in or working in the installation would be at great risk.

Residual current devices (RCDs)

To test a RCD, first press the integral test button on the device. Use an RCD tester to determine the disconnection time. This should be within 200 milliseconds when tested at the tripping current of the device.

An RCD is fitted to the circuit for safety and protection. If this device is not working then the installation is unprotected and the danger of electric shock exists. Failure to operate to the required standard demands that the device should be replaced immediately.

Advising the customer

On completion of this topic area the candidate will be able to state the necessity for advising the customer/client of the need for additional restoration work to the fabric of the building.

Additional restoration work

As electricians, we are more often than not being asked to locate and repair faults on existing installations. Consequently, depending on the fault found and the method of rectification needed, we may need to disturb the fabric of the building. It is therefore vitally important to fully discuss with the client all of the implications of the fault diagnosis/rectification process. As part of investigating a fault, we may have to do nothing more than look inside an item of equipment. However, if the fault has been caused by someone putting a nail through a cable, then the repair will involve replacing the cable and that in turn may involve having to damage the building fabric. You should therefore explain these aspects to the client and agree what repair work (e.g. plastering, brickwork, decorating) will be required and which, if any, of these can be carried out satisfactorily by yourself.

Obviously, as with any aspect of fault finding and rectification, always leave the site in a clean and safe condition.

Disposal of waste

On completion of this topic area the candidate will be able to list the procedures for the disposal of waste materials.

Waste

Any substance or object that you discard, intend to discard or are required to discard is waste and as such is subject to a number of regulatory requirements. The term 'discard' has a special meaning. Even if material is sent for recycling or undergoes treatment in-house, it can still be waste.

Under the Controlled Waste Regulations 1998, the duty of care requires that you ensure all waste is handled, recovered or disposed of responsibly, that it is only handled, recovered or disposed of by individuals or businesses that are authorised to do so, and that a record is kept of all wastes received or transferred through a system of signed Waste Transfer Notes.

If you work as a subcontractor and the main contractor arranges for the recovery or disposal of waste that you produce, you are still responsible for those wastes under the duty of care.

Be aware that materials requiring recycling (such as copper cable), either on your premises or elsewhere, are likely to be waste and will be subject to the waste management regime and the duty of care. If in any doubt, seek advice from the Environmental Regulator.

Special waste

If the material you are handling has hazardous properties, it may need to be dealt with as **special waste**.

Covered by the Special Waste Regulations 1996, special waste is waste that is potentially hazardous or dangerous and which may therefore require extra precautions during handling, storage, treatment or disposal. In reality this means that most businesses are likely to produce some special wastes. The following are examples:

- asbestos
- lead-acid batteries
- used engine oils and oil filters
- oily sludges
- solvent-based paint and ink
- solvents
- chemical wastes
- pesticides.

Did you know?

If you leave materials on site when your work is complete, you may be discarding them. If they are discarded they will be 'waste' and, as the producer of that waste, you will be responsible for it

Waste disposal

If you intend to discard containers, an assessment must be made as to whether they are special waste. Containers may be special waste if they contain residues of hazardous or dangerous substances/materials. If the residue is 'special' then the whole container is special waste.

Below are some guidelines for good waste disposal practice.

- Do not burn cable or cable drums on site; find another method of disposal.

- Before allowing any waste haulier or contractor to remove a waste material from your site, ask where the material will be taken and for a copy of the waste management licence or evidence of exemption for that facility.

- Segregate the different types of waste that arise from your works. This will make it easier to supply an accurate description of the waste for waste transfer purposes.

- Minimise the quantity of waste you produce to save you money on raw materials and disposal costs.

- Label all waste skips – make it clear to everyone which waste types should be disposed of in that skip.

Waste disposal is commonly segregated into different types

Completion

On completion of this topic area the candidate will be able to describe the process of completion after rectification of faults in an electro-technical system.

On completion of the fault rectification process the client will need to be:

- shown the circuit working
- given the completed certification, including schedule of test results and schedule of inspection
- given any recommendations that may result in the prevention of the fault reoccurring
- given a contact number for any future emergencies
- given a date for future re-inspection
- encouraged to ask questions.

It must always be kept in mind that most clients will be non-technical. Therefore, if they are to be provided with sufficient information to enable the installation to be used safely, this will have to be provided in the most direct and simple way.

The customer will need to be shown the circuit working and the method of control/switching as well as the location of the protective devices. They will need to know how to reset the device, especially if it is a RCD, and what to do should it fail to reset. Reference should always be made to any manufacturer's information that is available, particularly if there is a choice of control settings and machine speeds to be set.

The client should be provided with an electrical installation certificate or minor works certificate – depending on precisely how much work had to be completed to effect successful rectification of the fault.

Recommendations may need to be made if a repeat of the fault is to be avoided; typically these may include regular maintenance and inspection. If equipment is in constant use then suggestions on periods for replacement of the equipment can be made, or recommendations on how equipment should be used referring to the manufacturer's literature, or recommendations about changing the type of fittings/equipment to be more robust and suitable for the environment that it has been installed in.

Always ensure that the client has up-to-date contact details for your organisation, and always leave a label by the distribution board recommending the period for the next inspection. Answer fully any questions asked by the client.

Remember

It is not sufficient to just pass clients the test sheets as you leave the building

FAQ

Q Most installations don't have any drawings. What do I do?

A BS 7671 states that: 'Suitable diagrams, instructions and similar information must be left within or adjacent to each distribution board.' If the information isn't at the board, then you should ask the owner of the installation or the facility's management company for copies of the relevant paperwork. If the diagrams are not available, then you should take time to familiarise yourself with the installation before commencing work. If necessary, involve colleagues or clients' staff who have worked on the installation before or are familiar with it. Do not proceed if you are unsure.

Q My client won't pay me because I haven't given him a certificate. Can they do that?

A Well, it's normal to be paid at the end of the job, but the job's not finished until the paperwork is done! BS 7671 states that a certificate should be issued after every new installation or addition or alteration of an existing installation. If not, the installation work does not conform to the Regs, and you haven't finished!

Q I've filled in the certificate but the main contractor says it's wrong. Do I have to do it again?

A There could be various reasons for this. It may just be a simple misinterpretation of the results or your results may be incorrect and they may have spotted this. Always remember that inspection and testing is not just about filling in the certificate. The main purpose of inspection and testing is to confirm that the installation conforms to BS 7671 and is safe to be put into service.

Activity

The next time you do a job, have a look at the waste you have produced to see if this could either be reduced or recycled. Keep a particular look-out for hazardous or restricted wastes, such as oil and batteries, that should not find their way into ordinary landfill rubbish.

Knowledge check

1. List five objectives of the re-commissioning process.

2. Does an installation which was installed in accordance with an earlier edition of BS 7671 have to conform to the current edition of BS 7671?

3. How often should electrical test instruments be calibrated?

4. List six items of an electrical installation that may require functional testing.

5. Describe the potential safety hazards arising from reversed polarity.

6. List four possible remedial actions to correct a circuit with an earth loop impedance which was too high.

7. Why is it important to consider carefully where you put any waste you produce?

8. Explain why it is important to talk to the customer about the completed electrical installation?

Index

A

a.c. motors 233
   single-phase synchronous 248–9
   speed control 258
   synchronous 248
   three-phase synchronous 248
accidents 16
   prevention 21–2
   reporting 11–12, 23–4, 58–9
adiabatic equation 306–10
alarm procedures 60–2
antistatic precautions 430
Approved Codes of Practice (ACOPs) 16
arcing 225, 401
armoured cables 207–10
   conductor colour identification 210
   installation 209–10
   terminating 208–9
asphyxiation 57–8
action 58
assembly points 61
auto transformer 92

B

barriers 14, 262, 334
bending conduit 178–81
   machines 177
bipolar transistor 124, 146
block diagrams 272
book of rates 266
breathing apparatus 51–2
BS 3036 fuses 301, 319-20
BS 7671 Regulations 2, 271
   access 347
   conduit requirements 312
   inspection and testing 324, 341–2, 376–8
   recommissioning 436
   and test results 440
BS EN 61010 355

building fabric 263
building regulations 269–70
busbar trunking 191
bystander safety requirements 14

C

cable ladder 291
cable selection 296–311
   ambient temperature 300
   BS 3036 fuse 301
correction factors 299–302
   diversity 298–9
   grouping factors 300
   mineral-insulated cable 301
   over-current protection 296–7
   shock protection 297, 304–6
   thermal constraints 297, 306–10
   thermal insulation 301
   voltage drop 297, 302–4
cable size calculations 298–9
   diversity 298–9, 310–11
   rating of protective device 299
cable tray 201–7, 291–2
   advantages/disadvantages 292
   bending 202–5
   fabricating on site 202–6
   fixing 207
   installing 201
   reduction 206
cables
   advantages and limitations 286–8
   armoured 207–10
   Cat 5 cable 211
   faults in interconnections 410–12
   faults in seals and entries 413–14
   faults in terminations 412
   fibre-optic 212
   FP 200 Gold 217–18
   joints and connections 294–5
   lighting circuits 396

   MICC 212–16, 286–7
   multicore 397
   size of 284
   trailing 211
capacitance and resistance (RC) 106–8
   see also resistance
capacitive circuits: dangers 431
capacitor smoothing 123
capacitor start-capacitor run motor 242–3
capacitor-start motor 241–2
capacitors 95–102, 140–4, 416
   basic principles 95–6
   calculations with 98
capacitance of 95–6
   charging and discharging 101–2
   coding 140–3
   in combination 99–100
   electrolytic 96
   fixed 96–7
   non-electrolytic 97
   polarity 144
   types of 96–8
   variable 96, 98
carcinogens 7, 29
career development 33–4, 68
Cat 5 cable 211
ceramic capacitor 97
certification for installations 379–87
choke smoothing 123
circuit breakers 225–7
   high voltage 227
   low voltage 226–7
   testing 338, 340
circuit diagrams 272
circuit protective conductors
   inspection checklists 351
   testing 361–2
clamp meters 354
Clean Air Act (1993) 28, 270
clients *see* customers
CMRs 28

collector columns 211
communication 276–7
    in writing 276–7
completion dates 34, 35
completion process 444
compound motor 232, 233
compression joints 295
conductor resistance 83–6
conductor size 296–311
    see also cables
conduit 176–89
    bending 178–81
    bending machines 177
    cable capacities 186
    cutting/screwing/terminating 183
    drawing in cables 185–6
    fixings 182
    flexible 188
    inspection checklists 349–50
    non-inspection elbows and tees 185
    plastic 187–8, 189, 289–90
    running coupling 184
    screwed 176
    size determination 312–13
    steel 289
    termination 183, 184
    types of bend 177
    wiring 185
confined spaces 42, 44-5, 57–8
Construction (Design and Management)
    Regulations (1994) 270
contactors: faults 415
continuity testing 361–4, 376, 385, 407
contracts 37–8, 272–4
    breach of 273
    monitoring progress 274
Control of Major Accidents and Hazards
    Regulations (1999) 9–10
Control of Substances Hazardous to
    Health Regulations (2002) 2, 7–8, 9–10
control systems 398–9
Controlled Waste Regulations (1998) 29,
    270, 437, 442
customers 34–8, 267
    advising on restoration work 442
    and building regulations 270
    communication 276
    and completion 444
    handover to 391, 438
    reports for 391, 438

D

dangerous occurrences 11, 12, 58–9
Dangerous Substances Regulations 28,
    29, 270
dangers 280–3, 429–32

Data Protection Act (1998) 23, 73
d.c. motors 232–3
    armature resistance control 257
    field control 257
    pulse width modulation 257–8
    speed control 257–8
degree qualifications 33–4
delta connection 169, 170
demand, changes in 30–3
diac 119, 120
diodes 147–8
    light-emitting 115–16
    testing 116
    zener 148
Disability Discrimination Act 73
discharge lighting 152, 154-8
discrimination
    direct and indirect 74
    legislation 72–4
    positive 74
disease reporting 11, 12
distribution diagrams 436
distribution of electricity 168-75
    measuring power 174-5
    neutral currents 173-4
    three phase supplies 169-70
    voltages 278
DOL (direct-on-line) starter 250–3
doping 113
dosimeter pocket card 28–9
drawing in cables 185–6
dual in-line ICs 148
dynamic induction 91

E

earth electrode resistance
    measuring 368–9
    test equipment 390
earth-fault loop impedance
    measured values 371–3, 374
    Tables 41.2, 41.3, 41.4 371
    test equipment 389
    testing 337, 340, 361–2, 370–4, 378,
        385, 408–9, 441
earth-loop impedance testers 353
earthing 228
    and bonding 313
    BS 7671 compliance 324
EEBADS (Earthed Equipotential
    Bonding and Automatic
    Disconnection) 280, 313
electric lamps 152–4
    tungsten halogen lamps 153–4
electric shock 53
    and cable selection 304–6
    hazards 54

precautions against 54–6
    protection 280–1, 297
    reducing voltage 54–5
Electrical Installation Certificate 328,
    330, 379–80
    schedule of test results 384–5
    to customer 391
    Type 1 381
    Type 2 382–3
electrical rotating machines 232–43
    see also a.c. motor; d.c motor; motor
Electricity at Work Regulations 2, 4–6, 9,
    271, 332, 335, 342
    absolute/reasonably practicable 4–5
    access 347
    rectification 437
Electricity, Safety, Quality and Continuity
    Regulations (2002) 2, 3–4, 271, 343, 437
Electricity Supply Regulations (1988) 332
electromagnetic induction–89–91
electromagnetism 88–94
    on current-carrying conductor 92–4
electron flow 82
electrostatic discharge 430–1
employers' duties 49
    health and safety 2–3, 15, 16, 18
    PPE 13
employment legislation 72–4
Employment Rights Act (1996) 72
environmental conditions 284, 347
environmental health officers 20
environmental legislation 27–9
Environmental Management Systems 26
Environmental Protection Act (1990) 27,
    270
environmental regulations 270
equipment
    faults 422
    Regulations 8–9
equipotential bonding conductors 361–2
erection methods 347
evacuation procedure 60–2
escape routes 60
    reporting in 61–2
excavation work 46–7
    planning 46
exposure monitoring 8
eye protection 51

F

Faraday, Michael/Law 90, 140, 237
fault diagnosis 395, 400
    basic principles 402–4
    information required 402, 405, 406
    logical approach 404
    non-live tests 406–8

and rectification 404–9
safe working practices 400–1
stages 405–6
use of other's expertise 403
faults 323
abuse, misuse and negligence 423
accessories 414
cable interconnections 410–12
cable seals and entries 413–14
cable terminations 412
contactors 415
control equipment 414–15
downtime 425-6
general categories 418
instrumentation 416
insulation failure 422
loss of supply 420–1
luminaires 417
position of 418–21
prevention by maintenance 424
protective devices 416–17
repair or replacement 424–7
solid-state and electronic
devices 415–16
see also rectification
fences/barriers 14, 53, 262, 334
fibre optic cable 150–2, 211, 212
dangers 429
field effect transistors 146–7
film resistors 134
filter circuits 123–4
fire
drills 64
evacuation 60–2
fighting 63–4
legislation 62–3
prevention 63
protection 281–2
risk 48
safety 62–4
Fire Officer 25
first aid 49-50
fixed resistors 133–4
Fleming's left-hand rule 93
flexible cables and cord faults 417
floor trunking 190
fluorescent lighting 154–7
glow-type starter 154–5
quick start 157
thermal starter circuit 157
forward biased connection 113–14
FP 200 Gold cable 217–18
full-wave bridge rectifier 122
full-wave rectification 122
fume inhalation 58
functional testing 408–9

fuses 301, 314, 319–20

G
generation of electricity 168–9
voltages 277
generators 232–3
GES lamp 154
glow-type starter circuit 154–5
GLS lamp: dimmer circuit 120

H
half-wave rectification 120
hand protection 50
hazard reporting 24–6
communication 24–5
writing process 25–6
hazardous malfunction 58–9
hazardous substances 7–8
hazards 22
COMAH 9–10
COSHH 7–8, 9–10
electric shock 54
guarding 22
health and safety 17–20, 270–1
implementing and controlling 14–15
policy statement 16
precautions 262–3
responsibility for 15, 16
sources of information 21
training 22
Health and Safety at Work Act (1974) 1,
2-3, 9, 270
bystander safety 14
penalties for offences 20
rectification 437
risk assessments 429
written policy 16
Health and Safety Commission 18, 19
Health and Safety Executive 11, 19
inspectors 19
health surveillance 8
height, work at 45–6
high-frequency circuits 156
Human Rights Act (1998) 73

I
impedance triangle 105
improvement notice 19
Incident Contact Centre 11–12
inductance 88–93
examples of use 92
and resistance (RL) 103–6
see also resistance
induction motors see motor
industrial distribution
systems/equipment 176–218

infrared source and sensors 116
injury 283
reporting 11, 12
inspection checklists 347–51
inspection and commissioning 324–8
anticipation of danger 343
basic and fault protection 344, 345
BS 7671 compliance 324, 341–2
certification 438
competence of inspector 326
conductors 343
contract specification/drawings 342
detailed requirements 343–5
enclosures/mechanical protection 345
flexible cables and cords 344
information sources 328, 341–5
initial verification procedures 324
instruments 352–60
isolating and switching devices 346
isolation 326, 327
joints and connections 343
legislation 341–3
manufacturer's instructions 342
marking and labelling 328
mutual detrimental influence 346
operating devices 327
planning and disruption 328
and project specification 325
protective devices 345
requirements 437–8
safe operation 325–6
switches 344
systems and equipment 327–8
test instruments 326–7
thermal effects protection 344
visual inspection 345–7
see also periodic inspection; testing
inspection schedule 383–4
installations
activities 262–3
certification 379–87
documentation, drawings and
specifications 271–2
parties concerned 267–8
planning 263–2
programming 275
regulatory requirements 269–71
schedule 388–9
see also planning
instrumentation faults 416
rectification 439
instruments see test instruments
insulation failure 422
insulation resistance testing 337, 340,
364–7, 377–8, 385, 390, 407–8, 441
ohmmeters 352–3

integrated circuits 148–9
Investors in People 76
ISO 9001 76–7, 78
isolation 56, 262, 282–3
    and fault diagnosis 400
    and inadvertent switching on 327
    lock-off procedures 43
    rotating machinery 283
isolators 227
IT equipment 343, 421, 431

J

joints and connections 294–5
junction boxes 295

K

Kirchhoff's Law 173

L

labelling test instruments 355
ladder rack 291
lamp dimmer circuit 120
learning and performance 68
legislation 2–12
    confined spaces 42, 44–5
    employment 72–4
    environmental 27–9, 270
    fire 62–3
    first aid 49
    health and safety 270–1
    inspection and testing 341–3
    pollution and waste 27–9
    recommissioning 437
    safe working 42
    see also individual Acts
letter writing 277
light emitting diodes (LEDs) 115–16
light-dependent resistor 140, 149
lighting 152–63
    cosine law 161–2
    discharge 152, 154–8
    illumination calculations 159–63
    inverse square law 160
    lumen method of calculation 162–3
lighting circuits 396–7, 400–1
    cabling arrangements 396, 397
    regulations 158
    switching arrangements 396–7
lighting controls checklist 349
lighting points checklist 349
load balancing 279
location diagrams 272
lock-off procedures 43
logic gate circuits 130–1
low-resistance ohmmeters 352

luminaires: faults 417
    see also fluorescent; lighting

M

machinery, unguarded 48
magnetism 87–94, 231
magnetic flux 87–8
    screw rule 88
Management of Health and Safety at
    Work Regulations (1999) 2, 7, 9, 15, 437
marking and labelling 328
MCBs 226–7, 314, 318
    combined tripping 315–16
    magnetic tripping 315
    selection of 316
    thermal tripping 314–15
MCCBs 227
measuring power 174–5
mica capacitors 97
MICC cable 212–16, 286–7
    running 216
    stripping 213–15
    terminating 212–13
    testing and fault-finding 216
Minor Works Certificate 380
    contents 386–7
    guidance notes 387
misuse of equipment 423
motor construction 244–58
motor designs 244–5
motor windings 249
    remote stop/start control 253
    speed and slip calculations 245–7
motor starters 250–6, 283
    auto-transformer starter 256
    automatic star-delta starter 255
    DOL starter 250–3
    hand-operated star-delta 25–-4
    rotor-resistance starter 256
    soft starters 255
multi-lock system 43
multicore cables 397
multimode fibre optic cable 151
mutagens 28, 29
mutual induction 91

N

n-type material 113
neutral currents 173–4
Noise Act (1996) 10
Noise and Statutory Nuisance Act
    (1993) 10
non-live tests 406–8
npn transistor 125, 132

O

off-load device 401
ohmmeters
    insulation-resistance 352–3
    low-resistance 352
Ohm's Law 82–3
on-load device 401
operational requirements 13
opto-coupler 150
overcurrent protection 314
    and cable size 296–7
    see also earth-fault loop
    impedance

P

p-n junction 113–14
p-type material 113
penalty clauses 37–8
periodic inspection and testing 329–40
    circuit breakers 338, 340
    continuity 336, 340
    earth-fault loop impedance 337, 340
    frequency 329–31
    inspection process 332–4
    insulation resistance 337, 340
    items covered 332–4
    main switches and isolators 338, 340
polarity tests 337, 340
preparing 335
purpose 329
    RCD testing 338, 340
    report 338–9, 379-80
    routine checks 331–2
    summary 340
    testing procedures 336-8
    see also inspection; testing
permanent split capacitor motor 242
permit to work 42
personal hygiene 50–3
personal protective equipment 22, 50-1
    employers' duties 8
    Regulations (1992) 13
phasor diagrams 103–4, 107, 109
photo cell 149
photodiode 149
phototransistor 150
PIR (passive infrared detector) 116
planning an installation 263–2
    estimating time 266
    external influences 264
    fabric of building 263
    factors to consider 263
    minimising disruption 265–6
    parts and materials storage 264–5
    programme of work 265
    tools and equipment 265

plant failure 422
plastic conduit 187–8, 189, 289–90
pnp transistor 124, 132
polarity testing 337, 340, 367, 376–7, 408, 441
Pollution Prevention and Control Act (1999) 27
Pollution Prevention and Control Regulations (2000) 27–8, 270
polyester capacitor 97
Portable Appliance Testing (PAT) 9
portable appliances 418
potential divider rule 358–9
power circuits 397–8
prohibition notice 19
project management 75, 265, 275
project specification 325
protection systems 228
protective conductors continuity 407
protective devices 285, 314–20
    disconnection times 304–6
    discrimination 320
    faults 416–17
    fuse and circuit breaker 318–20
    fusing factor 312
    testing 345
Provision and Use of Work Equipment Regulations (1999) 2, 8–9, 437
PVC conduit 187–8, 189, 289–90
PVC insulated and sheathed flat wiring cables 288
PVC/GSWB/PVC cable 210
PVC/SWA/PVC cable 207, 208–9, 287

Q

qualifications 33–4, 68
quality systems 76–7, 78

R

Race Relations Acts 73, 74
Radioactive Substances Act 28, 270
RCDs
    functional testing 378, 385
    test equipment 353–4, 390
    testing 338, 340, 374–5, 441
    recommissioning 436–7
    statutory regulations 437
    rectification 120, 122–4, 147–8
    contingency plans 427–9
    emergency supply 429
    factors affecting 427–9
    instrument accuracy 439
    minimising disruption 427
    see also faults
regulatory requirements 2
    absolute/reasonably practicable 4–5

see also BS 7671; legislation
relationships with colleagues 69–71
repair or replacement 424–7
    costs 424–5
    downtime 425–6
    legal responsibility 426–7
    see also rectification
report writing 277
reporting
    accidents 11–12, 23–4, 58–9
    hazards 24–6
Reporting of Injuries, Diseases and Dangerous Occurrences Regulations (1995) 11–12, 58–9
resistance 82–6
    and capacitance in series (RC) 106–8
    and inductance 103–6
    inductance and capacitance in parallel 110–11
    inductance and capacitance in series (RLC) 108–10
    and inductance in series 103–6
    Ohm's Law 82–3
resistivity 83–6
    conductor resistance 83
resistors 133–40
    coding 136, 137–9
    as current limiters 356–7
    fixed 133–4
    light-dependent 140, 149
    markings 135–6
    power ratings 359–60
    preferred values 135
    testing 356–60
    variable 134–5
    for voltage control 357–9
respiratory protective equipment 13, 51–2
restoration building work 442
restoration of supply 401
reverse biased connection 113–14
RIDDOR 11–12, 58–9
ring final circuit conductors 362–4, 407
risk assessment 9, 21, 24, 44, 48, 429
routine checks 331–2

S

safe working practices 16–17, 42, 325–6, 400–1
    procedures 50–3, 262–3
    see also health and safety
safety guards 52
safety officer 17–18
safety representative 18–19
safety screens 52
safety signs 17
safety training 55–6

schedule of test results 384–5
screw rule 88
screwed conduit 176
screwits 295
self-induction 92
semi-resonant starting 155–6
semiconductor devices 112–32
series motor 232, 233
series-wound (universal) motor 233–4
Sex Discrimination Act (1975) 72, 74
shaded-pole motors 243
shunt motor 232, 233
silicon controlled rectifier 116–19
single-core PVC insulated sheathed/ unsheathed cable 288
single-phase a.c. synchronous motor 248-9
single-phase induction motors 238–43, 245
    capacitor start-capacitor run 242–3
    capacitor-start motor 241–2
    permanent split capacitor 242
shaded-pole motor 243
split-phase motor 239–40
site plans 272
slip-ring motor 237
smoke alarms 64
smoothing 123–4
snubber circuit 120
socket outlet inspection 348
solenoid 89
solid state temperature device 152
special waste 29, 443
split-phase motor 239–40
sprinkler systems 63–4
squirrel-cage rotor 236–7
star connection 169, 170
starters see motor starters
static charges 430–1
static induction 91
steel conduit 289
storage battery dangers 431–2
stroboscopic effect 156–7
supply, restoration of 401
switchgear 225–7
    access to 347
    faults 414–15
    inspection checklist 347–8
switching arrangements 396–7
synchronous motors 248–9
synchronous speed 245, 246–7

T

tantalum capacitor 96, 144
team development stages 70–1
teamwork 69–71

termination 293–5
  connecting to terminals 293–4
  crimp terminals 294
  flexible cords 293
test instruments 352–60, 409
  accuracy 355–60
  applied voltage testers 389
  calibration accuracy 355–6
  checking 326-7
  earth electrode resistance 390
  earth-fault loop impedance 389
  for inspection and commissioning
    352–60
  labelling 355–6
  requirements 389–90
testing resistors 356–60
testing 323, 361–75, 439–40
  BS 7671 compliance 376–8
  continuity 361–4, 376, 385, 407
  earth-fault loop impedance 361–2,
    370–4, 378, 385, 408–9
  functional 408–9, 440
  insulation resistance 364–7, 377-8,
    385, 407–8, 441
  polarity 367, 376–7, 408, 441
  RCDs 374–5, 378, 385, 441
  requirements 388–90
  sequence of tests 388, 437
types of test equipment 437
  see also inspection
thermistors 139–40, 152
three-phase a.c. synchronous motor 248
three-phase induction motor 234–8
  production of rotating field 235
  rotor construction 236–8
  squirrel cage 244, 253
  stator construction 236
three-phase power 170–3
three-phase rectifier circuits 124
three-phase supplies 169–70
thyristors 116–19
timers 398–9
tong testers 354
toxic substances 29
trade unions 18
training
  opportunities 68
  safety 55–6
transferable skills 31
transformers 91–2, 218–24
  calculations 219–20, 221–4
  copper losses 219
  iron losses 220
  ratings 221
  step-up and step-down 218
transient voltages 421

transistors 124–32, 144–7
  codes 145–6
  connecting 144
  FETs 146–7
  operation 125–9
  as switch 129–31
  testing 132
transmission voltages 278
triac 119, 120
trimmer capacitor 98
trunking 189–200
  advantages/disadvantages 290–1
  busbar 191
  capacities 198–200
  fabricating on site 192–7
  floor 190
  flush cable 190
  inspection checklists 350–1
  multi-compartment 190
  PVC 200
  Regulations 197–8
  rising mains 191
  skirting 190
  steel 190
tungsten halogen lamps 153–4
tuning capacitor 98

U

under voltage protection 346
unguarded machinery 48
UPS (uninterruptible power supplies) 431
UTP cable 211

V

valence electrons 112
variable resistors 134–5
variable-frequency drive 258
visual inspection 345–7
voltage drop 168
  and cable selection 297, 302–4
voltage indicating devices 326–7
  applied voltage testers 389
voltages, standard 277–9
  distribution 278
  generation 277
  single/three-phase four-wire
    systems 278-9
  transmission 278

W

warning signs 14, 17, 56
warning tape 262
waste
  disposal 442-3
  regulations 29
  special 443

Waste Transfer Notes 29, 442
wire-wound resistors 133
wiring accessories checklist 348
wiring systems and enclosures
  advantages and limitations 286–92
  appropriate 283–5
  current demand 284
  environmental conditions 284
  overcurrent protection 285
  and use of building 283
Work at Height Regulations (2005) 45–6
work plan 75
working alone 44–8
working practices 67
  benefits of improving 78
  see also safe working
Works Rescue Team 25
wound rotor motor 237–8
writing reports 25–6

X

XLPE/LSF/SWA/LSF cable 208

Z

zener diode 114–15, 148